QUANTUM MECHANICS
For Electrical Engineers

QUANTUM MECHANICS
MECHANICS
For Electrical Engineers

Isaak Mayergoyz
University of Maryland, USA

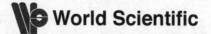

World Scientific

NEW JERSEY · LONDON · SINGAPORE · BEIJING · SHANGHAI · HONG KONG · TAIPEI · CHENNAI · TOKYO

Published by

World Scientific Publishing Co. Pte. Ltd.
5 Toh Tuck Link, Singapore 596224
USA office: 27 Warren Street, Suite 401-402, Hackensack, NJ 07601
UK office: 57 Shelton Street, Covent Garden, London WC2H 9HE

Library of Congress Cataloging-in-Publication Data
Names: Mayergoyz, I. D., author.
Title: Quantum mechanics for electrical engineers / Isaak Mayergoyz
    (University of Maryland, USA).
Description: Hackensack, New Jersey : World Scientific, 2016. |
    Includes bibliographical references and index.
Identifiers: LCCN 2016038481 | ISBN 9789813146907 (hardcover) | ISBN 9789813148017 (pbk)
Subjects: LCSH: Quantum theory.
Classification: LCC QC174.12 .M376 2016 | DDC 530.12--dc23
LC record available at https://lccn.loc.gov/2016038481

British Library Cataloguing-in-Publication Data
A catalogue record for this book is available from the British Library.

Printed in Singapore

To my grandson Jacob

Preface

You have in your hands a book on quantum mechanics. This book is designed as a graduate course for electrical engineers, and it reflects the experience of the author in teaching quantum mechanics to graduate students in the Electrical and Computer Engineering Department of the University of Maryland, College Park. This book can also be used for teaching of quantum mechanics to graduate students in materials science and engineering departments as well as to applied physicists.

The development of quantum mechanics in the late 1920s has been universally acclaimed as one of the greatest achievements of modern physics. This development eventually resulted in the emergence of new technologies such as lasers, nuclear power, semiconductor electronics, superconducting devices, spintronics, quantum computing, etc. This, in turn, led to the penetration of quantum mechanics into electrical and computer engineering curriculum in order to provide the theoretical foundation for graduate courses dealing with the above-mentioned technologies. This state of affairs is similar to the penetration of electromagnetic field theory into electrical engineering curriculum that occurred about eighty years ago.

There are many excellent books on quantum mechanics. However, most of these books are written for physicists, and this is reflected in the selection of the topics covered in those books. The selection of many topics covered in this book is based on their relevance to engineering applications. In other words, this book is designed to serve as the theoretical foundation for graduate courses in quantum optics and lasers, semiconductor electronics, applied superconductivity and quantum computing, which are currently being taught in many electrical and computer engineering departments. To achieve the above goal, the decision was made to cover (along with traditional subjects) the following topics: tunneling, including resonant tunneling and Josephson tunneling in superconductors; Landau levels and their relation to the integer quantum Hall effect; Bloch waves, band structures, effective mass Schrödinger equation and semiclassical transport; energy gap formation in superconductors; quantum transitions in two-level systems and Rabi flopping frequency; Berry phase and Berry curvature; density matrix and optical Bloch equation for two-level systems; Wigner function and quantum transport; and exchange interaction and spintronics. The book also contains a short chapter covering the very basic facts related to Lagrangian and Hamiltonian mechanics. The reason is that electrical engineers are usually unfamiliar with this form of classical mechanics, whereas the Lagrangian and Hamiltonian mechanics have many intimate connections with the mathematical formalism of quantum mechanics. The bibliography contains a short list of books on quantum mechanics and related topics. This list is not exhaustive but rather suggestive.

A special effort has been made to produce a relatively short book that can be used for one semester course in quantum mechanics. Naturally, this could only be achieved by omitting many topics. In particular, many important facts related to atomic, molecular and nuclear physics are not discussed in the book.

In undertaking this project, the intention was to produce a student-friendly textbook on quantum mechanics. It has been realized that this is a very difficult task because many facts and concepts of quantum mechanics are very remote from (and contradictory to) our everyday macroscopic experience. Furthermore, the mathematical formalism of quantum mechanics is quite sophisticated. For these reasons, every effort was made to introduce quantum mechanical concepts and related mathematical formalism in a straightforward way and to strive to achieve clarity and precision in their exposition. It is for students to judge to what extent these efforts have been successful.

I wish to express my sincere gratitude to Dr. Nathan A. Moody who carefully assembled and edited the first version of my lecture notes on quantum mechanics. I am greatly indebted to Dr. Patrick C. McAvoy for reading the manuscript and providing the valuable suggestions for its improvement. I am also very grateful to Priscilla Tang for all her help in the preparation of the book manuscript. My thanks to Chelsea Chin from World Scientific Publishing Company for her assistance and patience. I gratefully acknowledge the financial support derived from the Alford L. Ward Professorship that made this project possible.

Contents

Chapter 1

Basics of Lagrangian and Hamiltonian Mechanics

1.1 What is Quantum Mechanics

Classical mechanics was developed as a result of observations of motion of macroscopic objects. When geometric dimensions of such objects can be neglected in describing their dynamics, we can talk about their motion along trajectories. In studying the motion of macro-objects along trajectories, such physical quantities as momentum, angular momentum and energy were introduced, and it was observed that these quantities assume continuous sets of values.

At the beginning of the last century, numerous experimental facts were established which have been conceptually inconsistent with the basic principles of classical mechanics and electrodynamics. These experimental facts are related to the motion of microscopic particles (electrons, for instance), and they can be roughly subdivided into two distinct groups. The first group consists of facts which clearly reveal that microparticles do not move along trajectories. Examples of these facts are stability of atoms and electron diffraction discovered by C. Davisson, L. Germer and G.P. Thomson. The second group consists of experimental facts which clearly established that on the microscopic level the possible values of energy and angular momentum are quantized. Examples of these facts include atomic radiation spectra which imply discrete energy levels in atoms and Stern–Gerlach type experiments that revealed the quantization of magnetic and angular momenta.

Quantum mechanics is the physical theory of motion of microscopic particles which is consistent with experimental facts. In this theory, the motion of microparticles is described in terms of macroscopic observables such as momentum, angular momentum and energy. The following question can be

1

immediately asked: "Why do we study the motion of microscopic particles in terms of macroscopic quantities?" The reason is that the motion of microscopic particles is not directly observable. In other words, this motion is not directly tangible to human beings. As a consequence, the motion and properties of microparticles are studied through their interactions with macroscopic objects and by observing changes in the motion (and in states) of macroscopic objects caused by these interactions. Practically, all experimental studies of microscopic particles are based on their interactions with macroscopic objects. These interactions are called **measurements**.

Here, however, lies the intrinsic difficulty. As discussed before, experiments show that microscopic particles do not move along trajectories. In other words, there is no such a thing as an electron path. Since microscopic particles do not move along trajectories, they do not have certain velocities $v(t)$ at any instant of time. Consequently, we cannot talk about their momentum **p**, angular momentum **L** and energy $\mathcal{E}$ defined by the following formulas, respectively,

$$\mathbf{p} = m\boldsymbol{v}, \tag{1.1.1}$$

$$\mathbf{L} = \mathbf{r} \times \mathbf{p} = m\left(\mathbf{r} \times \boldsymbol{v}\right), \tag{1.1.2}$$

$$\mathcal{E} = \frac{mv^2}{2} + U_{potential}, \tag{1.1.3}$$

where all the notations have their usual meanings.

This leads to the question: "How can we describe the motion of microscopic particles in terms of macroscopical observables **p**, **L** and $\mathcal{E}$ if microparticles do not move along trajectories and the notion of velocity $v(t)$ is not applicable to their motion?" It turns out that there is a way out of this predicament which is based on the fact that macroscopic observables **p**, **L** and $\mathcal{E}$ can be defined without any reference to the motion along trajectories, but rather on the basis of general properties of space and time for closed systems (i.e. systems which are not subject to any external forces).

It is shown in this chapter, that in classical mechanics, **p**, **L** and $\mathcal{E}$ can be defined as the physical quantities whose conservations for a closed system follow from homogeneity of space, isotropicity of space and homogeneity of time, respectively. It is shown in the next chapter, that the same definitions can be given for **p**, **L** and $\mathcal{E}$ in quantum mechanics. These common definitions of **p**, **L** and $\mathcal{E}$ form a very important bridge between classical and quantum mechanics. Indeed, measurements are usually performed in such a way that a microscopic particle and a measuring macroscopic device form a closed system. Then, observing changes in momentum, angular momentum and energy of macroscopic devices allows one to infer (due to the conservation of these quantities) the changes in $\mathbf{p}$, $\mathbf{L}$ and $\mathcal{E}$ of microscopic particles. These inferences are the essence of measurements.

It turns out that the definitions of $\mathbf{p}$, $\mathbf{L}$ and $\mathcal{E}$ based on general properties of space and time can be most conveniently and generally introduced in classical mechanics if this mechanics is formulated in Lagrangian and Hamiltonian forms. As will be discussed later in this text, there are also other reasons why Lagrangian and Hamiltonian forms of classical mechanics are important in quantum mechanics.

1.2   The Principle of Least Action and Lagrange Equations

Consider a macroscopic object whose geometric dimensions can be neglected in describing its mechanical motion. Naturally, such an object can be called a macroparticle. Consider a system of N macroparticles. Their instantaneous positions in space with respect to an arbitrary chosen origin of a Cartesian coordinate system can be fully specified by N vectors $\mathbf{r}_i(t)$, $(i = 1, 2, \ldots N)$. It is apparent that in these Cartesian coordinates the mechanical system of macroparticles has $3N$ degrees of freedom. Velocities $\boldsymbol{v}_i(t)$ and accelerations $\mathbf{a}_i(t)$ of the macroparticles are defined by the formulas

$$\boldsymbol{v}_i(t) = \frac{d\mathbf{r}_i(t)}{dt}, \quad (i = 1, 2, \ldots N) \tag{1.2.1}$$

$$\mathbf{a}_i(t) = \frac{d\boldsymbol{v}_i(t)}{dt} = \frac{d^2\mathbf{r}_i(t)}{dt^2}, \quad (i = 1, 2, \ldots N). \tag{1.2.2}$$

It turns out that in many problems it may be more convenient to specify the instantaneous positions of macroparticles in space by using non-Cartesian coordinates. Namely, any n functions $q_i(t)$, $(i = 1, 2, \ldots n)$ which uniquely specify the positions of macroparticles in space at any instant of time t can be used for this purpose. They are usually called **generalized** coordinates, and their time-derivatives $\dot{q}_i(t) = \frac{dq_i(t)}{dt}$, $(i = 1, 2, \ldots n)$ are called **generalized** velocities. According to the principle of least action, any mechanical system of macroparticles can be characterized by a Lagrangian function (or Lagrangian) which depends on generalized coordinates, generalized velocities and time. The following concise notation will be used for the Lagrangian

$$L\left[q_i(t), \dot{q}_i(t), t\right] \tag{1.2.3}$$

where it is assumed that L is a function of all generalized coordinates and generalized velocities.

Next, suppose that at time instants t_1 and t_2 the spatial positions of the mechanical system are specified by two sets of generalized coordinates

$$\{q_1(t_1), q_2(t_1), \ldots q_n(t_1)\} \quad \text{and} \quad \{q_1(t_2), q_2(t_2), \ldots q_n(t_2)\}. \quad (1.2.4)$$

Now, the principle of least action can be stated as follows: the mechanical system evolves between times t_1 and t_2 in such a way that the following integral, called "action S"

$$S = \int_{t_1}^{t_2} L[q_i(t), \dot{q}_i(t), t]\, dt \quad (1.2.5)$$

assumes the least possible value.

It is clear from the stated principle that the least action is what distinguishes an actual trajectory of a mechanical system from infinitely many other trajectories that connect two positions specified in (1.2.4).

It turns out that differential equations of motion can be derived from the stated principle of least action. The derivation proceeds as follows.

Let $q_i(t)$ be functions for which S assumes the least possible value. This implies that S is increased when $q_i(t)$ are replaced by

$$\widetilde{q}_i(t) = q_i(t) + \alpha_i \delta q_i(t), \quad (i = 1, 2, \ldots n) \quad (1.2.6)$$

where α_i and $\delta q_i(t)$ are some numbers and functions of time, respectively. Since positions of the mechanical system at t_1 and t_2 are fixed, this requires that

$$\delta q_i(t_1) = 0 \quad \text{and} \quad \delta q_i(t_2) = 0, \quad (i = 1, 2, \ldots n). \quad (1.2.7)$$

By replacing $q_i(t)$ by $\widetilde{q}_i(t)$ in formula (1.2.5), we find

$$S(\alpha_1, \alpha_2, \ldots \alpha_n) = \int_{t_1}^{t_2} L\left[q_i(t) + \alpha_i \delta q_i(t), \dot{q}_i(t) + \alpha_i \delta \dot{q}_i(t), t\right] dt \quad (1.2.8)$$

where

$$\delta \dot{q}_i(t) = \frac{d}{dt}\left(\delta q_i(t)\right). \quad (1.2.9)$$

According to the principle of least action, S assumes its minimum value for

$$\alpha_1 = \alpha_2 = \cdots = \alpha_n = 0. \quad (1.2.10)$$

This means that

$$\left.\frac{\partial S}{\partial \alpha_i}\right|_{\alpha_i = 0} = 0, \quad (i = 1, 2, \ldots n). \quad (1.2.11)$$

By performing the differentiation in formula (1.2.8) and taking into account (1.2.11), we obtain

$$\frac{\partial S}{\partial \alpha_i}\bigg|_{\alpha_i=0} = \int_{t_1}^{t_2} \left[\frac{\partial L}{\partial q_i}\delta q_i + \frac{\partial L}{\partial \dot{q}_i}\delta \dot{q}_i\right] dt = 0, \quad (i=1,2,\ldots n). \quad (1.2.12)$$

By using integration by parts and (1.2.7), we derive

$$\int_{t_1}^{t_2} \frac{\partial L}{\partial \dot{q}_i}\delta \dot{q}_i \, dt = \frac{\partial L}{\partial \dot{q}_i}\delta q_i\bigg|_{t_1}^{t_2} - \int_{t_1}^{t_2} \frac{d}{dt}\left(\frac{\partial L}{\partial \dot{q}_i}\right)\delta q_i \, dt = -\int_{t_1}^{t_2} \frac{d}{dt}\left(\frac{\partial L}{\partial \dot{q}_i}\right)\delta q_i \, dt. \quad (1.2.13)$$

By using the last formula, the equalities (1.2.12) can be transformed as follows

$$\int_{t_1}^{t_2} \left[\frac{\partial L}{\partial q_i}\delta q_i - \frac{d}{dt}\frac{\partial L}{\partial \dot{q}_i}\right]\delta q_i(t) \, dt = 0, \quad (i=1,2,\ldots n). \quad (1.2.14)$$

The last equalities imply that

$$\frac{d}{dt}\left(\frac{\partial L}{\partial \dot{q}_i}\right) - \frac{\partial L}{\partial q_i} = 0, \quad (i=1,2,\ldots n), \quad (1.2.15)$$

or

$$\boxed{\frac{d}{dt}\left(\frac{\partial L}{\partial \dot{q}_i}\right) = \frac{\partial L}{\partial q_i}, \quad (i=1,2,\ldots n).} \quad (1.2.16)$$

Indeed, suppose that for some $i=i_0$ and $t=t_0$

$$\frac{d}{dt}\left(\frac{\partial L}{\partial \dot{q}_{i_0}}\right) - \frac{\partial L}{\partial q_{i_0}} > 0. \quad (1.2.17)$$

Then, according to the continuity argument, the last inequality is also valid for any t within some time interval $(t_0 - \Delta, t_0 + \Delta)$ for sufficiently small Δ. By choosing $\delta q_{i_0}(t) = 0$ outside the interval $(t_0 - \Delta, t_0 + \Delta)$ and $\delta q_{i_0}(t) > 0$ inside the above time interval, we find that

$$\int_{t_1}^{t_2} \left[\frac{\partial L}{\partial q_{i_0}} - \frac{d}{dt}\frac{\partial L}{\partial \dot{q}_{i_0}}\right]\delta q_{i_0}(t) \, dt > 0, \quad (1.2.18)$$

which contradicts equalities (1.2.14). This contradiction establishes the validity of equations (1.2.16). These equations are the Lagrange equations of motion of a mechanical system of macroparticles. In Cartesian coordinates, the Lagrangian is defined as

$$L\left(\mathbf{r}_i(t), \boldsymbol{v}_i(t), t\right) \quad (1.2.19)$$

and the Lagrangian equations (1.2.16) assume the form

$$\frac{d}{dt}\left(\frac{\partial L}{\partial \boldsymbol{v}_i}\right) = \frac{\partial L}{\partial \mathbf{r}_i}, \quad (i=1,2,\ldots N), \quad (1.2.20)$$

where $\frac{\partial L}{\partial v_i}$ and $\frac{\partial L}{\partial r_i}$ are the three-dimensional vectors:

$$\frac{\partial L}{\partial \boldsymbol{v}_i} = \mathbf{e}_x \frac{\partial L}{\partial v_{x_i}} + \mathbf{e}_y \frac{\partial L}{\partial v_{y_i}} + \mathbf{e}_z \frac{\partial L}{\partial v_{z_i}}, \qquad (1.2.21)$$

$$\frac{\partial L}{\partial \boldsymbol{r}_i} = \mathbf{e}_x \frac{\partial L}{\partial x_i} + \mathbf{e}_y \frac{\partial L}{\partial y_i} + \mathbf{e}_z \frac{\partial L}{\partial z_i}. \qquad (1.2.22)$$

Now, the following important remark is in order. The Lagrange equations (1.2.16) are valid for any choice of generalized coordinates q_i, that is for any "coordinate representation" of Lagrangian L. In other words, the mathematical form of the Lagrange equations is invariant with respect to the choice of coordinate representation of the Lagrangian. It will be shown later in this text that this important property of description of laws of nature in mathematical form invariant with respect to the choice of representation is replicated in the mathematical structure of quantum mechanics, where by using the Dirac notations the Schrödinger equation can be written for wave functions from abstract Hilbert space, i.e., for wave function in any representation.

Next, we consider the analytical formulas for the Lagrangian function L. First, it must be noted that the Lagrangian L is not unique, and it is determined up to an additive term which is a total time derivative of an arbitrary function depending on generalized coordinates and time. In other words, the Lagrangian L' defined as

$$\boxed{L' [q_i(t), \dot{q}_i(t), t] = L [q_i(t), \dot{q}_i(t), t] + \frac{d}{dt} f (q_i(t), t)} \qquad (1.2.23)$$

is equivalent to the original Lagrangian L. Indeed, from the last formula follows that the action functional S' defined for the Lagrangian L' is related to the action functional S by the expression

$$S' = \int_{t_1}^{t_2} L' [q_i(t), \dot{q}_i(t), t] \, dt = S + f [q_i(t_2), t_2] - f [q_i(t_1), t_1] , \qquad (1.2.24)$$

which is equivalent to

$$S' = S + const. \qquad (1.2.25)$$

It is apparent from the last formula that the least possible values for actions S' and S are achieved for the same trajectory of a mechanical system and that the same Lagrangian equation of motion (1.2.16) can be derived for the Lagrangian L'. Thus, it can be concluded that for any mechanical system there is a set (a class) of equivalent Lagrangians defined by formula (1.2.23).

Now, we shall derive the formula for a Lagrangian for a free (i.e., subject to no forces) macroparticle in an inertial frame of reference. In this reference frame, time is homogeneous, while space is homogeneous and isotropic. Homogeneity of time implies that the Lagrangian of a free particle does not depend on time explicitly. Homogeneity of space implies that the Lagrangian does not depend explicitly on coordinates of the free particle. Finally, isotropicity of space implies that the Lagrangian does not depend on the direction of free particle velocity. Thus, it can be concluded that the Lagrangian depends only on the magnitude of velocity $v = |\boldsymbol{v}|$, that is on its square. This means that

$$L = F(v^2). \tag{1.2.26}$$

Now, consider another inertial reference frame moving with respect to the previous inertial frame with velocity $\mathbf{b}$. Consequently, the velocity $\boldsymbol{v}'$ of the free particle in the new reference frame is

$$\boldsymbol{v}' = \boldsymbol{v} + \mathbf{b}. \tag{1.2.27}$$

According to formula (1.2.26), the Lagrangian L' of the free particle in the new inertial reference frame is

$$L' = F\left[(\boldsymbol{v}')^2\right] = F\left(v^2 + 2\boldsymbol{v} \cdot \mathbf{b} + b^2\right). \tag{1.2.28}$$

It is clear that

$$2\boldsymbol{v} \cdot \mathbf{b} + b^2 = \frac{d}{dt}\left[2\mathbf{r}(t) \cdot \mathbf{b} + b^2 t\right]. \tag{1.2.29}$$

From the last two formulas, we find

$$L' = F\left(v^2 + \frac{d}{dt}\left[2\mathbf{r}(t) \cdot \mathbf{b} + b^2 t\right]\right). \tag{1.2.30}$$

According to the Galilean relativity principle, any two inertial reference frames are equivalent as far as the mathematical description of motion of the free particle is concerned. This implies that the two Lagrangians L and L' must be equivalent and, consequently, must be related by formula (1.2.23). This will be the case if function F in (1.2.26) and (1.2.30) is linear, that is, if

$$L = \alpha v^2. \tag{1.2.31}$$

Constant α is typically written as $\alpha = \frac{m}{2}$, where m is identified as the mass (inertial mass, to be precise) of the particle. Thus, we arrive at the following formula for the Lagrangian of a free particle

$$L = \frac{mv^2}{2}. \tag{1.2.32}$$

By using the same line of reasoning as before, it can be established that the Lagrangian of a system of N free (noninteracting) particles can be written as

$$L = \sum_{i=1}^{N} \frac{m_i v_i^2}{2}. \tag{1.2.33}$$

Next, we consider a closed system of N interacting particles. It is clear that time is homogeneous for such a system. This implies that the Lagrangian does not depend on time explicitly. We shall also assume that the forces of interaction between the particles depend only on their instantaneous positions in space. This interaction can be described by introducing an additional term in Lagrangian L which depends only on instantaneous position of the particles. Namely,

$$L = \sum_{i=1}^{N} \frac{m_i v_i^2}{2} - U(\mathbf{r}_1(t), \mathbf{r}_2(t), \dots \mathbf{r}_N(t)). \tag{1.2.34}$$

The Lagrangian equations (1.2.20) of motion of this closed system take the form

$$m_i \frac{d\boldsymbol{v}_i}{dt} = -\frac{\partial U}{\partial \mathbf{r}_i}, \quad (i = 1, 2, \dots N). \tag{1.2.35}$$

It is clear that the last equations coincide with the Newton's Second Law if the right-hand sides in those equations are understood as forces $\mathbf{F}_i$ acting on particles:

$$\mathbf{F}_i = -\frac{\partial U}{\partial \mathbf{r}_i}, \quad (i = 1, 2, \dots N). \tag{1.2.36}$$

In formula (1.2.34), the Lagrangian is written in Cartesian coordinates. By expressing these coordinates in terms of generalized coordinates

$$x_i(t) = f_{x_i}(q_1(t), q_2(t), \dots q_n(t)), \tag{1.2.37}$$

$$y_i(t) = f_{y_i}(q_1(t), q_2(t), \dots q_n(t)), \tag{1.2.38}$$

$$z_i(t) = f_{z_i}(q_1(t), q_2(t), \dots q_n(t)) \tag{1.2.39}$$

and taking into account that

$$v_i^2(t) = (\dot{x}_i(t))^2 + (\dot{y}_i(t))^2 + (\dot{z}_i(t))^2 \tag{1.2.40}$$

as well as that

$$\dot{x}_i(t) = \sum_{i=1}^{n} \frac{\partial f_{x_i}}{\partial q_i} \dot{q}_i(t), \tag{1.2.41}$$

$$\dot{y}_i(t) = \sum_{i=1}^{n} \frac{\partial f_{y_i}}{\partial q_i} \dot{q}_i(t), \tag{1.2.42}$$

$$\dot{z}_i(t) = \sum_{i=1}^{n} \frac{\partial f_{z_i}}{\partial q_i} \dot{q}_i(t), \tag{1.2.43}$$

formula (1.2.34) for the Lagrangian can be transformed into the following:

$$L = \sum_{i=1}^{n} \sum_{k=1}^{n} a_{ik} \left(q_1(t), q_2(t), \ldots q_n(t)\right) \dot{q}_i \dot{q}_k - U\left(q_1(t), q_2(t), \ldots q_n(t)\right).$$

$$(1.2.44)$$

This formula can also be written as

$$L = T - U, \qquad (1.2.45)$$

where

$$T = \sum_{i=1}^{n} \sum_{k=1}^{n} a_{ik} \left(q_1(t), q_2(t), \ldots q_n(t)\right) \dot{q}_i \dot{q}_k. \qquad (1.2.46)$$

It is apparent that T is a homogeneous function of order 2 with respect to generalized velocities $\dot{q}_i(t)$. The latter means that

$$\sum_{i=1}^{n} \sum_{k=1}^{n} a_{ik} \left(q_1(t), q_2(t), \ldots q_n(t)\right) (\beta \dot{q}_i)(\beta \dot{q}_k)$$

$$= \beta^2 \sum_{i=1}^{n} \sum_{k=1}^{n} a_{ik} \left(q_1(t), q_2(t), \ldots q_n(t)\right) \dot{q}_i \dot{q}_k. \qquad (1.2.47)$$

By differentiating both sides of the last formula with respect to β and then setting β to one, the following identity can be established:

$$\sum_{i=1}^{n} \frac{\partial T}{\partial \dot{q}_i} \dot{q}_i = 2T. \qquad (1.2.48)$$

This identity will be used in the discussion presented in the next section.

1.3 Conservation Laws

First, we consider the physical quantity whose conservation for closed mechanical systems follows from the homogeneity of time for such systems. The homogeneity of time implies that no instant of time is special. Mathematically, this homogeneity is reflected in the fact that the Lagrangian for a closed mechanical system does not depend explicitly on time. As is clear from formula (1.2.44), the Lagrangian depends only on generalized coordinates and generalized velocities that is,

$$L\left(q_1(t), q_2(t), \ldots q_n(t); \dot{q}_1(t), \dot{q}_2(t), \ldots \dot{q}_n(t)\right). \qquad (1.3.1)$$

By differentiating the function L in (1.3.1) with respect to time, we find

$$\frac{dL}{dt} = \sum_{i=1}^{n} \left[\frac{\partial L}{\partial q_i} \dot{q}_i + \frac{\partial L}{\partial \dot{q}_i} \ddot{q}_i(t) \right], \qquad (1.3.2)$$

where $\ddot{q}_i(t)$ are second-order time derivatives of $q_i(t), (i = 1, 2, \ldots n)$. The explicit independence of time of the Lagrangian results in the absence of the term $\frac{\partial L}{\partial t}$ in the right-hand side of formula (1.3.2).

According to the Lagrangian equations (1.2.16), we have

$$\frac{\partial L}{\partial q_i} = \frac{d}{dt}\left(\frac{\partial L}{\partial \dot{q}_i}\right), \quad (i = 1, 2, \ldots n). \tag{1.3.3}$$

By using the last formula, the relation (1.3.2) can be transformed as follows

$$\frac{dL}{dt} = \sum_{i=1}^{n}\left[\frac{d}{dt}\left(\frac{\partial L}{\partial \dot{q}_i}\right)\dot{q}_i + \frac{\partial L}{\partial \dot{q}_i}\frac{d\dot{q}_i(t)}{dt}\right] = \frac{d}{dt}\left[\sum_{i=1}^{n}\frac{\partial L}{\partial \dot{q}_i}\dot{q}_i\right], \tag{1.3.4}$$

which leads to:

$$\frac{d}{dt}\left[\sum_{i=1}^{n}\frac{\partial L}{\partial \dot{q}_i}\dot{q}_i - L\right] = 0. \tag{1.3.5}$$

This implies that the quantity

$$\mathcal{E} = \sum_{i=1}^{n}\frac{\partial L}{\partial \dot{q}_i}\dot{q}_i - L = const \tag{1.3.6}$$

is conserved. This quantity is called **energy**, and it is apparent from the above derivation that its conservation is the consequence of homogeneity of time for a closed system. The expression (1.3.6) can be transformed in the mathematical form which is typically used for energy in classical mechanics. The transformation is based on formulas (1.2.45) and (1.2.48). By using these formulas in (1.3.6), we obtain

$$\mathcal{E} = T + U = const. \tag{1.3.7}$$

In Cartesian coordinates, the last formula can be written as

$$\mathcal{E} = \sum_{i=1}^{N}\frac{m_i v_i^2}{2} + U\left[\mathbf{r}_1(t), \mathbf{r}_2(t), \ldots \mathbf{r}_N(t)\right] = const. \tag{1.3.8}$$

This is the familiar expression for energy where the first term in the right-hand side is identified as kinetic energy, while the second term in the right-hand side represents the potential energy.

Next, we consider the physical quantity whose conservation for closed mechanical systems follows from the homogeneity of space for such systems. The homogeneity of space means that no position of an entire mechanical system in space is special. Mathematically, this homogeneity implies that the Lagrangian for a closed system is **unchanged** if the entire system is arbitrary displaced (shifted) in space. Accordingly, consider an arbitrary infinitesimally small displacement $\boldsymbol{\delta}$ of the mechanical system as a whole:

$$\mathbf{r}_i \Rightarrow \mathbf{r}_i + \boldsymbol{\delta}, \quad (i = 1, 2, \ldots N). \tag{1.3.9}$$

This results in the following mathematical change in the Lagrangian:

$$\sum_{i=1}^{N} \frac{\partial L}{\partial \mathbf{r}_i} \cdot \boldsymbol{\delta}. \tag{1.3.10}$$

Due to the homogeneity of space, this change must be equal to zero, which leads to:

$$\boldsymbol{\delta} \cdot \sum_{i=1}^{N} \frac{\partial L}{\partial \mathbf{r}_i} = 0. \tag{1.3.11}$$

Since $\boldsymbol{\delta}$ is arbitrary, we find

$$\sum_{i=1}^{N} \frac{\partial L}{\partial \mathbf{r}_i} = 0. \tag{1.3.12}$$

From the last formula and the Lagrangian equations of motion (1.2.20) follows that

$$\frac{d}{dt} \left(\sum_{i=1}^{N} \frac{\partial L}{\partial \mathbf{v}_i} \right) = 0. \tag{1.3.13}$$

This implies that

$$\mathbf{P} = \sum_{i=1}^{N} \frac{\partial L}{\partial \mathbf{v}_i} = const. \tag{1.3.14}$$

Thus, the conservation of vector $\mathbf{P}$ follows from the homogeneity of space. By using formula (1.2.34), the last equality can be written as

$$\mathbf{P} = \sum_{i=1}^{N} m_i \mathbf{v}_i = const. \tag{1.3.15}$$

It is easy to recognize now that $\mathbf{P}$ in (1.3.15) coincides with the momentum in classical mechanics. Thus, the classical momentum can be defined as the physical quantity whose conservation for a closed system follows from the homogeneity of space.

It is clear that

$$\mathbf{P} = \sum_{i=1}^{N} \mathbf{p}_i, \tag{1.3.16}$$

where

$$\mathbf{p}_i = m_i \boldsymbol{v}_i, \quad (i = 1, 2, \ldots N) \tag{1.3.17}$$

are momenta of individual particles.

Next, we consider the physical quantity whose conservation for closed mechanical systems follows from the isotropicity of space for such systems. The isotropicity of space means that no orientation of an entire mechanical system in space is special. In other words, the Lagrangian for a closed system is unchanged if the entire mechanical system is rotated in space as a whole through any angle around an arbitrary chosen axis. Accordingly, consider an infinitesimally small rotation around an arbitrary chosen axis. This rotation is characterized by a vector $\boldsymbol{\delta\varphi}$, whose magnitude is equal to the rotation angle $\delta\varphi$ and whose direction is along an arbitrary chosen axis (see Figure 1.1). From this figure we find

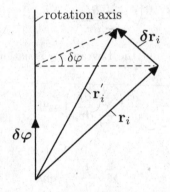

Fig. 1.1

$$\boldsymbol{\delta r}_i = \boldsymbol{\delta\varphi} \times \mathbf{r}_i, \quad (i = 1, 2, \ldots N), \tag{1.3.18}$$

$$\boldsymbol{\delta v}_i = \frac{d}{dt}(\boldsymbol{\delta r}_i) = \frac{d}{dt}(\boldsymbol{\delta\varphi} \times \mathbf{r}_i), \quad (i = 1, 2, \ldots N). \tag{1.3.19}$$

This results in the following mathematical change in the Lagrangian:

$$dL = \sum_{i=1}^{N} \left(\frac{\partial L}{\partial \mathbf{r}_i} \cdot \boldsymbol{\delta r}_i + \frac{\partial L}{\partial \boldsymbol{v}_i} \cdot \boldsymbol{\delta v}_i \right). \tag{1.3.20}$$

Due to the isotropicity of space, this change must be equal to zero:

$$\sum_{i=1}^{N} \left(\frac{\partial L}{\partial \mathbf{r}_i} \cdot \boldsymbol{\delta r}_i + \frac{\partial L}{\partial \boldsymbol{v}_i} \cdot \boldsymbol{\delta v}_i \right) = 0. \tag{1.3.21}$$

According to the Lagrange equations of motion (1.2.20), we have

$$\frac{\partial L}{\partial \mathbf{r}_i} = \frac{d}{dt}\left(\frac{\partial L}{\partial \boldsymbol{v}_i}\right), \quad (i = 1, 2, \ldots N). \tag{1.3.22}$$

By using formulas (1.3.18), (1.3.19) and (1.3.22) in the equality (1.3.21), we obtain

$$\sum_{i=1}^{N}\left[\frac{d}{dt}\left(\frac{\partial L}{\partial \boldsymbol{v}_i}\right) \cdot (\boldsymbol{\delta\varphi} \times \mathbf{r}_i) + \frac{\partial L}{\partial \boldsymbol{v}_i} \cdot \frac{d}{dt}(\boldsymbol{\delta\varphi} \times \mathbf{r}_i)\right] = 0, \tag{1.3.23}$$

that can be further transformed as follows

$$\sum_{i=1}^{N}\frac{d}{dt}\left[\frac{\partial L}{\partial \boldsymbol{v}_i} \cdot (\boldsymbol{\delta\varphi} \times \mathbf{r}_i)\right] = \boldsymbol{\delta\varphi} \cdot \frac{d}{dt}\left(\sum_{i=1}^{N}\mathbf{r}_i \times \frac{\partial L}{\partial \boldsymbol{v}_i}\right) = 0. \tag{1.3.24}$$

Since $\boldsymbol{\delta\varphi}$ is arbitrary, we find

$$\frac{d}{dt}\left(\sum_{i=1}^{N}\mathbf{r}_i \times \frac{\partial L}{\partial \boldsymbol{v}_i}\right) = 0. \tag{1.3.25}$$

This implies that

$$\mathbf{L} = \sum_{i=1}^{N}\mathbf{r}_i \times \frac{\partial L}{\partial \boldsymbol{v}_i} = const. \tag{1.3.26}$$

Thus, the conservation of vector $\mathbf{L}$ follows from the isotropicity of space.

By using formulas (1.2.34) and (1.3.17), the last equality can be written as

$$\mathbf{L} = \sum_{i=1}^{N}\mathbf{r}_i \times (m_i\boldsymbol{v}_i) = const, \tag{1.3.27}$$

which leads to

$$\mathbf{L} = \sum_{i=1}^{N}(\mathbf{r}_i \times \mathbf{p}_i) = const. \tag{1.3.28}$$

Now, it is easy to recognize that vector $\mathbf{L}$ in formulas (1.3.27) and (1.3.28) coincides with the angular momentum of classical mechanics. Thus, the classical angular momentum can be defined as the physical quantity whose conservation for a closed system follows from the isotropicity of space.

It is apparent that

$$\mathbf{L} = \sum_{i=1}^{N}\mathbf{l}_i, \tag{1.3.29}$$

where

$$\mathbf{l}_i = \mathbf{r}_i \times \mathbf{p}_i, \quad (i = 1, 2, \ldots N) \tag{1.3.30}$$

are angular momenta of individual particles.

1.4   Hamiltonian Equations and Poisson Brackets

The Lagrange equations of motion (1.2.16) can be represented in another equivalent and more symmetric form if a state of a mechanical system is specified by generalized coordinates and generalized momenta instead of generalized coordinates and generalized velocities. This leads to Hamiltonian equations of motion which have many intimate connections with the mathematical formalism of quantum mechanics. The derivation of Hamiltonian equations proceeds as follows.

First, we introduce the generalized momenta $p_i(t)$ by the formula

$$p_i(t) = \frac{\partial L}{\partial \dot{q}_i}, \quad (i = 1, 2, \ldots n). \tag{1.4.1}$$

It is clear from formulas (1.2.3) and (1.4.1) that generalized momenta are functions of $q_i(t)$, $\dot{q}_i(t)$ and t:

$$p_i(t) = g_i\left[q_i(t), \dot{q}_i(t), t\right], \quad (i = 1, 2, \ldots n). \tag{1.4.2}$$

It is also clear from (1.4.2) that $\dot{q}_i(t)$ can be viewed as functions of $q_i(t)$, $p_i(t)$, and t:

$$\dot{q}_i(t) = f_i\left[q_i(t), p_i(t), t\right], \quad (i = 1, 2, \ldots n). \tag{1.4.3}$$

By using equation (1.3.6) for energy and by substituting into this equation formulas (1.4.1) and (1.4.3), we arrive at the expression for energy as a function of generalized coordinates and generalized momenta:

$$\mathcal{E} = H\left[p_i(t), q_i(t), t\right]. \tag{1.4.4}$$

Such a function is called a Hamiltonian function. We shall next derive the equations of motion in terms of this function. To this end, we represent equation (1.3.6) in the form

$$H = \sum_{i=1}^{n} \left[p_i \dot{q}_i - L(q_i, \dot{q}_i, t)\right], \tag{1.4.5}$$

from which we find:

$$dH = \sum_{i=1}^{n} \left[p_i dq_i + \dot{q}_i dp_i - \frac{\partial L}{\partial q_i} dq_i - \frac{\partial L}{\partial \dot{q}_i} d\dot{q}_i\right] - \frac{\partial L}{\partial t} dt. \tag{1.4.6}$$

From the Lagrange equations (1.2.16) follows that

$$\frac{\partial L}{\partial q_i} = \frac{d}{dt}\left(\frac{\partial L}{\partial \dot{q}_i}\right) = \dot{p}_i, \quad (i = 1, 2, \ldots n). \tag{1.4.7}$$

By substituting formulas (1.4.1) and (1.4.7) into (1.4.6), we find

$$dH = \sum_{i=1}^{n} [p_i d\dot{q}_i + \dot{q}_i dp_i - \dot{p}_i dq_i - p_i d\dot{q}_i] - \frac{\partial L}{\partial t} dt, \qquad (1.4.8)$$

which after cancellations leads to

$$dH = \sum_{i=1}^{n} [\dot{q}_i dp_i - \dot{p}_i dq_i] - \frac{\partial L}{\partial t} dt. \qquad (1.4.9)$$

On the other hand, from formula (1.4.4) we obtain

$$dH = \sum_{i=1}^{n} \left[\frac{\partial H}{\partial p_i} dp_i + \frac{\partial H}{\partial q_i} dq_i \right] + \frac{\partial H}{\partial t} dt. \qquad (1.4.10)$$

By comparing formulas (1.4.9) and (1.4.10) we conclude that

$$\boxed{\dot{q}_i = \frac{\partial H}{\partial p_i}, \quad \dot{p}_i = -\frac{\partial H}{\partial q_i}, \quad (i = 1, 2, \ldots n),} \qquad (1.4.11)$$

and

$$\frac{\partial H}{\partial t} = -\frac{\partial L}{\partial t}. \qquad (1.4.12)$$

Equations (1.4.11) are Hamiltonian equations. They play a central role in modern physics because they reveal that the dynamics is controlled by energy, that is by the Hamiltonian function H. As discussed in the next chapter, this fundamental physical feature is replicated in the mathematical structure of quantum mechanics. The Hamiltonian equations (1.4.11) are highly symmetric. This aspect is widely exploited in the development of the extensive mathematical theory of Hamiltonian equations. One of the most salient results of this theory is the conservation of phase volumes in Hamiltonian dynamics. Hamiltonian equations are also instrumental in the derivation of the condition under which some physical quantity

$$G\left[p_i(t), q_i(t), t\right] \qquad (1.4.13)$$

is conserved when $p_i(t)$ and $q_i(t)$ are the solution of the Hamiltonian equations (1.4.11). To find this condition, we first differentiate function G in (1.4.13) with respect to time:

$$\frac{dG}{dt} = \sum_{i=1}^{n} \left[\frac{\partial G}{\partial p_i} \dot{p}_i + \frac{\partial G}{\partial q_i} \dot{q}_i \right] + \frac{\partial G}{\partial t}. \qquad (1.4.14)$$

By using equations (1.4.11), the last formula can be transformed as follows

$$\frac{dG}{dt} = \sum_{i=1}^{n} \left[-\frac{\partial H}{\partial q_i} \frac{\partial G}{\partial p_i} + \frac{\partial H}{\partial p_i} \frac{\partial G}{\partial q_i} \right] + \frac{\partial G}{\partial t}. \qquad (1.4.15)$$

Now, we introduce the Poisson bracket

$$\{H, G\} = \sum_{i=1}^{n} \left[\frac{\partial H}{\partial p_i} \frac{\partial G}{\partial q_i} - \frac{\partial H}{\partial q_i} \frac{\partial G}{\partial p_i} \right] \tag{1.4.16}$$

and rewrite the relation (1.4.15) in the form

$$\frac{dG}{dt} = \frac{\partial G}{\partial t} + \{H, G\}. \tag{1.4.17}$$

It is clear that the conservation of the physical quantity G takes place if

$$\boxed{\frac{\partial G}{\partial t} + \{H, G\} = 0.} \tag{1.4.18}$$

In the case when G does not depend on time explicitly, that is when

$$\frac{\partial G}{\partial t} = 0, \tag{1.4.19}$$

the conservation condition for G is

$$\boxed{\{H, G\} = 0.} \tag{1.4.20}$$

Thus, the Poisson bracket is quite instrumental in expressing the conservation of a physical quantity under Hamiltonian dynamics. In other words, this bracket is instrumental in checking if a specific physical quantity is an integral of motion for Hamiltonian dynamics. As discussed in the next chapter, commutators of two operators play in quantum mechanics the role analogous to Poisson brackets and, historically, this analogy was central in the development of the mathematical formalism of quantum mechanics. It is also worthwhile to mention that the Hamiltonian equation (1.4.11) can be written in terms of Poisson brackets. Indeed, it is easy to check by using the definition (1.4.16) of the Poisson bracket that the Hamiltonian equations (1.4.11) can be written as follows:

$$\dot{p}_i = \{H, p_i\}, \quad \dot{q}_i = \{H, q_i\}, \quad (i = 1, 2, \ldots n). \tag{1.4.21}$$

Hamiltonian equations (1.4.11) are usually called canonical Hamiltonian equations, while p_i and q_i are called canonical momenta and coordinates. It is interesting to point out that there exist noncanonical Hamiltonian equations that cannot be written in the form (1.4.11) but can be represented in the form (1.4.21) for specially defined Poisson brackets. We shall not consider this matter further, but instead we shall rather turn to the discussion of another topic which will be of importance in the subsequent exposition of quantum mechanics.

In our previous discussion, we considered mechanical systems in which interactions (i.e. forces) between particles depended only on their instantaneous positions. However, this is not the case for a charged particle moving in an external electromagnetic field. In this case, the force acting on the particle is the Lorentz force

$$\mathbf{F} = q\mathbf{E}(\mathbf{r}, t) + q\boldsymbol{v} \times \mathbf{B}(\mathbf{r}, t), \tag{1.4.22}$$

where q is the charge of the particle, $\boldsymbol{v}$ is its velocity, while $\mathbf{E}(\mathbf{r}, t)$ and $\mathbf{B}(\mathbf{r}, t)$ are electric field and magnetic flux density, respectively.

It is apparent from the last formula that the force depends not only on spatial position of the charged particle, but on its velocity as well.

Our immediate goal is to find such a Lagrangian that the Lagrange equations (1.2.20) coincide with the Newton dynamic equation

$$m\frac{d\boldsymbol{v}}{dt} = q\mathbf{E}(\mathbf{r}, t) + q\boldsymbol{v} \times \mathbf{B}(\mathbf{r}, t). \tag{1.4.23}$$

It is shown below that such a Lagrangian is given by the formula

$$\boxed{L = \frac{mv^2}{2} - q\left[\varphi(\mathbf{r}, t) - \boldsymbol{v} \cdot \mathbf{A}(\mathbf{r}, t)\right],} \tag{1.4.24}$$

where $\varphi(\mathbf{r}, t)$ and $\mathbf{A}(\mathbf{r}, t)$ are electric scalar and magnetic vector potentials, respectively, which are related to $\mathbf{E}(\mathbf{r}, t)$ and $\mathbf{B}(\mathbf{r}, t)$ by the following well-known formulas

$$\mathbf{E}(\mathbf{r}, t) = -\nabla\varphi(\mathbf{r}, t) - \frac{\partial\mathbf{A}(\mathbf{r}, t)}{\partial t}, \tag{1.4.25}$$

$$\mathbf{B}(\mathbf{r}, t) = \nabla \times \mathbf{A}(\mathbf{r}, t). \tag{1.4.26}$$

First, from formula (1.4.24) we find

$$\frac{\partial L}{\partial \boldsymbol{v}} = m\boldsymbol{v} + q\mathbf{A}(\mathbf{r}, t), \tag{1.4.27}$$

and, consequently,

$$\frac{d}{dt}\left[\frac{\partial L}{\partial \boldsymbol{v}}\right] = \frac{d}{dt}\left[m\boldsymbol{v} + q\mathbf{A}(\mathbf{r}, t)\right]. \tag{1.4.28}$$

Second,

$$\frac{\partial L}{\partial \mathbf{r}} = -q\nabla\varphi(\mathbf{r}, t) + q\nabla\left[\boldsymbol{v} \cdot \mathbf{A}(\mathbf{r}, t)\right]. \tag{1.4.29}$$

From the last two formulas follows that the Lagrange equation of particle motion

$$\frac{d}{dt}\left[\frac{\partial L}{\partial \boldsymbol{v}}\right] = \frac{\partial L}{\partial \mathbf{r}} \tag{1.4.30}$$

can be written as

$$m\frac{d\boldsymbol{v}}{dt} = -q\nabla\varphi(\mathbf{r},t) - q\frac{d\mathbf{A}(\mathbf{r},t)}{dt} + q\nabla\left[\boldsymbol{v}\cdot\mathbf{A}(\mathbf{r},t)\right]. \qquad (1.4.31)$$

Since a spatial position of the charged particle changes with time, $\mathbf{r}$ in $\mathbf{A}(\mathbf{r},t)$ is a function of time. Consequently,

$$\frac{d\mathbf{A}(\mathbf{r},t)}{dt} = \frac{\partial\mathbf{A}(\mathbf{r},t)}{\partial t} + (\boldsymbol{v}\cdot\nabla)\,\mathbf{A}(\mathbf{r},t). \qquad (1.4.32)$$

By substituting the last equation into formula (1.4.31), we find

$$m\frac{d\boldsymbol{v}}{dt} = -q\left[\nabla\varphi(\mathbf{r},t) + \frac{\partial\mathbf{A}(\mathbf{r},t)}{\partial t}\right] + q\{\nabla\left[\boldsymbol{v}\cdot\mathbf{A}(\mathbf{r},t)\right] - (\boldsymbol{v}\cdot\nabla)\,\mathbf{A}(\mathbf{r},t)\}. $$
$$(1.4.33)$$

According to the well-known identity from vector calculus, we have

$$\nabla\left[\boldsymbol{v}\cdot\mathbf{A}(\mathbf{r},t)\right] - (\boldsymbol{v}\cdot\nabla)\,\mathbf{A}(\mathbf{r},t) = \boldsymbol{v}\times\left[\nabla\times\mathbf{A}(\mathbf{r},t)\right]. \qquad (1.4.34)$$

By substituting the last formula into equation (1.4.33), we obtain

$$m\frac{d\boldsymbol{v}}{dt} = -q\left[\nabla\varphi(\mathbf{r},t) + \frac{\partial\mathbf{A}(\mathbf{r},t)}{\partial t}\right] + q\boldsymbol{v}\times\left[\nabla\times\mathbf{A}(\mathbf{r},t)\right]. \qquad (1.4.35)$$

Now, recalling formulas (1.4.25) and (1.4.26), we conclude that the Lagrange equation of motion (1.4.35) coincides with the Newton equation (1.4.23). Thus, formula (1.4.24) is an appropriate expression for a Lagrangian of a charged particle moving in external electromagnetic field.

It is known that electromagnetic potentials are not unique and that there exists a class (a set) of equivalent electromagnetic potentials which are related to one another by the following gauge transformation:

$$\mathbf{A}'(\mathbf{r},t) = \mathbf{A}(\mathbf{r},t) + \nabla\gamma(\mathbf{r},t), \qquad (1.4.36)$$

$$\varphi'(\mathbf{r},t) = \varphi(\mathbf{r},t) - \frac{\partial\gamma(\mathbf{r},t)}{\partial t}, \qquad (1.4.37)$$

where $\gamma(\mathbf{r},t)$ is an arbitrary differentiable function of spatial position and time.

By using formulas (1.4.24), (1.4.36) and (1.4.37), it can be demonstrated (and it is suggested as a useful exercise) that the gauge transformation of electromagnetic potentials results in the following equivalent transformation of Lagrangian:

$$L' = L + q\frac{d}{dt}\left[\gamma(\mathbf{r},t)\right]. \qquad (1.4.38)$$

Thus, there is one to one correspondence between the equivalent class of Lagrangians and the equivalent class of electromagnetic potentials.

We conclude this section by deriving the expression for the Hamiltonian of a charged particle moving in external magnetic field. To this end, we recall that the Hamiltonian is the energy expressed in terms of generalized (canonical) momenta and coordinates. According to the definition of the canonical momentum $\mathbf{p} = \frac{\partial L}{\partial \boldsymbol{v}}$ and formula (1.4.27), we have

$$\mathbf{p} = m\boldsymbol{v} + q\mathbf{A}(\mathbf{r}, t). \tag{1.4.39}$$

Furthermore, in the case of a single particle, the expression (1.3.6) for energy can be written as

$$\mathcal{E} = \boldsymbol{v} \cdot \frac{\partial L}{\partial \boldsymbol{v}} - L. \tag{1.4.40}$$

By using formulas (1.4.24) and (1.4.40), we derive

$$\mathcal{E} = \frac{mv^2}{2} + q\varphi(\mathbf{r}, t). \tag{1.4.41}$$

It is worthwhile to point out that the energy of a charged particle does not depend on applied magnetic field. This is because the component of the force caused by the magnetic field is normal to the trajectory of the particle (see formula (1.4.22)) and, for this reason, does not change its energy. One practical implication of this fact is that in particle accelerators electric fields are used for particle acceleration, while magnetic fields are utilized for bending and focusing of particle beams.

Now, we shall complete the derivation of the Hamiltonian. From formula (1.4.39) we find

$$\boldsymbol{v} = \frac{\mathbf{p} - q\mathbf{A}(\mathbf{r}, t)}{m}. \tag{1.4.42}$$

By substituting the last formula into equation (1.4.41), we derive

$$\boxed{H(\mathbf{p}, \mathbf{r}, t) = \frac{[\mathbf{p} - q\mathbf{A}(\mathbf{r}, t)]^2}{2m} + q\varphi(\mathbf{r}, t).} \tag{1.4.43}$$

In a particular case when the charged particle moves in electric field, the last formula is reduced to

$$\boxed{H(\mathbf{p}, \mathbf{r}, t) = \frac{p^2}{2m} + q\varphi(\mathbf{r}, t).} \tag{1.4.44}$$

It is important to stress that electromagnetic potentials rather than electric and magnetic fields are used in the Lagrangian and Hamiltonian formulations of classical mechanics of charged particles. This feature is replicated in the mathematical structure of quantum mechanics.

Problems

(1) Prove formula (1.2.48).

(2) Prove that:
$$\{q_i, q_k\} = 0, \quad \{p_i, p_k\} = 0, \quad \{p_i, q_k\} = \delta_{ik} \tag{P.1.1}$$
and compare with commutation relations (2.3.9).

(3) Prove that:
$$\{f, g\} = -\{g, f\}, \tag{P.1.2}$$
$$\{f_1 + f_2, g\} = \{f_1, g\} + \{f_2, g\}, \tag{P.1.3}$$
$$\{f_1 f_2, g\} = f_1\{f_2, g\} + f_2\{f_1, g\}. \tag{P.1.4}$$

(4) Prove that
$$\{f, \{g, h\}\} + \{g, \{h, f\}\} + \{h, \{f, g\}\} = 0. \tag{P.1.5}$$

(5) Prove that if f and g are integrals of motion, which do not depend on time explicitly, then
$$\{f, g\} = const, \tag{P.1.6}$$
and it is an integral of motion as well.

(6) Prove (P.1.6) when the integrals of motion f and g depend on time explicitly.

(7) Prove formulas (1.4.21).

(8) Consider the precessional dynamics of magnetization (see Sections 7.1 and 7.2)
$$\frac{d\mathbf{M}}{dt} = -\gamma \left(\mathbf{M} \times \mathbf{H}_{eff}(\mathbf{M})\right), \tag{P.1.7}$$
where
$$\mathbf{H}_{eff} = -\frac{1}{\mu_0} \frac{\partial w(\mathbf{M})}{\partial \mathbf{M}}, \tag{P.1.7}$$
where $w(\mathbf{M})$ is the micromagnetic energy, which can be viewed as the Hamiltonian function. Prove that this precessional dynamics has noncanonical Hamiltonian structure and can be represented in the form
$$\frac{dM_x}{dt} = \{w, M_x\}, \quad \frac{dM_y}{dt} = \{w, M_y\}, \quad \frac{dM_z}{dt} = \{w, M_z\}, \tag{P.1.8}$$
where the Poisson bracket for $f(\mathbf{M})$ and $g(\mathbf{M})$ is defined by the formula
$$\{f, g\} = \mathbf{M} \cdot \left(\frac{\partial f}{\partial \mathbf{M}} \times \frac{\partial g}{\partial \mathbf{M}}\right). \tag{P.1.9}$$

(9) Prove formula (1.4.34).

(10) Prove formula (1.4.38).

(11) Find out (through literature search) the statement of Liouville's phase space volume conservation theorem for Hamiltonian dynamics and provide its proof.

(12) Find out (through literature search) the statement of the Poincaré recurrence theorem and provide its proof.

Chapter 2

Fundamentals of Quantum Mechanics

2.1 Hilbert Space and Hermitian Operators

Hilbert space and the theory of Hermitian operators in this space form the mathematical foundation of quantum mechanics. For this reason, the basic facts related to Hermitian operators in Hilbert spaces are summarized in this section.

First, we start with the definition of Hilbert space. Actually, we initially introduce one specific Hilbert space which is denoted as L_2. Consider a complex-valued function $\psi(\mathbf{r}, t)$ which is square-integrable with respect to spatial coordinates $\mathbf{r}$. The latter means that the following integral is finite:

$$\int \psi^*(\mathbf{r}, t)\psi(\mathbf{r}, t) \, dV = \int |\psi(\mathbf{r}, t)|^2 \, dV. \qquad (2.1.1)$$

In the last formula $\psi^*(\mathbf{r}, t)$ is the complex conjugate of $\psi(\mathbf{r}, t)$ and the integration is carried out over the entire three-dimensional space.

It is clear that the set of square-integrable functions is a linear function space in the sense that if $\psi_1(\mathbf{r}, t)$ and $\psi_2(\mathbf{r}, t)$ are square-integrable, then their linear combination

$$\psi(\mathbf{r}, t) = \alpha_1\psi_1(\mathbf{r}, t) + \alpha_2\psi_2(\mathbf{r}, t) \qquad (2.1.2)$$

is square-integrable as well for any complex numbers α_1 and α_2. It is also apparent now that linear combinations of any finite number of square-integrable functions are square-integrable.

In the set of square-integrable functions, one can introduce the inner product $\langle \varphi | \psi \rangle$ by the formula

$$\langle \varphi | \psi \rangle = \int \varphi^*(\mathbf{r}, t)\psi(\mathbf{r}, t) \, dV. \qquad (2.1.3)$$

It is easy to check that the following formulas are valid for the inner product:

$$\langle \varphi | \psi \rangle = \langle \psi | \varphi \rangle^*, \qquad (2.1.4)$$

$$\langle \varphi | \alpha_1 \psi_1 + \alpha_2 \psi_2 \rangle = \alpha_1 \langle \varphi | \psi_1 \rangle + \alpha_2 \langle \varphi | \psi_2 \rangle , \qquad (2.1.5)$$

$$\langle \alpha_1 \varphi_1 + \alpha_2 \varphi_2 | \psi \rangle = \alpha_1^* \langle \varphi_1 | \psi \rangle + \alpha_2^* \langle \varphi_2 | \psi \rangle , \qquad (2.1.6)$$

$$\langle \psi | \psi \rangle = \| \psi \|^2 > 0. \qquad (2.1.7)$$

The quantity

$$\| \psi \| = \left(\int |\psi(\mathbf{r}, t)|^2 \, dV \right)^{\frac{1}{2}} \qquad (2.1.8)$$

is called the L_2-norm of function ψ.

There are mathematically technical reasons (such as "completeness" of function space with respect to limits of function sequences) that the integration in all above formulas should be understood in the Lebesgue sense, and this is reflected by letter "L" used in the notation L_2 for the function space. Square-integrable in the Lebesgue-sense functions are usually called square-summable functions. The above mathematical details will not be important for our subsequent discussion of quantum mechanics and the conventional integration (in the Riemann sense) will be sufficient for the understanding of this discussion.

Next, consider an infinite set $\{\psi_k\}$ of orthonormal functions which satisfy the relations

$$\langle \psi_i | \psi_k \rangle = \delta_{ik} = \begin{cases} 1, & \text{if } i = k \\ 0, & \text{if } i \neq k \end{cases}. \qquad (2.1.9)$$

This set is called complete if any function ψ from L_2 can be represented as

$$\psi = \sum_{k=1}^{\infty} a_k \psi_k. \qquad (2.1.10)$$

It is easy to derive from the last two formulas that

$$a_k = \langle \psi_k | \psi \rangle , \quad (k = 1, 2, \ldots \infty). \qquad (2.1.11)$$

Consequently, formula (2.1.10) can be written as follows

$$\psi = \sum_{k=1}^{\infty} \langle \psi_k | \psi \rangle \psi_k. \qquad (2.1.12)$$

It also can be concluded from formula (2.1.10) that

$$|\psi|^2 = \psi^* \psi = \sum_{i=1}^{\infty} \sum_{k=1}^{\infty} a_i^* a_k \psi_i^* \psi_k. \qquad (2.1.13)$$

By integrating all parts of the last equality and taking into account formulas (2.1.9), we derive that

$$\|\psi\|^2 = \langle \psi^* | \psi \rangle = \sum_{k=1}^{\infty} |a_k|^2. \tag{2.1.14}$$

Now, we introduce another Hilbert space, denoted as ℓ_2, of infinite-dimensional vectors

$$\mathbf{a} = \begin{pmatrix} a_1 \\ a_2 \\ \vdots \\ a_k \\ \vdots \end{pmatrix}. \tag{2.1.15}$$

The inner product $\langle \mathbf{a}, \mathbf{b} \rangle$ for two of such vectors is defined as follows:

$$\langle \mathbf{a} | \mathbf{b} \rangle = \sum_{k=1}^{\infty} a_k^* b_k. \tag{2.1.16}$$

It is easy to check that formulas similar to formulas (2.1.4)–(2.1.6) are valid for the above inner product. Furthermore, it is apparent that

$$\|\mathbf{a}\|^2 = \langle \mathbf{a}, \mathbf{a} \rangle = \sum_{k=1}^{\infty} |a_k|^2. \tag{2.1.17}$$

Of course, it has been tacitly assumed that components a_k of vector $\mathbf{a}$ are such that all above series converge.

If the set (the basis) of orthonormal functions $\{\psi_k\}$ is chosen, then for any function ψ from L_2 there exists a vector $\mathbf{a}$ from ℓ_2 with components a_k defined by formulas (2.1.11). Conversely, for any vector $\mathbf{a}$ from ℓ_2 there exists a function ψ defined by the formula (2.1.10). In other words, there is one-to-one correspondence between L_2 and ℓ_2. This correspondence is isomorphic in the sense that if vectors $\mathbf{a}$ and $\mathbf{b}$ from ℓ_2 correspond to functions ψ and φ from L_2, respectively, then there exists one-to-one correspondence between their linear combinations $\alpha_1 \mathbf{a} + \alpha_2 \mathbf{b}$ and $\alpha_1 \psi + \alpha_2 \psi$. This one-to-one correspondence between L_2 and ℓ_2 is isometric in the sense that it is norm preserving. Indeed, from formulas (2.1.14) and (2.1.17) follows that

$$\|\psi\| = \|\mathbf{a}\|. \tag{2.1.18}$$

Mathematically, this means that there is isometric isomorphism between L_2 and ℓ_2. The presented discussion suggests that there is some equivalence between Hilbert spaces L_2 and ℓ_2. One may say that in the case of L_2 we deal with coordinate representation of ψ, while in the case of ℓ_2 we deal with the representation of ψ in terms of expansion coefficients in the chosen basis $\{\psi_k\}$. By choosing different orthonormal bases, we obtain different representations of ψ in ℓ_2. In quantum mechanics, different representations of the state of microscopic systems (microparticles) are used. Historically, two equivalent forms (the wave form and matrix form) of quantum mechanics were almost simultaneously developed which correspond to equivalent L_2 and ℓ_2 representations of quantum mechanical states, respectively.

The existence of different equivalent representations led to the development of the concept of abstract Hilbert space. This space consists of elements ψ whose mathematical nature is not explicitly specified. This is a linear space in the sense that if ψ and φ are its elements, then linear combinations $\alpha_1\psi + \alpha_2\varphi$ are elements of this space as well. For each two elements φ and ψ the inner product $\langle\varphi|\psi\rangle$ is introduced as a complex number and it is postulated that this inner product has the properties expressed by the formulas (2.1.4)–(2.1.7). Spaces L_2 and ℓ_2 are particular realizations of this abstract Hilbert space. The laws of quantum mechanics can be stated in terms of this abstract Hilbert space, and these laws will be applicable for any representations of quantum mechanical states of microparticles or microsystems. Such a formulation of quantum mechanics was developed by P. Dirac and J. von Neumann. Dirac also introduced the notations which are now ubiquitous in quantum mechanical literature. Namely, he suggested to split the inner product $\langle\varphi|\psi\rangle$ in two parts: $|\psi\rangle$ and $\langle\varphi|$, which he called ket–vector and bra–vector, respectively. Ket–vectors are the elements of abstract Hilbert space and they describe the quantum mechanical states without specifying their representations. Bra–vectors are used to form inner products and they can be viewed as belonging to dual space, i.e., the space of linear functional defined on the abstract Hilbert space. Indeed, for a fixed ket–vector $\langle\varphi|$ the inner product $\langle\varphi|\psi\rangle$ maps any element ψ of abstract Hilbert space into a complex number. That is why $\langle\psi|\varphi\rangle$ is called a functional. This is a linear functional according to the property (2.1.5) of the inner product. According to the Riesz representation theorem from functional analysis, any linear functional can be represented as the inner product of ket–vectors with a specific (for a given functional) bra–vector.

Next, we proceed to the discussion of operators in Hilbert space. An operator $\hat{f}$ in Hilbert space is a specific mapping of one element ψ of this space into another element φ of the same space. This is mathematically written in the form

$$\varphi = \hat{f}\psi, \qquad (2.1.19)$$

or in terms of Dirac notations as

$$|\varphi\rangle = \hat{f}|\psi\rangle. \tag{2.1.20}$$

Below are two examples of operators: a) integral operator

$$\varphi(\mathbf{r}) = \int K(\mathbf{r}, \mathbf{r}')\psi(\mathbf{r}')\,dV' \tag{2.1.21}$$

where $K(\mathbf{r}, \mathbf{r}')$ is called the kernel of the operator and integration is performed with respect to $\mathbf{r}'$; b) differential operator

$$\varphi(\mathbf{r}) = \frac{\partial \psi(\mathbf{r})}{\partial x}. \tag{2.1.22}$$

In quantum mechanics, one deals with linear operators, which have the following property

$$\hat{f}(\alpha_1\psi_1 + \alpha_2\psi_2) = \alpha_1\hat{f}\psi_1 + \alpha_2\hat{f}\psi_2, \tag{2.1.23}$$

that can be written in Dirac notations as follows

$$\hat{f}|\alpha_1\psi_1 + \alpha_2\psi_2\rangle = \alpha_1\hat{f}|\psi_1\rangle + \alpha_2\hat{f}|\psi_2\rangle. \tag{2.1.24}$$

A linear operator $\hat{f}$ is called Hermitian if for **any** elements φ and ψ of the Hilbert space the following equality is valid:

$$\boxed{\langle\varphi|\hat{f}\psi\rangle = \langle\hat{f}\varphi|\psi\rangle.} \tag{2.1.25}$$

Thus, in the case of a Hermitian operator the value of the inner product does not depend to which element (ψ or φ) operator is applied. To reflect this fact, Dirac introduced the following notation:

$$\langle\varphi|\hat{f}\psi\rangle = \langle\varphi|\hat{f}|\psi\rangle. \tag{2.1.26}$$

Hermitian operators are very special, because their eigenvalues and eigenfunctions have unique properties. Eigenvalues and eigenfunctions of operator $\hat{f}$ are defined by the equation

$$\hat{f}\psi_\lambda = \lambda\psi_\lambda. \tag{2.1.27}$$

In other words, eigenfunctions are mapped into their scaled versions and the scaling coefficients are eigenvalues. The set of all eigenvalues is called the spectrum of the operator. An important case is the case of the discrete spectrum, when eigenvalues are separated from one another and can be numerated. In this case, formula (2.1.27) can be written as follows

$$\hat{f}\psi_k = \lambda_k\psi_k. \tag{2.1.28}$$

Proposition 2.1. *The spectrum of a Hermitian operator is real.*

Proof is given below for a discrete spectrum, although it is applicable to the case of a continuous spectrum as well.

First, by using formula (2.1.28) we find

$$\langle \psi_k | \hat{f}\psi_k \rangle = \langle \psi_k | \lambda_k \psi_k \rangle = \lambda_k \|\psi_k\|^2. \qquad (2.1.29)$$

On the other hand, by taking into account that $\hat{f}$ is Hermitian, we derive

$$\langle \psi_k | \hat{f}\psi_k \rangle = \langle \hat{f}\psi_k | \psi_k \rangle = \langle \lambda_k \psi_k, \psi_k \rangle = \lambda_k^* \|\psi_k\|^2. \qquad (2.1.30)$$

From the last two formulas follows that

$$\lambda_k = \lambda_k^*, \qquad (2.1.31)$$

which is only possible when eigenvalues are real numbers. □

Proposition 2.2. *Eigenfunctions ψ_i and ψ_k of Hermitian operators corresponding to different eigenvalues λ_i and λ_k, $(\lambda_i \neq \lambda_k)$, are orthogonal*

$$\langle \psi_i | \psi_k \rangle = 0. \qquad (2.1.32)$$

Proof. It is stated that

$$\hat{f}\psi_i = \lambda_i \psi_i, \quad \hat{f}\psi_k = \lambda_k \psi_k. \qquad (2.1.33)$$

Consequently,

$$\langle \psi_i | \hat{f}\psi_k \rangle = \langle \psi_i | \lambda_k \psi_k \rangle = \lambda_k \langle \psi_i | \psi_k \rangle. \qquad (2.1.34)$$

On the other hand, taking into account that $\hat{f}$ is Hermitian, we find

$$\langle \psi_i | \hat{f}\psi_k \rangle = \langle \hat{f}\psi_i | \psi_k \rangle = \langle \lambda_i \psi_i | \psi_k \rangle = \lambda_i^* \langle \psi_i | \psi_k \rangle = \lambda_i \langle \psi_i | \psi_k \rangle \qquad (2.1.35)$$

because the spectrum is real.

From the last two formulas, we conclude that

$$(\lambda_k - \lambda_i) \langle \psi_i | \psi_k \rangle = 0. \qquad (2.1.36)$$

Since λ_k and λ_i are different, the last equality is only possible if formula (2.1.32) is valid. Thus, ψ_i and ψ_k are orthogonal. □

Remark 2.1. Hermitian operators may have degenerate eigenvalues. This is the case when there are several linearly independent eigenfunctions corresponding to the same eigenvalue:

$$\hat{f}\psi_k^{(n)} = \lambda_k \psi_k^{(n)}, \quad (n = 1, 2, \ldots N). \qquad (2.1.37)$$

In this case, eigenfunctions $\psi_k^{(n)}$ can be chosen to be orthogonal to one another, i.e.,

$$\langle \psi_k^{(m)} | \psi_k^{(n)} \rangle = 0 \quad \text{if } m \neq n. \qquad (2.1.38)$$

Remark 2.2. By using the appropriate scaling of eigenfunctions, they can be normalized, resulting in the following relations:

$$\langle \psi_i | \psi_k \rangle = \delta_{ik}. \tag{2.1.39}$$

It will be tacitly assumed in our subsequent discussion that such a normalization is performed and the eigenfunctions form the orthonormal set $\{\psi_k\}$ of functions.

Proposition 2.3. *The set of eigenfunctions $\{\psi_k\}$ of the Hermitian operator $\hat{f}$ is complete in the sense that any element of Hilbert space can be expanded in terms of these eigenfunctions,*

$$\psi = \sum_{k=1}^{n} a_k \psi_k, \tag{2.1.40}$$

where

$$a_k = \langle \psi_k | \psi \rangle. \tag{2.1.41}$$

It is clear that the expansion coefficients a_k can be viewed as components of an infinite-dimensional vector **a** which can be interpreted as the f-representation of ψ, because ψ_k are eigenfunctions of the $\hat{f}$-operator.

As far as the proof of Proposition 2.3 is concerned, the matter is quite complicated. It turns out that the completeness of eigenfunctions can be rigorously proven for so-called "compact" Hermitian operators. This mathematical subtlety is usually ignored in the exposition of quantum mechanics and eigenfunction expansions (2.1.40) are commonly assumed to be valid.

Next, we shall discuss the multiplication of operators which has important implications in quantum mechanics. Consider two operators $\hat{f}$ and $\hat{g}$. The product $\hat{g}\hat{f}$ of these operators is the operator which consists of two consecutive mappings performed first by operator $\hat{f}$ and subsequently by operator $\hat{g}$. Mathematically, this is expressed as follows:

$$\varphi_1 = \hat{g}(\hat{f}\psi) = \hat{g}\hat{f}\psi. \tag{2.1.42}$$

Similarly, the product $\hat{f}\hat{g}$ is defined as follows:

$$\varphi_2 = \hat{f}(\hat{g}\psi) = \hat{f}\hat{g}\psi. \tag{2.1.43}$$

In general,

$$\varphi_1 \neq \varphi_2, \tag{2.1.44}$$

and

$$\hat{g}\hat{f} \neq \hat{f}\hat{g}. \qquad (2.1.45)$$

Thus, in general operators $\hat{f}$ and $\hat{g}$ do not commute. If

$$\hat{g}\hat{f} = \hat{f}\hat{g}, \qquad (2.1.46)$$

then it is said that operators $\hat{f}$ and $\hat{g}$ commute. The commutator of two operators

$$\boxed{\left[\hat{f},\hat{g}\right] = \hat{f}\hat{g} - \hat{g}\hat{f}} \qquad (2.1.47)$$

is often used to check the commutativity of operators. It is clear that two operators $\hat{f}$ and $\hat{g}$ commute if and only if

$$\left[\hat{f},\hat{g}\right] = 0. \qquad (2.1.48)$$

Proposition 2.4. *Two Hermitian operators $\hat{f}$ and $\hat{g}$ commute if and only if they have the same set of eigenfunctions.*

Proof.

A) Consider the case when two operators $\hat{f}$ and $\hat{g}$ have the same set of eigenfunctions $\{\psi_n\}$. This implies that

$$\hat{f}\psi_n = f_n\psi_n \quad \text{and} \quad \hat{g}\psi_n = g_n\psi_n, \qquad (2.1.49)$$

where f_n and g_n are used as notations for eigenvalues of operators $\hat{f}$ and $\hat{g}$, respectively. Consider an arbitrary element (vector) ψ of Hilbert space and expand it in terms of eigenfunctions $\{\psi_n\}$:

$$\psi = \sum_{n=1}^{\infty} a_n\psi_n. \qquad (2.1.50)$$

Then,

$$\hat{f}\psi = \sum_{n=1}^{\infty} a_n\hat{f}\psi_n = \sum_{n=1}^{\infty} a_n f_n\psi_n, \qquad (2.1.51)$$

and

$$\hat{g}(\hat{f}\psi) = \sum_{n=1}^{\infty} a_n f_n\hat{g}\psi_n = \sum_{n=1}^{\infty} a_n f_n g_n\psi_n. \qquad (2.1.52)$$

Similarly, we find

$$\hat{g}\psi = \sum_{n=1}^{\infty} a_n\hat{g}\psi_n = \sum_{n=1}^{\infty} a_n g_n\psi_n, \qquad (2.1.53)$$

and

$$\hat{f}(\hat{g}\psi) = \sum_{n=1}^{\infty} a_n g_n \hat{f}\psi_n = \sum_{n=1}^{\infty} a_n g_n f_n \psi_n = \sum_{n=1}^{\infty} a_n f_n g_n \psi_n. \qquad (2.1.54)$$

It is clear from formulas (2.1.52) and (2.1.54) that

$$\hat{g}(\hat{f}\psi) = \hat{f}(\hat{g}\psi). \qquad (2.1.55)$$

Thus, it is proven that operators $\hat{f}$ and $\hat{g}$ commute.

B) Suppose that operators $\hat{f}$ and $\hat{g}$ commute, and we intend to prove that they have the same set of eigenfunctions. For the sake of simplicity, we shall provide the proof for the case of simple (nondegenerate) eigenvalues, i.e., for the case when there exists only one eigenfunction ψ_k for any eigenvalues g_k and f_k. The proof proceeds as follows.

From equation

$$\hat{f}\psi_k = f_k \psi_k \qquad (2.1.56)$$

follows that

$$\hat{g}(\hat{f}\psi_k) = f_k \hat{g}\psi_k. \qquad (2.1.57)$$

Since it is assumed that operators $\hat{f}$ and $\hat{g}$ commute, we have

$$\hat{g}(\hat{f}\psi_k) = \hat{f}(\hat{g}\psi_k). \qquad (2.1.58)$$

From the last two formulas, we find

$$\hat{f}(\hat{g}\psi_k) = f_k \hat{g}\psi_k, \qquad (2.1.59)$$

which means that $\hat{g}\psi_k$ is the eigenfunction of operator $\hat{f}$ corresponding to the eigenvalue f_k. Since for eigenvalue f_k there exists only one eigenfunction ψ_k, formula (2.1.59) implies that $\hat{g}\psi_k$ is a scaled version of ψ_k:

$$\hat{g}\psi_k = g_k \psi_k. \qquad (2.1.60)$$

Thus, it is proven that operators $\hat{f}$ and $\hat{g}$ have the same set of eigenfunctions.

$$\square$$

We conclude this section with the brief discussion of the case of the continuous spectrum when eigenvalues are not separated from one another and cannot be numerated. In this case, it is convenient to write the eigenvalue equation (2.1.27) in the form

$$\hat{f}\psi_f = f\psi_f, \qquad (2.1.61)$$

where f stands for a particular eigenvalue of operator $\hat{f}$, while ψ_f is an eigenfunction corresponding to this eigenvalue.

In the case of the continuous spectrum, summation in eigenfunction expansion (2.1.40) is replaced by integration

$$\psi(\mathbf{r}) = \int a(f)\psi_f(\mathbf{r})\,df, \tag{2.1.62}$$

and the expansion coefficient (function) $a(f')$ is given by the formula

$$a(f') = \int \psi_{f'}^*(\mathbf{r})\psi(\mathbf{r})\,dV, \tag{2.1.63}$$

which is similar to formula (2.1.41). By substituting formula (2.1.62) into the last equation, we find

$$a(f') = \int \psi_{f'}^*(\mathbf{r})\left[\int a(f)\psi_f(\mathbf{r})\,df\right]dV. \tag{2.1.64}$$

By changing the order of integration in the last formula, we arrive at

$$a(f') = \int a(f)\left[\int \psi_{f'}^*(\mathbf{r})\psi_f(\mathbf{r})\,dV\right]df. \tag{2.1.65}$$

It is clear from formula (2.1.65) that

$$\int \psi_{f'}^*(\mathbf{r})\psi_f(\mathbf{r})\,dV = \delta(f - f'), \tag{2.1.66}$$

where $\delta(f - f')$ is the Dirac delta function.

It follows from the last equation that

$$\int |\psi_f(\mathbf{r})|^2\,dV = \infty, \tag{2.1.67}$$

which means that the eigenfunctions of the continuous spectrum are not square-integrable and do not belong to the Hilbert space L_2. In most physical applications, this mathematical difficulty is usually circumvented by considering square-integrable superpositions of eigenfunctions of the continuous spectrum, i.e., the superpositions which are physically realizable.

2.2 Postulates of Quantum Mechanics

In this section, the postulates of quantum mechanics are stated and their immediate implications are discussed.

Postulate 1 (Wave Function Postulate).

According to this postulate, the physical state of a microparticle (electron, for instance) is fully described by a wave function which is a complex-valued function from Hilbert space. In the case of coordinate representation, the wave function is a function $\psi(\mathbf{r}, t)$ of spatial coordinates and time from L_2. The physical meaning of the wave function $\psi(\mathbf{r}, t)$ can be stated as follows. Consider an infinitesimally small volume dV "centered" at $\mathbf{r}$. Then,

$$|\psi(\mathbf{r}, t)|^2 \, dV \tag{2.2.1}$$

is the probability of finding the particle in volume dV at time t by measuring its position. In other words, $|\psi(\mathbf{r}, t)|^2$ can be viewed as the probability density of particle location at time t. This physical meaning of $\psi(\mathbf{r}, t)$ implies that

$$\int |\psi(\mathbf{r}, t)|^2 \, dV = 1, \tag{2.2.2}$$

where the integration is carried out over the entire three-dimensional space. The stated postulate is consistent with electron diffraction experiments which reveal the randomness of measured electron locations.

In the case of a system of N microparticles, the physical state of this system is described by a wave function $\psi(\mathbf{r}_1, \mathbf{r}_2, \ldots \mathbf{r}_N, t)$ in $3N$-dimensional configuration space and

$$|\psi(\mathbf{r}_1, \mathbf{r}_2, \ldots \mathbf{r}_N, t)|^2 \, dV_{3N} \tag{2.2.3}$$

is the probability of the measured system localization in the infinitesimally small volume dV_{3N} of $3N$-dimensional space. Thus, $|\psi(\mathbf{r}_1, \mathbf{r}_2, \ldots \mathbf{r}_N, t)|^2$ can be interpreted as the probability density of the microsystem localization in $3N$-dimensional configuration space. As before, it is clear that the physical meaning of $\psi(\mathbf{r}_1, \mathbf{r}_2, \ldots, \mathbf{r}_N, t)$ implies that

$$\int |\psi(\mathbf{r}_1, \mathbf{r}_2, \ldots, \mathbf{r}_N, t)|^2 \, dV_{3N} = 1, \tag{2.2.4}$$

where the integration is performed over the entire configuration space.

Postulate 2 (Superposition Postulate).

According to this postulate, if any two wave functions ψ_1 and ψ_2 describe some possible states of a microparticle (or a system of microparticles), then any linear combination

$$\psi = \alpha_1 \psi_1 + \alpha_2 \psi_2 \tag{2.2.5}$$

normalized according to formula (2.2.2) (or formula (2.2.4)) also describes a possible state of microparticle (or microsystem).

This postulate implies that the set of all possible wave functions is a linear space. It also suggests that the time-evolution of the wave function should be described by a linear equation.

Postulate 3 (Causality Postulate).

This postulate states that the wave function at any instant of time t uniquely defines the time-evolution of the wave function at subsequent (future) instants of time. This means that there is one-to-one correspondence between the wave function at the given instant of time t and its time derivative at the same instant of time. Mathematically, this can be expressed by the equation

$$\frac{\partial \psi}{\partial t} = \hat{K}\psi, \tag{2.2.6}$$

where $\hat{K}$ is the operator realizing the above-mentioned one-to-one correspondence.

It is convenient to represent operator $\hat{K}$ in the form

$$\hat{K} = \frac{1}{i\hbar}\hat{H}, \tag{2.2.7}$$

where $i = \sqrt{-1}$, $\hbar$ is normalized by 2π Planck constant

$$\hbar = 1.05 \cdot 10^{-34} \text{J} \cdot \text{s} \tag{2.2.8}$$

and the operator $\hat{H}$ is called the Hamiltonian operator. It is shown later that the use of "i" in formula (2.2.7) results in the operator $\hat{H}$ being Hermitian, while $\hbar$ is introduced in formula (2.2.7) to properly reflect the scale of energy on the microscopic level.

By substituting formula (2.2.7) into equation (2.2.6) we end up with

$$\boxed{i\hbar\frac{\partial \psi}{\partial t} = \hat{H}\psi.} \tag{2.2.9}$$

This is the time-dependent Schrödinger equation which describes the evolution of the wave function in time.

Postulate 4 (Quantization as an Eigenvalue Problem).

According to this postulate, each physical quantity f is associated with a linear Hermitian operator $\hat{f}$ and the only possible values of quantity f are the eigenvalues f_n of operator $\hat{f}$:

$$\hat{f}\psi_n = f_n\psi_n, \qquad (2.2.10)$$

and each eigenfunction ψ_n describes the state in which the measurement of f results only in value f_n. This postulate is consistent with experimental facts such as observed discrete energy levels in atoms (and other microsystems) as well as observed quantized values of angular momentum.

Next, consider another physical quantity g and its corresponding Hermitian operator $\hat{g}$. As discussed in the previous section of this chapter, two Hermitian operators $\hat{f}$ and $\hat{g}$ commute if and only if they have the same set of eigenfunctions. These eigenfunctions describe the states in which the simultaneous measurements of f and g produce **certain** results equal to the eigenvalues of operators $\hat{f}$ and $\hat{g}$, respectively. In this sense, the physical quantities f and g are simultaneously measurable. Thus, **we conclude that the commutativity of operators of physical quantities implies their simultaneous measurability**.

Usually, several (more than two) physical quantities can be simultaneously measurable. These simultaneously measurable physical quantities form a complete set if the common eigenfunctions of their operators are **uniquely** defined. In other words, the certain results of simultaneous measurements of these quantities uniquely define the quantum mechanical states which are common eigenfunctions of operators of measured physical quantities.

Postulate 5 (Measurement Postulate).

Consider a physical quantity f and some state (i.e., wave function) ψ, which is not an eigenfunction of operator $\hat{f}$. This wave function can be expanded in terms of eigenfunctions ψ_n of operator $\hat{f}$:

$$\psi = \sum_n c_n\psi_n. \qquad (2.2.11)$$

According to the Measurement Postulate, the result of measurement of physical quantity f in the state described by the wave function ψ cannot be predicted with certainty. It is possible to predict only the probabilities of different results of measurement. Namely, the probability of the result $f = f_n$ is equal to $|c_n|^2$, that is:

$$\boxed{|c_n|^2 = \text{Prob(measurement result of } f \text{ is } f_n).} \qquad (2.2.12)$$

The measurement process in quantum mechanics also irreversibly affects the state of the microscopic particle (or microsystem). Namely, the Measurement Postulate states that after the measurement a new state of the microscopic particle emerges whose wave function coincides with the eigenfunction ψ_n corresponding to the measurement result $f = f_n$, that is

$$\boxed{\psi \Rightarrow \psi_n.} \qquad (2.2.13)$$

It is apparent that the emergence of new state ψ_n is not deterministic but rather random in nature to be consistent with the randomness of the measurement result $f = f_n$. This physical phenomenon is coined as "random collapse" of the wave function.

It is clear from the previous discussion that the result of measurement of physical quantity f in the state ψ defined by formula (2.2.11) is not predictable with certainty. In other words, the physical quantity f does not have a definite value in the state ψ. However, it is possible to speak about average (mean) value $\bar{f}$ of the physical quantity f in the state ψ. According to formula (2.2.12), this mean value can be computed as:

$$\boxed{\bar{f} = \sum_n f_n |c_n|^2.} \qquad (2.2.14)$$

The last formula can be transformed to compute $\bar{f}$ without invoking the eigenfunction expansion (2.2.11) but using directly the state wave function ψ. Namely, we intend to demonstrate that $\bar{f}$ can be computed by using the formula

$$\boxed{\bar{f} = \langle \psi | \hat{f} \psi \rangle = \langle \psi | \hat{f} | \psi \rangle.} \qquad (2.2.15)$$

The demonstration proceeds as follows

$$\bar{f} = \langle \psi | \hat{f} \psi \rangle = \Big\langle \sum_i c_i \psi_i \Big| \hat{f} \Big| \sum_n c_n \psi_n \Big\rangle$$

$$= \Big\langle \sum_i c_i \psi_i \Big| \sum_n c_n \hat{f} \psi_n \Big\rangle = \Big\langle \sum_i c_i \psi_i \Big| \sum_n f_n c_n \psi_n \Big\rangle$$

$$= \sum_n \sum_i f_n c_n c_i^* \langle \psi_i | \psi_n \rangle. \qquad (2.2.16)$$

Now, by taking into account the orthonormality of eigenfunctions ψ_n (see formula (2.1.39)), we conclude that formulas (2.2.14) and (2.2.15) are indeed equivalent.

Previously, we discussed the Measurement Postulate for the case of the discrete spectrum. If the spectrum of physical quantity f is continuous, then the series expansion (2.2.11) must be replaced by the integral expansion

$$\psi(\mathbf{r}) = \int c(f)\psi_f(\mathbf{r})\,df. \tag{2.2.17}$$

In this case $|c(f_0)|^2 df$ is the probability that the measurement result will be within the interval

$$f_0 \le f \le f_0 + df. \tag{2.2.18}$$

Furthermore, formula (2.2.14) is replaced by the following one:

$$\bar{f} = \int f|c(f)|^2\,df. \tag{2.2.19}$$

This concludes the description of the main postulates of quantum mechanics. It has been demonstrated over the years that the quantum mechanics based on these postulates has the remarkable predictive power, that is, the ability to produce experimentally verifiable predictions. Thus, it can be concluded that the validity of these postulates has been well established. However, it must be noted that from the physical and logical points of view the Measurement Postulate is the source of ongoing controversy and scientific debate. The origin of the controversy is the fact that the Causality Postulate and equation (2.2.9) indicate that the time-evolution of a state wave function is purely deterministic, while the Measurement Postulate prescribes that during the measurement process the state wave function undergoes a "random collapse".

Now, we consider some immediate consequences of the stated postulates.

Proposition 2.5. *Operator $\hat{H}$ in the dynamic equation (2.2.9) is linear.*

Proof. Consider the possible states ψ_1 and ψ_2 of the quantum mechanical system (or particle). This means that these states evolve in time according to equation (2.2.9). Consequently,

$$i\hbar\frac{\partial\psi_1}{\partial t} = \hat{H}\psi_1 \quad \text{and} \quad i\hbar\frac{\partial\psi_2}{\partial t} = \hat{H}\psi_2 \tag{2.2.20}$$

Then, according to the Superposition Postulate, any normalized linear combination

$$\psi = \alpha_1\psi_1 + \alpha_2\psi_2 \tag{2.2.21}$$

is also a possible state wave function which evolves in time according to equation

$$i\hbar\frac{\partial\psi}{\partial t} = \hat{H}\psi. \tag{2.2.22}$$

From the last two formulas follows that

$$\alpha_1 i\hbar \frac{\partial \psi_1}{\partial t} + \alpha_2 i\hbar \frac{\partial \psi_2}{\partial t} = \hat{H}(\alpha_1 \psi_1 + \alpha_2 \psi_2). \tag{2.2.23}$$

Finally, by using formulas (2.2.20) in the left hand side of equation (2.2.23), we arrive at

$$\alpha_1 \hat{H}\psi_1 + \alpha_2 \hat{H}\psi_2 = \hat{H}(\alpha_1 \psi_1 + \alpha_2 \psi_2), \tag{2.2.24}$$

which proves that $\hat{H}$ is a linear operator.   □

Proposition 2.6. *Operator $\hat{H}$ is Hermitian.*

Proof consists of two distinct steps.

Step 1. According to the probabilistic interpretation of the wave function, its L_2-norm does not change with time (see formula (2.2.2)). This means that the time evolution of the wave function is unitary. From formula (2.2.2), we find

$$\frac{d}{dt} \int |\psi(\mathbf{r}, t)|^2 \, dV = \frac{d}{dt} \int \psi^*(\mathbf{r}, t)\psi(\mathbf{r}, t) \, dV = 0. \tag{2.2.25}$$

The last equation implies that

$$\int \frac{\partial \psi^*}{\partial t} \psi \, dV + \int \psi^* \frac{\partial \psi}{\partial t} \, dV = 0. \tag{2.2.26}$$

By using equation (2.2.9) in the last formula, we find

$$\frac{i}{\hbar} \int \left(\hat{H}\psi\right)^* \psi \, dV - \frac{i}{\hbar} \int \psi^* \hat{H}\psi \, dV = 0, \tag{2.2.27}$$

which leads to

$$\langle \hat{H}\psi | \psi \rangle = \langle \psi | \hat{H}\psi \rangle. \tag{2.2.28}$$

This implies that for any ψ, the inner product $\langle \hat{H}\psi | \psi \rangle$ is a real number.

Step 2. The completion of the proof is based on the following mathematical identity which is valid for any ψ_1 and ψ_2:

$$\langle \hat{H}\psi_1 | \psi_2 \rangle$$
$$= \frac{1}{4}\{\langle \hat{H}(\psi_1 + \psi_2) | \psi_1 + \psi_2 \rangle - \langle \hat{H}(\psi_1 - \psi_2) | \psi_1 - \psi_2 \rangle$$
$$- i[\langle \hat{H}(\psi_1 + i\psi_2) | \psi_1 + i\psi_2 \rangle - \langle \hat{H}(\psi_1 - i\psi_2) | \psi_1 - i\psi_2 \rangle]\}. \tag{2.2.29}$$

By using the properties of the inner product in Hilbert space, the right-hand side of this identity can be simplified as follows:

$$\langle \hat{H}\psi_1 | \psi_2 \rangle = \frac{1}{2}\{\langle \hat{H}\psi_1 | \psi_2 \rangle + \langle \hat{H}\psi_2 | \psi_1 \rangle - i[i\langle \hat{H}\psi_1 | \psi_2 \rangle - i\langle \hat{H}\psi_2 | \psi_1 \rangle]\}. \tag{2.2.30}$$

From formula (2.2.30) follows the validity of (2.2.29). Furthermore, by using the last formula, it is easy to confirm the validity of the following mathematical identity:

$$\langle \hat{H}\psi_2|\psi_1 \rangle$$
$$= \frac{1}{4}\{\langle \hat{H}(\psi_1 + \psi_2)|\psi_1 + \psi_2 \rangle - \langle \hat{H}(\psi_1 - \psi_2)|\psi_1 - \psi_2 \rangle$$
$$+ i[\langle \hat{H}(\psi_1 + i\psi_2)|\psi_1 + i\psi_2 \rangle - \langle \hat{H}(\psi_1 - i\psi_2)|\psi_1 - i\psi_2 \rangle]\}. \quad (2.2.31)$$

According to the result of Step 1, all inner products in the right hand sides of identities (2.2.29) and (2.2.31) are real. This implies that

$$\langle \hat{H}\psi_1|\psi_2 \rangle = \langle \hat{H}\psi_2|\psi_1 \rangle^*. \quad (2.2.32)$$

Consequently,

$$\langle \hat{H}\psi_1|\psi_2 \rangle = \langle \psi_1|\hat{H}\psi_2 \rangle. \quad (2.2.33)$$

Thus, the operator $\hat{H}$ is Hermitian. □

Next, we consider the conservation of a physical quantity in quantum mechanics. In classical mechanics, a physical quantity is conserved if its value does not change (i.e., remains constant) during the dynamics of the mechanical system. This notion of conservation is not literally applicable in quantum mechanics. This is because, if some physical quantity has a definite value at some instant of time, it may not have definite values in subsequent (future) instants of time. The reason is that if the wave function is an eigenfunction of the operator of a physical quantity at some instant of time, it may not be the eigenfunction of the same operator at subsequent instants of time. Thus, the notion of conservation of a physical quantity in quantum mechanics must be modified. It is said that the physical quantity is conserved in quantum mechanics if its mean (average) value $\overline{f(t)}$ does not change with time,

$$\overline{f(t)} = const, \quad (2.2.34)$$

which is equivalent to

$$\frac{d\overline{f(t)}}{dt} = 0. \quad (2.2.35)$$

To derive the condition for the conservation of a physical quantity in quantum mechanics, we now invoke formula (2.2.15). From that formula and (2.2.35), we find

$$\frac{d\overline{f(t)}}{dt} = \langle \frac{\partial \psi}{\partial t}|\hat{f}\psi \rangle + \langle \psi|\frac{\partial \hat{f}}{\partial t}\psi \rangle + \langle \psi|\hat{f}\frac{\partial \psi}{\partial t} \rangle = 0, \quad (2.2.36)$$

where the second term in the right-hand side of the last formula appears because operator $\hat{f}$ may explicitly depend on time.

By using equation (2.2.9), the last formula can be transformed as follows:

$$\frac{i}{\hbar}\langle \hat{H}\psi|\hat{f}\psi\rangle + \langle\psi|\frac{\partial \hat{f}}{\partial t}\psi\rangle - \frac{i}{\hbar}\langle\psi|\hat{f}\hat{H}\psi\rangle = 0. \qquad (2.2.37)$$

Taking into account that $\hat{H}$ is Hermitian and combining the three inner products into one, we obtain

$$\langle\psi|\left[\frac{\partial \hat{f}}{\partial t} - \frac{i}{\hbar}\left(\hat{f}\hat{H} - \hat{H}\hat{f}\right)\right]\psi\rangle = 0. \qquad (2.2.38)$$

Since the last equality is valid for any ψ, this means that

$$\frac{\partial \hat{f}}{\partial t} - \frac{i}{\hbar}\left(\hat{f}\hat{H} - \hat{H}\hat{f}\right) = 0. \qquad (2.2.39)$$

By recalling (see previous section) that

$$\hat{f}\hat{H} - \hat{H}\hat{f} = \left[\hat{f}, \hat{H}\right] \qquad (2.2.40)$$

is the commutator, the formula (2.2.39) can be written as follows:

$$\boxed{\frac{\partial \hat{f}}{\partial t} - \frac{i}{\hbar}\left[\hat{f}, \hat{H}\right] = 0.} \qquad (2.2.41)$$

This is the condition for the conservation of physical quantity f in quantum mechanics. The mathematical form of this condition is analogous to the condition of conservation of a physical quantity in classical mechanics discussed in the last section of the previous chapter. It is clear that the commutator in the last formula plays the role analogous to the Poisson bracket in classical Hamiltonian mechanics.

In the case when the operator $\hat{f}$ does not depend on time explicitly, i.e., when $\frac{\partial \hat{f}}{\partial t} = 0$, the conservation condition (2.2.41) is simplified and it is reduced to

$$\boxed{\left[\hat{f}, \hat{H}\right] = 0.} \qquad (2.2.42)$$

Thus, in this case **the commutativity of operator $\hat{f}$ with the Hamiltonian guarantees the conservation of f.**

Now, we shall demonstrate that the Hamiltonian operator can be viewed as the operator of energy. To this end, consider a closed quantum mechanical system. It is clear that for a closed system time is homogeneous. This means that the form of the evolution equation (2.2.9) for wave function ψ should not depend on the choice of origin of time. In other words, this equation must be invariant with respect to arbitrary translations in time. This will be the case when the Hamiltonian does not depend on time explicitly, that is, when

$$\frac{\partial \hat{H}}{\partial t} = 0. \tag{2.2.43}$$

Since the Hamiltonian commutes with itself,

$$\left[\hat{H}, \hat{H}\right] = 0, \tag{2.2.44}$$

we find that the conservation condition (2.2.41) is fulfilled for the Hamiltonian. Thus, if the Hamiltonian operator is the operator of some physical quantity, the conservation of this physical quantity for a closed system follows from the homogeneity of time. Like in classical mechanics, this physical quantity is energy $\mathcal{E}$, and the Hamiltonian is the energy operator. Now, equation (2.2.9) reveals that the time evolution of the wave function is controlled by the energy operator. This situation is very analogous to the Hamiltonian equations in classical mechanics.

Since the Hamiltonian $\hat{H}$ is the energy operator, the possible (measurable) values of energy $\mathcal{E}_n$ are eigenvalues of this operator:

$$\hat{H}\psi_n(\mathbf{r}, t) = \mathcal{E}_n \psi_n(\mathbf{r}, t). \tag{2.2.45}$$

These eigenvalues form the energy spectrum.

According to equation (2.2.9), we have

$$i\hbar \frac{\partial \psi_n(\mathbf{r}, t)}{\partial t} = \hat{H}\psi_n(\mathbf{r}, t). \tag{2.2.46}$$

From the last two formulas, we find

$$i\hbar \frac{\partial \psi_n(\mathbf{r}, t)}{\partial t} = \mathcal{E}_n \psi_n(\mathbf{r}, t). \tag{2.2.47}$$

By integrating the last equation, we obtain

$$\boxed{\psi_n(\mathbf{r}, t) = \Phi_n(\mathbf{r}) e^{-\frac{i}{\hbar}\mathcal{E}_n t}.} \tag{2.2.48}$$

Thus, the eigenstates of the energy operator are time-harmonic functions of time.

By substituting the last formula into equation (2.2.45), we end up with

$$\boxed{\hat{H}\Phi_n(\mathbf{r}) = \mathcal{E}_n \Phi_n(\mathbf{r}).} \tag{2.2.49}$$

This is the so-called time-independent Schrödinger equation, which is used for the calculation of the energy spectrum.

From formula (2.2.48) follows that

$$|\psi_n(\mathbf{r}, t)|^2 = |\Phi_n(\mathbf{r})|^2. \tag{2.2.50}$$

Similarly, if operator $\hat{f}$ does not depend on time, we find

$$\bar{f} = \langle \hat{f}\psi_n | \psi_n \rangle = \langle \hat{f}\Phi_n | \Phi_n \rangle. \tag{2.2.51}$$

Formulas (2.2.50) and (2.2.51) reveal that in **energy eigenstates the probability density of coordinates as well as average value $\bar{f}$ are constant in time. For this reason, the energy eigenstates are called stationary states**.

In this section, the postulates of quantum mechanics and their immediate consequences have been discussed. In the next three sections, the expressions for operators of coordinates, momentum, angular momentum and energy are studied.

2.3   Operators of Coordinates and Momentum

Consider a microscopic particle whose quantum mechanical state is described by the wave function $\psi(\mathbf{r}, t)$. According to the physical meaning of $|\psi(\mathbf{r}, t)|^2$ as the probability density of (measured) particle location, the mean (average) location $\bar{\mathbf{r}}$ of the microparticle can be computed by using the formula

$$\bar{\mathbf{r}} = \int \mathbf{r}|\psi(\mathbf{r}, t)|^2 \, dV. \tag{2.3.1}$$

This formula can be further transformed as follows:

$$\bar{\mathbf{r}} = \int \psi^*(\mathbf{r}, t)\mathbf{r}\psi(\mathbf{r}, t) \, dV = \langle \psi | \mathbf{r}\psi \rangle. \tag{2.3.2}$$

Let $\hat{\mathbf{r}}$ be the notation for the operator of particle coordinates. Then, according to formula (2.2.15) from the previous section, we have

$$\bar{\mathbf{r}} = \langle \psi | \hat{\mathbf{r}}\psi \rangle. \tag{2.3.3}$$

The comparison between the last two formulas leads to the conclusion that

$$\boxed{\hat{\mathbf{r}}\psi = \mathbf{r}\psi.} \tag{2.3.4}$$

Thus, $\hat{\mathbf{r}}$ is the operator of multiplication by $\mathbf{r}$. It is easy to verify that $\hat{\mathbf{r}}$ is a Hermitian operator.

Operator $\hat{\mathbf{r}}$ can be represented as

$$\hat{\mathbf{r}} = \hat{x}\mathbf{e}_x + \hat{y}\mathbf{e}_y + \hat{z}\mathbf{e}_z, \tag{2.3.5}$$

where $\hat{x}$, $\hat{y}$ and $\hat{z}$ are the operators of x, y and z coordinates, respectively. On the other hand,

$$\mathbf{r} = x\mathbf{e}_x + y\mathbf{e}_y + z\mathbf{e}_z. \tag{2.3.6}$$

By substituting the last two formulas into (2.3.4), we obtain

$$\mathbf{e}_x(\hat{x}\psi) + \mathbf{e}_y(\hat{y}\psi) + \mathbf{e}_z(\hat{z}\psi) = \mathbf{e}_x(x\psi) + \mathbf{e}_y(y\psi) + \mathbf{e}_z(z\psi). \tag{2.3.7}$$

Consequently,

$$\boxed{\hat{x}\psi = x\psi, \quad \hat{y}\psi = y\psi, \quad \hat{z}\psi = z\psi.} \tag{2.3.8}$$

Thus, $\hat{x}$, $\hat{y}$ and $\hat{z}$ are the operators of multiplication by x, y and z, respectively. It is apparent that operators $\hat{x}$, $\hat{y}$ and $\hat{z}$ commute. This means that all three particle coordinates are simultaneously measurable.

Next, we consider the spectrum of the $\hat{\mathbf{r}}$-operator. Let $\mathbf{r}_0$ be an eigenvalue of this operator and $\psi_{\mathbf{r}_0}(\mathbf{r})$ be the corresponding eigenfunction. Then,

$$\hat{\mathbf{r}}\psi_{\mathbf{r}_0}(\mathbf{r}) = \mathbf{r}_0\psi_{\mathbf{r}_0}(\mathbf{r}). \tag{2.3.9}$$

By using formula (2.3.4) in the last equation, we find

$$\mathbf{r}\psi_{\mathbf{r}_0}(\mathbf{r}) = \mathbf{r}_0\psi_{\mathbf{r}_0}(\mathbf{r}), \tag{2.3.10}$$

which leads to

$$(\mathbf{r} - \mathbf{r}_0)\,\psi_{\mathbf{r}_0}(\mathbf{r}) = 0. \tag{2.3.11}$$

The last formula implies that $\psi_{\mathbf{r}_0}(\mathbf{r}) = 0$ if $\mathbf{r} \neq \mathbf{r}_0$ and $\psi_{\mathbf{r}_0}(\mathbf{r})$ must be infinite at $\mathbf{r} = \mathbf{r}_0$ in order for the integral of $\psi_{\mathbf{r}_0}(\mathbf{r})$ to be finite. This suggests that

$$\boxed{\psi_{\mathbf{r}_0}(\mathbf{r}) = \delta(\mathbf{r}_0 - \mathbf{r}),} \tag{2.3.12}$$

where $\delta(\mathbf{r}_0 - \mathbf{r}) = \delta(\mathbf{r} - \mathbf{r}_0)$ is the Dirac delta function.

The last formula can be justified as follows. Consider an arbitrary wave function $\psi(\mathbf{r}, t)$ and its integral expansion in terms of eigenfunctions $\psi_{\mathbf{r}_0}(\mathbf{r})$:

$$\psi(\mathbf{r}, t) = \int c(\mathbf{r}_0, t)\psi_{\mathbf{r}_0}(\mathbf{r})\,dV_{\mathbf{r}_0}, \tag{2.3.13}$$

where the integration is performed with respect to $\mathbf{r}_0$-coordinates.

It is clear that $|c(\mathbf{r}_0, t)|^2$ and $|\psi(\mathbf{r}_0, t)|^2$ have the same physical meaning of probability density of the microparticle location. Consequently, the choice can be made that

$$c(\mathbf{r}_0, t) = \psi(\mathbf{r}_0, t). \tag{2.3.14}$$

By substituting the last formula into equation (2.3.13), we find

$$\psi(\mathbf{r}, t) = \int \psi(\mathbf{r}_0, t)\psi_{\mathbf{r}_0}(\mathbf{r})\, dV_{\mathbf{r}_0}. \tag{2.3.15}$$

The last equality implies the validity formula (2.3.12).

The previous discussion reveals that the **spectrum of microparticle coordinates is continuous. In other words, coordinates are not quantized.**

Next, we turn to the discussion of the momentum operator. This discussion will be based on the specific commutation relation associated with symmetry transformations.

Consider some change of spatial coordinates:

$$\mathbf{r} = \boldsymbol{\lambda}(\mathbf{r}'). \tag{2.3.16}$$

This coordinate change transforms a wave function $\psi(\mathbf{r}, t)$ into another wave function $\psi'(\mathbf{r}', t)$:

$$\psi(\mathbf{r}, t) = \psi(\boldsymbol{\lambda}(\mathbf{r}'), t) = \psi'(\mathbf{r}', t). \tag{2.3.17}$$

This transformation of wave functions generates some operator $\hat{T}$ which maps ψ into ψ':

$$\psi' = \hat{T}\psi. \tag{2.3.18}$$

This coordinate transformation is called a symmetry transformation if it does not change the mathematical form of the time-dependent Schrödinger equation (2.2.9). In other words, Schrödinger equations for wave functions ψ' and ψ have the same mathematical form. Namely,

$$i\hbar \frac{\partial \psi'}{\partial t} = \hat{H}\psi', \tag{2.3.19}$$

$$i\hbar \frac{\partial \psi}{\partial t} = \hat{H}\psi. \tag{2.3.20}$$

Proposition 2.7. *If operator $\hat{T}$ corresponds to the symmetry transformation, then the Hamiltonian $\hat{H}$ and $\hat{T}$ commute.*

Proof. By using formulas (2.3.18) and (2.3.20), we derive

$$i\hbar \frac{\partial \psi'}{\partial t} = i\hbar \frac{\partial}{\partial t}\left(\hat{T}\psi\right) = i\hbar \hat{T}\frac{\partial \psi}{\partial t} = \hat{T}\hat{H}\psi. \tag{2.3.21}$$

On the other hand,

$$\hat{H}\psi' = \hat{H}\hat{T}\psi. \tag{2.3.22}$$

By substituting formulas (2.3.21) and (2.3.22) into equation (2.3.19), we obtain

$$\hat{T}\hat{H}\psi = \hat{H}\hat{T}\psi. \tag{2.3.23}$$

This means that operators $\hat{T}$ and $\hat{H}$ commute and their commutator is equal to zero:

$$\left[\hat{T}, \hat{H}\right] = 0. \tag{2.3.24}$$

□

Now, we shall use the commutativity relation (2.3.24) to introduce the momentum operator. Consider a **closed** system of N microparticles. This system is described by a wave function $\psi(\mathbf{r}_1, \mathbf{r}_2, \ldots \mathbf{r}_N, t)$. For a closed system, space is homogeneous. This means that the mathematical form of the time-dependent Schrödinger equation (2.2.9) should not depend on the choice of the origin of spatial coordinates. In other words, this equation must be invariant with respect to arbitrary translations (shifts) in space. This implies that spatial translations are symmetry transformations for closed systems and the operators of these translations must commute with the Hamiltonian.

Consider an infinitesimally small translation $\boldsymbol{\delta}$:

$$\mathbf{r}'_k = \mathbf{r}_k + \boldsymbol{\delta}, \quad (k = 1, 2, \ldots N). \tag{2.3.25}$$

Then,

$$\psi'(\mathbf{r}'_1, \mathbf{r}'_2, \ldots \mathbf{r}'_N, t) = \psi(\mathbf{r}_1 + \boldsymbol{\delta}, \mathbf{r}_2 + \boldsymbol{\delta}, \ldots \mathbf{r}_N + \boldsymbol{\delta}, t)$$

$$= \psi(\mathbf{r}_1, \mathbf{r}_2, \ldots \mathbf{r}_N, t) + \boldsymbol{\delta} \cdot \sum_{k=1}^{N} \nabla_k \psi(\mathbf{r}_1, \mathbf{r}_2, \ldots \mathbf{r}_N, t), \tag{2.3.26}$$

where subscript "k" in ∇_k implies that the differentiation is performed with respect to spatial coordinates of $\mathbf{r}_k$.

The last formula can be written as follows:

$$\psi' = \hat{T}\psi, \tag{2.3.27}$$

where $\hat{T}$ is the operator of infinitesimally small spatial translation, which has the form

$$\hat{T} = \hat{I} + \boldsymbol{\delta} \cdot \sum_{k=1}^{N} \nabla_k, \tag{2.3.28}$$

with $\hat{I}$ being the identity operator ($\hat{I}\psi = \psi$).

Since translations are symmetry transformations, operator $\hat{T}$ commutes with the Hamiltonian. Taking into account that the identity operator $\hat{I}$ commutes with any operator, we conclude that operator $\sum_{k=1}^{N} \nabla_k$ commutes with the Hamiltonian:

$$\hat{H}\left(\sum_{k=1}^{N} \nabla_k\right) - \left(\sum_{k=1}^{N} \nabla_k\right)\hat{H} = 0. \qquad (2.3.29)$$

Consequently, if some physical quantity is associated with operator $\sum_{k=1}^{N} \nabla_k$, this physical quantity is conserved and its conservation follows from the homogeneity of space for a closed system. As in classical mechanics (see previous chapter), this physical quantity is momentum. For the operator $\sum_{k=1}^{N} \nabla_k$ to be the operator of a physical quantity, it must be Hermitian. This can be achieved by multiplying $\sum_{k=1}^{N} \nabla_k$ by $(-i)$. Furthermore, the factor $\hbar$ is introduced to achieve the proper dimension of momentum as well as to properly reflect the scale of momentum at the microscopic level. Thus, the momentum operator $\hat{\mathbf{P}}$ of a microsystem of particles is given by the formula

$$\hat{\mathbf{P}} = -i\hbar \sum_{k=1}^{N} \nabla_k. \qquad (2.3.30)$$

For a single microparticle, the momentum operator is

$$\hat{\mathbf{p}} = -i\hbar\nabla. \qquad (2.3.31)$$

The last formula can be written as follows:

$$\hat{\mathbf{p}} = \mathbf{e}_x\hat{p}_x + \mathbf{e}_y\hat{p}_y + \mathbf{e}_z\hat{p}_z = -i\hbar\left(\mathbf{e}_x\frac{\partial}{\partial x} + \mathbf{e}_y\frac{\partial}{\partial y} + \mathbf{e}_z\frac{\partial}{\partial z}\right). \qquad (2.3.32)$$

Consequently,

$$\hat{p}_x = -i\hbar\frac{\partial}{\partial x}, \quad \hat{p}_y = -i\hbar\frac{\partial}{\partial y}, \quad \hat{p}_z = -i\hbar\frac{\partial}{\partial z}. \qquad (2.3.33)$$

As a simple exercise, it is suggested to prove that operators $\hat{p}_x$, $\hat{p}_y$ and $\hat{p}_z$ are Hermitian:

$$\langle\hat{p}_x\psi_1|\psi_2\rangle = \langle\psi_1|\hat{p}_x\psi_2\rangle, \quad \langle\hat{p}_y\psi_1|\psi_2\rangle = \langle\psi_1|\hat{p}_y\psi_2\rangle, \quad \langle\hat{p}_z\psi_1|\psi_2\rangle = \langle\psi_1|\hat{p}_z\psi_2\rangle. \qquad (2.3.34)$$

The proof reveals the importance of factor "i" in achieving hermiticity of the above operators.

It is clear from formulas (2.3.33) that operators $\hat{p}_x$, $\hat{p}_y$ and $\hat{p}_z$ commute:

$$\hat{p}_x\hat{p}_y = \hat{p}_y\hat{p}_x, \quad \hat{p}_x\hat{p}_z = \hat{p}_z\hat{p}_x, \quad \hat{p}_y\hat{p}_z = \hat{p}_z\hat{p}_y. \qquad (2.3.35)$$

This means that all three components of momentum are simultaneously measurable.

Next, consider the spectrum of the momentum operator $\hat{\mathbf{p}}$. Let $\mathbf{p}$ be an eigenvalue of this operator and $\psi_{\mathbf{p}}(\mathbf{r})$ be the corresponding eigenfunction. Then,

$$\hat{\mathbf{p}}\psi_{\mathbf{p}}(\mathbf{r}) = \mathbf{p}\psi_{\mathbf{p}}(\mathbf{r}). \tag{2.3.36}$$

By using formula (2.3.31), the last equation can be written as

$$-i\hbar\nabla\psi_{\mathbf{p}}(\mathbf{r}) = \mathbf{p}\psi_{\mathbf{p}}(\mathbf{r}). \tag{2.3.37}$$

It is easy to see that the solution of this equation is given by the formula

$$\psi_{\mathbf{p}}(\mathbf{r}) = \frac{1}{(2\pi\hbar)^{\frac{3}{2}}}e^{\frac{i}{\hbar}\mathbf{p}\cdot\mathbf{r}} = \frac{1}{(2\pi\hbar)^{\frac{3}{2}}}e^{\frac{i}{\hbar}(p_x x + p_y y + p_z z)}, \tag{2.3.38}$$

where the factor $\frac{1}{(2\pi\hbar)^{\frac{3}{2}}}$ is chosen to enforce the orthonormality condition (see (2.1.66)):

$$\int \psi_{\mathbf{p}'}^*(\mathbf{r})\psi_{\mathbf{p}}(\mathbf{r})\,dV = \delta(\mathbf{p} - \mathbf{p}'). \tag{2.3.39}$$

The presented discussion reveals that **the spectrum of momentum is continuous. In other words, momentum is not quantized.**

It is apparent from formula (2.3.38) that

$$|\psi_{\mathbf{p}}(\mathbf{r})|^2 = \frac{1}{(2\pi\hbar)^3} = const. \tag{2.3.40}$$

The latter means that, if a microparticle has a certain momentum $\mathbf{p}$, then the probability density of its localization anywhere in space is the same. In other words, if the particle is localized in $\mathbf{p}$-space, then it is completely spread out in $\mathbf{r}$-space. To demonstrate the converse statement, consider the integral expansion of an arbitrary wave function $\psi(\mathbf{r})$ in terms of eigenfunctions $\psi_{\mathbf{p}}(\mathbf{r})$:

$$\psi(\mathbf{r}) = \int c(\mathbf{p})\psi_{\mathbf{p}}(\mathbf{r})\,dV_{\mathbf{p}} = \frac{1}{(2\pi\hbar)^{\frac{3}{2}}}\int c(\mathbf{p})e^{\frac{i}{\hbar}\mathbf{p}\cdot\mathbf{r}}\,dV_{\mathbf{p}}, \tag{2.3.41}$$

where the function $c(\mathbf{p})$ is given by the formula

$$c(\mathbf{p}) = \int \psi_{\mathbf{p}}^*(\mathbf{r})\psi(\mathbf{r})\,dV_{\mathbf{r}} = \frac{1}{(2\pi\hbar)^{\frac{3}{2}}}\int \psi(\mathbf{r})e^{\frac{-i}{\hbar}\mathbf{p}\cdot\mathbf{r}}\,dV_{\mathbf{r}}. \tag{2.3.42}$$

It is clear that the function $c(\mathbf{p})$ can be regarded as the wave function of the microparticle in $\mathbf{p}$-representation. In this sense $|c(\mathbf{p})|^2$ has the physical

meaning of the probability density of the localization of the particle in
p-space.

Now, consider the particle which is localized in **r**-space and whose wave
function is given by the formula

$$\psi_{\mathbf{r}_0}(\mathbf{r}) = \delta(\mathbf{r} - \mathbf{r}_0). \tag{2.3.43}$$

Then, according to formula (2.3.42), the **p**-representation $c_{\mathbf{r}_0}(\mathbf{p})$ of this
wave function is

$$c_{\mathbf{r}_0}(\mathbf{p}) = \frac{1}{(2\pi\hbar)^{\frac{3}{2}}} e^{-\frac{i}{\hbar}\mathbf{p}\cdot\mathbf{r}_0}, \tag{2.3.44}$$

and

$$|c_{\mathbf{r}_0}(\mathbf{p})|^2 = \frac{1}{(2\pi\hbar)^3} = const. \tag{2.3.45}$$

This means that, if a particle has a certain spatial location $\mathbf{r}_0$, then the
probability density of its localization anywhere in **p**-space is the same. In
other words, if the particle is localized in **r**-space, it is completely spread
out in **p**-space. This clearly reveals that **r** and **p** are not simultaneously
measurable. It is clear from formulas (2.3.41) and (2.3.42) that the coordinate and momentum representations of microparticle wave functions
are related through the Fourier transform. It follows from mathematical
properties of the Fourier transform that it is impossible to achieve simultaneous coordinate and momentum localization of wave functions. In the
areas of control and communication, this phenomenon is well known for
time and frequency representations of signals. It is interesting to point out
that the momentum representation of the wave function $c(\mathbf{p})$ is isometric
to coordinate representation $\psi(\mathbf{r})$ of the same wave function. Namely:

$$\int |c(\mathbf{p})|^2 \, dV_{\mathbf{p}} = \int |\psi(\mathbf{r})|^2 \, dV_{\mathbf{r}}. \tag{2.3.46}$$

The last equality is well known in the theory of the Fourier transform as
the Plancherel identity. Due to the linearity of the Fourier transform, there
is isomorphism between the two L_2 spaces of wave functions in coordinate
and momentum representations. From the mathematical point of view,
momentum representation may be convenient because calculus operations
in coordinate representation are replaced by algebraic operations. In this
sense, the Fourier transform can be briefly summarized as isometric isomorphism of L_2 into L_2 with the algebraization of calculus.

The fact that **r** and **p** are not simultaneously measurable at the microscopic level is reflected in the commutation relations for $\hat{\mathbf{r}}$ and $\hat{\mathbf{p}}$ operators. We shall first demonstrate this for $\hat{x}$ and $\hat{p}_x$ operators. It is clear that

$$\hat{x}\hat{p}_x\psi = x\left(-i\hbar\frac{\partial\psi}{\partial x}\right) = -i\hbar x\frac{\partial\psi}{\partial x}. \tag{2.3.47}$$

On the other hand,

$$\hat{p}_x\hat{x}\psi = -i\hbar\frac{\partial}{\partial x}(x\psi) = -i\hbar x\frac{\partial\psi}{\partial x} - i\hbar\psi. \tag{2.3.48}$$

From the last two formulas follows that

$$\hat{p}_x\hat{x}\psi - \hat{x}\hat{p}_x\psi = -i\hbar\psi, \tag{2.3.49}$$

which is tantamount to

$$\boxed{\hat{p}_x\hat{x} - \hat{x}\hat{p}_x = -i\hbar\hat{I}.} \tag{2.3.50}$$

By using the same line of reasoning, similar commutation relations can be established:

$$\boxed{\hat{p}_y\hat{y} - \hat{y}\hat{p}_y = -i\hbar\hat{I},} \tag{2.3.51}$$

$$\boxed{\hat{p}_z\hat{z} - \hat{z}\hat{p}_z = -i\hbar\hat{I}.} \tag{2.3.52}$$

Thus, like components of p and r are not simultaneously measurable.

It is easy to demonstrate that

$$\hat{p}_x\hat{y} - \hat{y}\hat{p}_x = 0, \tag{2.3.53}$$

$$\hat{p}_x\hat{z} - \hat{z}\hat{p}_x = 0, \tag{2.3.54}$$

and similar commutation relations are valid for $\hat{p}_y$ and $\hat{p}_z$. These commutation relations imply **that unlike components of p and r are simultaneously measurable.**

All the above commutation relations can be summarized by one formula

$$\boxed{[\hat{p}_{x_i}, \hat{x}_j] = -i\hbar\delta_{ij}\hat{I},} \tag{2.3.55}$$

where δ_{ij} is the Kronecker delta, while x_1, x_2 and x_3 are notations for x, y and z, respectively.

As discussed before and evident from commutation relation (2.3.50), x and p_x cannot be simultaneously and accurately measured. It is interesting to establish a fundamental limit to the precision with which these (so-called complementary) physical quantities can be simultaneously known. This can be done by characterizing the uncertainties in measurements of x and p_x by standard deviations

$$\sigma_x = \sqrt{(x - \bar{x})^2}, \tag{2.3.56}$$

$$\sigma_{p_x} = \sqrt{(p_x - \bar{p}_x)^2} \tag{2.3.57}$$

and deriving the lower bound for their product $\sigma_x \sigma_{p_x}$.

For the sake of simplicity of computations, it is assumed below that

$$\bar{x} = 0, \quad \bar{p}_x = 0 \tag{2.3.58}$$

and

$$\sigma_x = \sqrt{\overline{x^2}}, \quad \sigma_{p_x} = \sqrt{\overline{p_x^2}}. \tag{2.3.59}$$

Now, consider the following integral:

$$\Lambda = \int_{-\infty}^{\infty} \left| \lambda x \psi + \frac{\partial \psi}{\partial x} \right|^2 dx \geq 0, \tag{2.3.60}$$

where λ is a real number.

The last formula can be transformed as follows:

$$\Lambda = \lambda^2 \int_{-\infty}^{\infty} x^2 |\psi|^2 \, dx + \lambda \int_{-\infty}^{\infty} x \left(\frac{d\psi^*}{dx} \psi + \psi^* \frac{d\psi}{dx} \right) dx + \int_{-\infty}^{\infty} \frac{d\psi^*}{dx} \frac{d\psi}{dx} \, dx \geq 0. \tag{2.3.61}$$

It is clear that

$$\int_{-\infty}^{\infty} x^2 |\psi|^2 \, dx = \sigma_x^2. \tag{2.3.62}$$

Furthermore,

$$\int_{-\infty}^{\infty} x \left(\frac{d\psi^*}{dx} \psi + \psi^* \frac{d\psi}{dx} \right) = \int_{-\infty}^{\infty} x \frac{d|\psi|^2}{dx} \, dx. \tag{2.3.63}$$

Performing the integration by parts in the last integral in (2.3.63), we find

$$\int_{-\infty}^{\infty} x \left(\frac{d\psi^*}{dx} \psi + \psi^* \frac{d\psi}{dx} \right) dx = - \int_{-\infty}^{\infty} |\psi|^2 \, dx = -1. \tag{2.3.64}$$

By using integration by parts in the last integral in formula (2.3.61), we derive

$$\int_{-\infty}^{\infty} \frac{d\psi^*}{dx} \cdot \frac{d\psi}{dx} \, dx = - \int_{-\infty}^{\infty} \psi^* \frac{d^2\psi}{dx^2} \, dx = \frac{1}{\hbar^2} \int_{-\infty}^{\infty} \psi^* \hat{p}_x^2 \psi \, dx = \frac{1}{\hbar^2} \sigma_{p_x}^2. \tag{2.3.65}$$

By using relations (2.3.62), (2.3.64) and (2.3.65), formula (2.3.61) can be transformed as follows:

$$\lambda^2 - \frac{\lambda}{\sigma_x^2} + \frac{\sigma_{p_x}^2}{\hbar^2 \sigma_x^2} \geq 0. \tag{2.3.66}$$

It is easy to see (by completing the square) that the quadratic form in the left-hand side of formula (2.3.66) is not negative for any real number λ if

$$\frac{\sigma_{p_x}^2}{\hbar^2 \sigma_x^2} - \frac{1}{4\sigma_x^4} \geq 0. \tag{2.3.67}$$

From the last inequality follows that

$$\boxed{\sigma_x \sigma_{p_x} \geq \frac{\hbar}{2}.} \tag{2.3.68}$$

This inequality establishes the fundamental limit for possible accuracy of simultaneous measurements of x and p_x. It is usually considered as the mathematical expression of the uncertainty principle of W. Heisenberg. The formal proof of the above inequality belongs to H. Weyl.

It is interesting to find the wave function for which the equality in formula (2.3.68) is achieved. For this wave function the product of uncertainties in simultaneous measurements of x and p_x is minimal. In this sense, this wave function describes the quantum mechanical state closest to the classical one. Such quantum states are called coherent states. It can be shown that the minimal uncertainty wave function is given by the formula

$$\psi(x) = \frac{1}{(2\pi)^{\frac{1}{4}} \sqrt{\sigma_x}} e^{\left(\frac{i}{\hbar} \bar{p}_x x - \frac{(x - \bar{x})^2}{4\sigma_x^2} \right)}, \tag{2.3.69}$$

where $\bar{p}_x$ and $\bar{x}$ are expected values of p_x and x, respectively.

It is easy to see from the last formula that probability density $|\psi|^2$ of coordinate is Gaussian:

$$|\psi|^2 = \frac{1}{\sqrt{2\pi}\sigma_x} e^{-\frac{(x - \bar{x})^2}{2\sigma_x^2}}. \tag{2.3.70}$$

It can be shown that the probability density $|c(p_x)|^2$ of momentum is Gaussian as well.

2.4 Angular Momentum Operators

Consider a **closed** system of N microparticles. The quantum mechanical state of this system is described in coordinate representation by a wave function $\psi(\mathbf{r}_1, \mathbf{r}_2, \ldots \mathbf{r}_N, t)$. For a closed system, space is isotropic. This means that the mathematical form of the time-dependent Schrödinger equation (2.2.9) should not depend on the chosen orientation of coordinate axes. In other words, this equation must be invariant with respect to arbitrary rotations of the Cartesian coordinate system in space. This implies that spatial rotations are symmetry transformations for closed systems and operators of these rotations must commute with the Hamiltonian.

Consider an infinitesimally small rotation through angle $d\varphi$ around an arbitrary chosen axis. As discussed in Chapter 1, this rotation results in the following transformation of particle coordinates:

$$\mathbf{r}'_k = \mathbf{r}_k + \mathbf{d}\boldsymbol{\varphi} \times \mathbf{r}_k, \qquad (2.4.1)$$

where the direction of vector $\mathbf{d}\boldsymbol{\varphi}$ is along the axis of rotation.

Thus,

$$\psi'(\mathbf{r}'_1, \mathbf{r}'_2, \dots \mathbf{r}'_N, t) = \psi(\mathbf{r}_1 + \mathbf{d}\boldsymbol{\varphi} \times \mathbf{r}_1, \mathbf{r}_2 + \mathbf{d}\boldsymbol{\varphi} \times \mathbf{r}_2, \dots \mathbf{r}_N + \mathbf{d}\boldsymbol{\varphi} \times \mathbf{r}_N, t)$$

$$= \psi(\mathbf{r}_1, \mathbf{r}_2, \dots \mathbf{r}_N, t) + \sum_{k=1}^{N} \nabla_k \psi \cdot (\mathbf{d}\boldsymbol{\varphi} \times \mathbf{r}_k),$$

$$(2.4.2)$$

which can be further transformed as follows:

$$\psi'(\mathbf{r}'_1, \mathbf{r}'_2, \dots \mathbf{r}'_N, t) = \psi(\mathbf{r}_1, \mathbf{r}_2, \dots \mathbf{r}_N, t) + \mathbf{d}\boldsymbol{\varphi} \cdot \sum_{k=1}^{N} \mathbf{r}_k \times \nabla_k \psi. \qquad (2.4.3)$$

The last formula can be written as

$$\psi' = \hat{T}\psi, \qquad (2.4.4)$$

where $\hat{T}$ is the operator of infinitesimally small spatial rotation, which has the form

$$\hat{T} = \hat{I} + \mathbf{d}\boldsymbol{\varphi} \cdot \sum_{k=1}^{N} \mathbf{r}_k \times \nabla_k. \qquad (2.4.5)$$

Since rotations are symmetry transformations for closed systems, operator $\hat{T}$ commutes with the Hamiltonian (see the Proposition from the last section). The identity operator commutes with any operator. Thus, it can be concluded that operator $\sum_{k=1}^{N} \mathbf{r}_k \times \nabla_k$ commutes with the Hamiltonian:

$$\hat{H}\left(\sum_{k=1}^{N} \mathbf{r}_k \times \nabla_k\right) - \left(\sum_{k=1}^{N} \mathbf{r}_k \times \nabla_k\right)\hat{H} = 0. \qquad (2.4.6)$$

Consequently, if some physical quantity is associated with operator $\sum_{k=1}^{N} \mathbf{r}_k \times \nabla_k$, this physical quantity is conserved and its conservation follows from the isotropicity of space for a closed system. As in classical mechanics (see previous chapter), this physical quantity is angular momentum. For the operator $\sum_{k=1}^{N} \mathbf{r}_k \times \nabla_k$ to be the operator of angular momentum, it must be Hermitian. The latter can be achieved by multiplying $\sum_{k=1}^{N} \mathbf{r}_k \times \nabla_k$ by $(-i)$. Furthermore, the factor $\hbar$ is introduced to achieve the proper dimension of angular momentum as well as to properly reflect the scale of angular momentum at the microscopic level. Thus, we arrive at the following expression for the angular momentum operator $\widehat{\mathbf{L}}$:

$$\boxed{\widehat{\mathbf{L}} = -i\hbar \sum_{k=1}^{N} \mathbf{r}_k \times \nabla_k.} \tag{2.4.7}$$

Recalling that $\hat{\mathbf{p}}_k = -i\hbar\nabla_k$, the last formula can also be written as

$$\boxed{\widehat{\mathbf{L}} = \sum_{k=1}^{N} \mathbf{r}_k \times \hat{\mathbf{p}}_k.} \tag{2.4.8}$$

Since the operator of multiplication by $\mathbf{r}_k$ coincides with operator $\mathbf{r}_k$, the last formula can be represented in the form

$$\widehat{\mathbf{L}} = \sum_{k=1}^{N} \hat{\mathbf{r}}_k \times \hat{\mathbf{p}}_k. \tag{2.4.9}$$

In the case of a single microparticle, the angular momentum operator is

$$\boxed{\widehat{\mathbf{L}} = \mathbf{r} \times \hat{\mathbf{p}}} \quad \text{or} \quad \boxed{\widehat{\mathbf{L}} = \hat{\mathbf{r}} \times \hat{\mathbf{p}}.} \tag{2.4.10}$$

Thus, the relation between operators $\widehat{\mathbf{L}}$, $\hat{\mathbf{r}}$ and $\hat{\mathbf{p}}$ in quantum mechanics is similar to the relation between physical quantities $\mathbf{L}$, $\mathbf{r}$ and $\mathbf{p}$ in classical mechanics.

Remark 2.3. We have introduced the momentum operator $\hat{\mathbf{p}}$ and angular momentum operator $\widehat{\mathbf{L}}$ by considering the invariance of the time-dependent Schrödinger equation (2.2.9) with respect to spatial translations and rotations of Cartesian coordinate systems, respectively. These translations and rotations have the mathematical structure of continuous (smooth) groups which are called Lie groups of translations and rotations. In a somewhat mathematically more sophisticated and rigorous treatment of the subject, operators $\hat{\mathbf{p}}$ and $\widehat{\mathbf{L}}$ are closely connected with infinitesimal operators of Lie groups of translations and rotations, respectively. The detailed discussion of this matter is beyond the scope of this text.

Now, we shall turn to the study of commutation properties of the operator of angular momentum. To do this, we shall first introduce the formulas for operators of Cartesian components $\hat{L}_x$, $\hat{L}_y$ and $\hat{L}_z$ of $\widehat{\mathbf{L}}$:

$$\widehat{\mathbf{L}} = \mathbf{e}_x \hat{L}_x + \mathbf{e}_y \hat{L}_y + \mathbf{e}_z \hat{L}_z. \tag{2.4.11}$$

According to formula (2.4.10), we have

$$\widehat{\mathbf{L}} = \begin{vmatrix} \mathbf{e}_x & \mathbf{e}_y & \mathbf{e}_z \\ x & y & z \\ \hat{p}_x & \hat{p}_y & \hat{p}_z \end{vmatrix}. \tag{2.4.12}$$

By comparing the last two formulas, we find

$$\boxed{\hat{L}_x = y\hat{p}_z - z\hat{p}_y, \quad \hat{L}_y = z\hat{p}_x - x\hat{p}_z, \quad \hat{L}_z = x\hat{p}_y - y\hat{p}_x.} \tag{2.4.13}$$

It is clear from formula (2.4.13) that $\hat{L}_x$ is the operator with respect to y and z coordinates. Consequently, $\hat{L}_x$ commutes with any operator acting with respect to x-coordinates. This implies that

$$\hat{x}\hat{L}_x - \hat{L}_x\hat{x} = 0 \tag{2.4.14}$$

as well as

$$\hat{p}_x\hat{L}_x - \hat{L}_x\hat{p}_x = 0. \tag{2.4.15}$$

By using the same line of reasoning, we also find that

$$\hat{y}\hat{L}_y - \hat{L}_y\hat{y} = 0, \quad \hat{z}\hat{L}_z - \hat{L}_z\hat{z} = 0, \tag{2.4.16}$$

$$\hat{p}_y\hat{L}_y - \hat{L}_y\hat{p}_y = 0, \quad \hat{p}_z\hat{L}_z - \hat{L}_z\hat{p}_z = 0. \tag{2.4.17}$$

Physically, the above commutation relations mean that coordinates and components of angular momentum along the same Cartesian axes are simultaneously measurable. Likewise, it is true that the same Cartesian components of momentum and angular momentum are simultaneously measurable.

Consider now the commutation relation for operators $\hat{L}_x$ and $\hat{y}$. From formula (2.4.13), we find

$$\hat{L}_x\hat{y} = (y\hat{p}_z - z\hat{p}_y)y = y\hat{p}_z y - z\hat{p}_y y. \tag{2.4.18}$$

According to formula (2.3.51), we have

$$\hat{p}_y y = y\hat{p}_y - i\hbar \hat{I}, \tag{2.4.19}$$

and, consequently, formula (2.4.18) can be further transformed as follows:

$$\hat{L}_x\hat{y} = y^2\hat{p}_z - zy\hat{p}_y + i\hbar z = y(y\hat{p}_z - z\hat{p}_y) + i\hbar z = y\hat{L}_x + i\hbar z. \tag{2.4.20}$$

Consequently,

$$\hat{L}_x \hat{y} - \hat{y}\hat{L}_x = i\hbar\hat{z}. \tag{2.4.21}$$

By using the same line of reasoning, it can be established that

$$\hat{L}_z \hat{x} - \hat{x}\hat{L}_z = i\hbar\hat{y}, \tag{2.4.22}$$

$$\hat{L}_y \hat{z} - \hat{z}\hat{L}_y = i\hbar\hat{x}. \tag{2.4.23}$$

It is also apparent that the last two formulas can be obtained from (2.4.21) by circular permutation of coordinates x, y and z.

It is suggested as an exercise to prove the commutation relations

$$\hat{L}_y \hat{x} - \hat{x}\hat{L}_y = -i\hbar\hat{z}, \tag{2.4.24}$$

$$\hat{L}_x \hat{z} - \hat{z}\hat{L}_x = -i\hbar\hat{y}, \tag{2.4.25}$$

$$\hat{L}_z \hat{y} - \hat{y}\hat{L}_z = -i\hbar\hat{x}. \tag{2.4.26}$$

Commutation relations (2.4.21)–(2.4.26) imply that coordinates and components of angular momentum along different Cartesian axes are not simultaneously measurable.

It is also suggested as an exercise to prove the following commutation relations:

$$\hat{L}_x \hat{p}_y - \hat{p}_y\hat{L}_x = i\hbar\hat{p}_z, \tag{2.4.27}$$

$$\hat{L}_z \hat{p}_x - \hat{p}_x\hat{L}_z = i\hbar\hat{p}_y, \tag{2.4.28}$$

$$\hat{L}_y \hat{p}_z - \hat{p}_z\hat{L}_y = i\hbar\hat{p}_x, \tag{2.4.29}$$

as well as

$$\hat{L}_y \hat{p}_x - \hat{p}_x\hat{L}_y = -i\hbar\hat{p}_z, \tag{2.4.30}$$

$$\hat{L}_x \hat{p}_z - \hat{p}_z\hat{L}_x = -i\hbar\hat{p}_y, \tag{2.4.31}$$

$$\hat{L}_z \hat{p}_y - \hat{p}_y\hat{L}_z = -i\hbar\hat{p}_x. \tag{2.4.32}$$

These commutation relations imply that components of momentum and angular momentum along different Cartesian axes are not simultaneously measurable.

We next proceed to the discussion of commutation relations for different components of angular momentum. According to formulas (2.4.13) and (2.4.14), we find

$$\hat{L}_x \hat{L}_y = \hat{L}_x \left(z\hat{p}_x - x\hat{p}_z \right) = \hat{L}_x z\hat{p}_x - x\hat{L}_x\hat{p}_z. \tag{2.4.33}$$

From commutation relations (2.4.25) and (2.4.31) follows that

$$\hat{L}_x z = z\hat{L}_x - i\hbar y, \tag{2.4.34}$$

$$\hat{L}_x \hat{p}_z = \hat{p}_z\hat{L}_x - i\hbar\hat{p}_y. \tag{2.4.35}$$

By using the last two relations, formula (2.4.33) can be further transformed as follows:

$$\hat{L}_x\hat{L}_y = z\hat{L}_x\hat{p}_x - i\hbar y\hat{p}_x - x\hat{p}_z\hat{L}_x + i\hbar x\hat{p}_y$$
$$= \left(z\hat{p}_x - x\hat{p}_z\right)\hat{L}_x + i\hbar\hat{L}_z = \hat{L}_y\hat{L}_x + i\hbar\hat{L}_z. \tag{2.4.36}$$

Consequently,

$$\boxed{\hat{L}_x\hat{L}_y - \hat{L}_y\hat{L}_x = i\hbar\hat{L}_z.} \tag{2.4.37}$$

By using the same line of reasoning, it can be established that

$$\boxed{\hat{L}_z\hat{L}_x - \hat{L}_x\hat{L}_z = i\hbar\hat{L}_y,} \tag{2.4.38}$$

$$\boxed{\hat{L}_y\hat{L}_z - \hat{L}_z\hat{L}_y = i\hbar\hat{L}_x.} \tag{2.4.39}$$

The last three commutation relations reveal that **no two components of angular momentum are simultaneously measurable.** In other words, angular momentum cannot be measured as a vector. This is in contrast with the position vector **r** and momentum vector **p**, for each of whom three components are simultaneously measurable.

Next, we shall show that while individual components of angular momentum are not simultaneously measurable, there exists another quantity related to angular momentum that does commute with its individual components and, consequently, can be simultaneously measured with each of them. This quantity is the squared magnitude of angular momentum, whose operator $\widehat{\mathbf{L}}^2$ is defined by the formula

$$\boxed{\widehat{\mathbf{L}}^2 = \hat{L}_x^2 + \hat{L}_y^2 + \hat{L}_z^2.} \tag{2.4.40}$$

We intend to prove the validity of commutation relations

$$\boxed{\widehat{\mathbf{L}}^2\hat{L}_x - \hat{L}_x\widehat{\mathbf{L}}^2 = 0,} \tag{2.4.41}$$

$$\boxed{\widehat{\mathbf{L}}^2\hat{L}_y - \hat{L}_y\widehat{\mathbf{L}}^2 = 0,} \tag{2.4.42}$$

$$\boxed{\widehat{\mathbf{L}}^2\hat{L}_z - \hat{L}_z\widehat{\mathbf{L}}^2 = 0.} \tag{2.4.43}$$

The proof of the commutation relation (2.4.41) proceeds as follows. From formula (2.4.40), we find

$$\widehat{\mathbf{L}}^2\hat{L}_x - \hat{L}_x\widehat{\mathbf{L}}^2 = \hat{L}_y^2\hat{L}_x - \hat{L}_x\hat{L}_y^2 + \hat{L}_z^2\hat{L}_x - \hat{L}_x\hat{L}_z^2. \tag{2.4.44}$$

The first two terms in the right hand side of the last formula are transformed as shown below:

$$\hat{L}_y^2\hat{L}_x - \hat{L}_x\hat{L}_y^2 = \hat{L}_y\hat{L}_y\hat{L}_x - \hat{L}_y\hat{L}_x\hat{L}_y + \hat{L}_y\hat{L}_x\hat{L}_y - \hat{L}_x\hat{L}_y\hat{L}_y$$
$$= \hat{L}_y\left(\hat{L}_y\hat{L}_x - \hat{L}_x\hat{L}_y\right) + \left(\hat{L}_y\hat{L}_x - \hat{L}_x\hat{L}_y\right)\hat{L}_y. \tag{2.4.45}$$

By recalling the commutation relation (2.4.37), the last formula can be further simplified as follows:

$$\hat{L}_y^2\hat{L}_x - \hat{L}_x\hat{L}_y^2 = -i\hbar\left(\hat{L}_y\hat{L}_z + \hat{L}_z\hat{L}_y\right). \tag{2.4.46}$$

By using similar transformations, the last two terms in the right-hand side of formula (2.4.44) can be reduced to

$$\hat{L}_z^2\hat{L}_x - \hat{L}_x\hat{L}_z^2 = \hat{L}_z\left(\hat{L}_z\hat{L}_x - \hat{L}_x\hat{L}_z\right) + \left(\hat{L}_z\hat{L}_x - \hat{L}_x\hat{L}_z\right)\hat{L}_z$$
$$= i\hbar\left(\hat{L}_z\hat{L}_y + \hat{L}_y\hat{L}_z\right). \tag{2.4.47}$$

By substituting the last two formulas into equation (2.4.44), we arrive at the commutation relation (2.4.41). The validity of commutation relations (2.4.42) and (2.4.43) can be established in a similar manner.

Since $\hat{\mathbf{L}}^2$ and $\hat{L}_z$ are simultaneously measurable, it is natural to find the states in which they simultaneously have certain values as well as to find their spectrum. The spectrum of $\hat{L}_z$ is easy to find. To this end, we shall use the spherical coordinates (r, θ, φ) which are related to the Cartesian coordinates by the formulas

$$x = r\sin\theta\cos\varphi, \tag{2.4.48}$$
$$y = r\sin\theta\sin\varphi, \tag{2.4.49}$$
$$z = \cos\theta. \tag{2.4.50}$$

By using these formulas, for any wave function $\psi(x, y, z, t)$ we derive

$$\frac{\partial\psi}{\partial\varphi} = \frac{\partial\psi}{\partial x}\frac{\partial x}{\partial\varphi} + \frac{\partial\psi}{\partial y}\frac{\partial y}{\partial\varphi} + \frac{\partial\psi}{\partial z}\frac{\partial z}{\partial\varphi}$$
$$= \frac{\partial\psi}{\partial x}\left(-r\sin\theta\sin\varphi\right) + \frac{\partial\psi}{\partial y}\left(r\sin\theta\cos\varphi\right) \tag{2.4.51}$$
$$= x\frac{\partial\psi}{\partial y} - y\frac{\partial\psi}{\partial x}.$$

On the other hand,

$$\hat{L}_z = x\hat{p}_y - y\hat{p}_x = -i\hbar\left(x\frac{\partial}{\partial y} - y\frac{\partial}{\partial x}\right). \tag{2.4.52}$$

From the last two formulas, we obtain

$$\hat{L}_z = -i\hbar\frac{\partial}{\partial\varphi}. \qquad (2.4.53)$$

The last equation implies that the $\hat{L}_z$ operator, as expected, is the operator of infinitesimally small rotation around the z-axis scaled by the factor $-i\hbar$. It is also apparent now that the equation for eigenvalues and eigenfunctions of operator $\hat{L}_z$ can be written as

$$-i\hbar\frac{\partial\psi(r,\theta,\varphi)}{\partial\varphi} = L_z\psi(r,\theta,\varphi). \qquad (2.4.54)$$

By integrating this equation, we find

$$\psi(r,\theta,\varphi) = \zeta(r,\theta)e^{\frac{i}{\hbar}L_z\varphi}, \qquad (2.4.55)$$

where $\zeta(r,\theta)$ is some function of r and θ.

Eigenfunction $\psi(r,\theta,\varphi)$ in the last equation must be continuous. This implies that the following equality must hold:

$$\psi(r,\theta,0) = \psi(r,\theta,2\pi). \qquad (2.4.56)$$

According to formula (2.4.55), the latter is only possible if

$$e^{\frac{i}{\hbar}L_z 2\pi} = 1. \qquad (2.4.57)$$

The last equation reveals that the eigenvalues L_z of operator $\hat{L}_z$ are

$$L_z = \hbar m, \quad (m = 0, \pm 1, \pm 2, \dots). \qquad (2.4.58)$$

Thus, the spectrum of $\hat{L}_z$ is discrete. In other words, the component of angular momentum along the z-axis is quantized. Since the z-axis can be arbitrarily chosen, it can be concluded that **a component of angular momentum along any axis is quantized** and its spectrum is given by formula (2.4.58). This is in contrast with the momentum operator whose spectrum is continuous for all its three components.

By using formula (2.4.58), equation (2.4.54) for eigenfunctions can be written as follows:

$$\psi_m(r,\theta,\varphi) = \zeta(r,\theta)e^{im\varphi}, \quad (m = 0, \pm 1, \pm 2, \dots), \qquad (2.4.59)$$

and the equation for eigenvalues of operator $\hat{L}_z$ can be represented in the form

$$\hat{L}_z\psi_m(r,\theta,\varphi) = \hbar m\psi_m(r,\theta,\varphi). \qquad (2.4.60)$$

Next, we consider the spectrum of the $\widehat{\mathbf{L}}^2$-operator. This can be done in two different ways. The first way is to cast this problem as the eigenvalue problem for the specific differential operator on the sphere. The other way to find the spectrum of the $\widehat{\mathbf{L}}^2$-operator is by using the commutation relations, that is without reducing $\widehat{\mathbf{L}}^2$ to the specific differential operator on the sphere. Below, we shall use the second way of computing the spectrum of the $\widehat{\mathbf{L}}^2$-operator. To do this, we introduce two operators

$$\hat{L}_+ = \hat{L}_x + i\hat{L}_y \tag{2.4.61}$$

and

$$\hat{L}_- = \hat{L}_x - i\hat{L}_y, \tag{2.4.62}$$

which are called the **promotion** and **demotion** operators, respectively. We next establish the following commutation relation:

$$\boxed{\hat{L}_z\hat{L}_+ - \hat{L}_+\hat{L}_z = \hbar\hat{L}_+.} \tag{2.4.63}$$

The proof proceeds as follows. First, it is clear that

$$\hat{L}_z\hat{L}_+ = \hat{L}_z\hat{L}_x + i\hat{L}_z\hat{L}_y, \tag{2.4.64}$$

while

$$\hat{L}_+\hat{L}_z = \hat{L}_x\hat{L}_z + i\hat{L}_y\hat{L}_z. \tag{2.4.65}$$

From the last two formulas, we derive

$$\hat{L}_z\hat{L}_+ - \hat{L}_+\hat{L}_z = \hat{L}_z\hat{L}_x - \hat{L}_x\hat{L}_z + i\left(\hat{L}_z\hat{L}_y - \hat{L}_y\hat{L}_z\right). \tag{2.4.66}$$

Now, by using formulas (2.4.38) and (2.4.39), we complete the proof:

$$\hat{L}_z\hat{L}_+ - \hat{L}_+\hat{L}_z = i\hbar\left(\hat{L}_y - i\hat{L}_x\right) = \hbar\left(\hat{L}_x + i\hat{L}_y\right) = \hbar\hat{L}_+. \tag{2.4.67}$$

We shall also need the following operator identity

$$\widehat{\mathbf{L}}^2 = \hat{L}_-\hat{L}_+ + \hat{L}_z^2 + \hbar\hat{L}_z, \tag{2.4.68}$$

which is established as follows

$$\hat{L}_-\hat{L}_+ + \hat{L}_z^2 + \hbar\hat{L}_z = \left(\hat{L}_x - i\hat{L}_y\right)\left(\hat{L}_x + i\hat{L}_y\right) + \hat{L}_z^2 + \hbar\hat{L}_z$$

$$= \hat{L}_x^2 + \hat{L}_y^2 + i\left(\hat{L}_x\hat{L}_y - \hat{L}_y\hat{L}_x\right) + \hat{L}_z^2 + \hbar\hat{L}_z = \hat{L}_x^2 + \hat{L}_y^2 + \hat{L}_z^2 = \widehat{\mathbf{L}}^2. \tag{2.4.69}$$

Since $\hat{L}_z$ and $\widehat{\mathbf{L}}^2$ are simultaneously measurable, there exists a quantum mechanical state $\psi_{\lambda m}$ in which L_z and L^2 have certain values $\hbar m$ and

λ, respectively. It is clear that $\psi_{\lambda m}$ is the solution of two simultaneous equations

$$\hat{L}_z \psi_{\lambda m} = \hbar m \psi_{\lambda m}, \tag{2.4.70}$$

$$\hat{\mathbf{L}}^2 \psi_{\lambda m} = \lambda \psi_{\lambda m}. \tag{2.4.71}$$

We intend to find the values of λ for which these two equations are valid. First, we note that

$$\langle \psi_{\lambda m} | \hat{\mathbf{L}}^2 \psi_{\lambda m} \rangle = \lambda \langle \psi_{\lambda m} | \psi_{\lambda m} \rangle = \lambda \| \psi_{\lambda m} \|^2. \tag{2.4.72}$$

On the other hand,

$$\langle \psi_{\lambda m} | \hat{\mathbf{L}}^2 \psi_{\lambda m} \rangle = \langle \psi_{\lambda m} | \hat{L}_x^2 \psi_{\lambda m} \rangle + \langle \psi_{\lambda m} | \hat{L}_y^2 \psi_{\lambda m} \rangle + \langle \psi_{\lambda m} | \hat{L}_z^2 \psi_{\lambda m} \rangle$$
$$= \langle \hat{L}_x \psi_{\lambda m} | \hat{L}_x \psi_{\lambda m} \rangle + \langle \hat{L}_y \psi_{\lambda m} | \hat{L}_y \psi_{\lambda m} \rangle + \hbar^2 m^2 \langle \psi_{\lambda m} | \psi_{\lambda m} \rangle$$
$$= \| \hat{L}_x \psi_{\lambda m} \|^2 + \| \hat{L}_y \psi_{\lambda m} \|^2 + \hbar^2 m^2 \| \psi_{\lambda m} \|^2.$$
$$\tag{2.4.73}$$

Comparing the last two formulas, we conclude that λ is positive and

$$\hbar^2 m^2 \leq \lambda. \tag{2.4.74}$$

This implies that, for a given λ, all possible values of m are bounded from above and below. Let l be the upper bound of $|m|$, i.e.,

$$l = \max|m| \text{ for a given } \lambda. \tag{2.4.75}$$

Next, by using identity (2.4.63) and equation (2.4.70), we find

$$\hat{L}_z \left(\hat{L}_+ \psi_{\lambda m} \right) = \hat{L}_+ \hat{L}_z \psi_{\lambda m} + \hbar \hat{L}_+ \psi_{\lambda m} = \hbar (m + 1) \hat{L}_+ \psi_{\lambda m}. \tag{2.4.76}$$

The last equation implies that $\hat{L}_+ \psi_{\lambda m}$ is the eigenfunction of operator $\hat{L}_z$ corresponding to the eigenvalue $\hbar(m + 1)$. This, according to equation (2.4.70) means that

$$\boxed{\hat{L}_+ \psi_{\lambda m} = \psi_{\lambda, m+1}.} \tag{2.4.77}$$

Thus, operator $\hat{L}_+$ maps any eigenfunction of operator $\hat{L}_z$ into the eigenfunction of the same operator with the corresponding eigenvalue increased by $\hbar$. This explains why the operator $\hat{L}_+$ is called the promotion operator. It is also clear that

$$\hat{L}_+ \psi_{\lambda l} = 0. \tag{2.4.78}$$

This is true because, according to formulas (2.4.74) and (2.4.75), there is no eigenvalue of operator $\hat{L}_z$ larger than $\hbar l$. Consequently, the equation (2.4.77) is satisfied only if equality (2.4.78) is valid.

Now, finally, we can compute the eigenvalues λ. To this end, we shall use the identity (2.4.68), formula (2.4.78) as well as the fact that, according to equation (2.4.70),

$$\hat{L}_z \psi_{\lambda l} = \hbar l \psi_{\lambda l}. \tag{2.4.79}$$

Consequently,

$$\begin{aligned}
\widehat{\mathbf{L}}^2 \psi_{\lambda l} &= \hat{L}_- \left(\hat{L}_+ \psi_{\lambda l} \right) + \hat{L}_z^2 \psi_{\lambda l} + \hbar \hat{L}_z \psi_{\lambda l} \\
&= \hbar^2 l^2 \psi_{\lambda l} + \hbar^2 l \psi_{\lambda l} \\
&= \hbar^2 l \, (l+1) \, \psi_{\lambda l}.
\end{aligned} \tag{2.4.80}$$

By comparing equations (2.4.71) and (2.4.80), we conclude that

$$\boxed{\lambda = \hbar^2 l \, (l+1), \quad (l = 0, 1, 2, \ldots).} \tag{2.4.81}$$

Thus, the spectrum of operator $\widehat{\mathbf{L}}^2$ is discrete. In other words, **the squared magnitude of angular momentum is quantized**.

Strictly speaking, formula (2.4.80) has so far been proved for wave function $\psi_{\lambda l}$, that is for the state when formula (2.4.79) is valid. It can also be proved that this formula is valid for $\psi_{\lambda m}$ with any value of m less than l. The proof is based on the following two formulas

$$\widehat{\mathbf{L}}^2 \hat{L}_- = \hat{L}_- \widehat{\mathbf{L}}^2, \tag{2.4.82}$$

$$\hat{L}_- \psi_{\lambda m} = \psi_{\lambda, m-1}. \tag{2.4.83}$$

Formula (2.4.82) immediately follows from the definition of the operator $\hat{L}_-$ and commutation relations (2.4.41) and (2.4.42), while formula (2.4.83) can be established in the same way as formula (2.4.77). The last formula explains why operator $\hat{L}_-$ is called the demotion operator. Now, from formulas (2.4.82) and (2.4.80) we find

$$\widehat{\mathbf{L}}^2 \hat{L}_- \psi_{\lambda l} = \hat{L}_- \widehat{\mathbf{L}}^2 \psi_{\lambda l} = \hbar^2 l \, (l+1) \, \hat{L}_- \psi_{\lambda l}. \tag{2.4.84}$$

On the other hand, according to formula (2.4.83) we have

$$\hat{L}_- \psi_{\lambda l} = \psi_{\lambda, l-1} \tag{2.4.85}$$

From the last two formulas follows that

$$\widehat{\mathbf{L}}^2 \psi_{\lambda, l-1} = \hbar^2 l \, (l+1) \, \psi_{\lambda, l-1}. \tag{2.4.86}$$

By repeating the same reasoning, it can be established that formula (2.4.86) is valid for any $|m| \leq l$, that is

$$\boxed{\widehat{\mathbf{L}}^2 \psi_{lm} = \hbar^2 l \, (l+1) \, \psi_{lm}.} \tag{2.4.87}$$

In the last formula, we introduce the conventional notation ψ_{lm} instead of $\psi_{\lambda m}$, because λ is fully defined by l according to formula (2.4.81). It is also clear that the last formula is valid for

$$m = 0, \pm 1, \pm 2, \cdots \pm l. \tag{2.4.88}$$

Thus, the eigenvalue $\hbar^2 l \, (l+1)$ of $\widehat{\mathbf{L}}^2$ is $(2l+1)$-times degenerate.

Finally, we consider the analytical expressions for eigenfunctions $\psi_{lm} \, (r, \theta, \varphi)$ of operator $\widehat{\mathbf{L}}^2$. By using spherical coordinates (2.4.48)–(2.4.50) and formulas (2.4.13) and (2.4.40), it can be shown that the following formula is valid

$$\widehat{\mathbf{L}}^2 = -\hbar^2 \left[\frac{1}{\sin\theta} \frac{\partial}{\partial\theta} \left(\sin\theta \frac{\partial}{\partial\theta} \right) + \frac{1}{\sin^2\theta} \frac{\partial^2}{\partial\varphi^2} \right]. \tag{2.4.89}$$

It is clear from the last equation that $\widehat{\mathbf{L}}^2$ is the Laplace operator on the sphere scaled by $\left(-\hbar^2\right)$. The Laplace operator on the sphere frequently appears in the area of mathematical physics in connection with the solution of the Laplace equation in spherical coordinates. It is proved in mathematical physics that the eigenfunctions of equation (2.4.87) can be written in the form

$$\psi_{lm}(r, \theta, \varphi) = R(r) Y_l^m(\theta, \varphi), \tag{2.4.90}$$

where $R(r)$ is some function of r, while $Y_l^m(\theta, \varphi)$ are spherical harmonics. These harmonics are given by the formula

$$Y_l^m(\theta, \varphi) = C P_l^{(m)}(\cos\theta) e^{im\varphi}, \tag{2.4.91}$$

where C is a normalization factor, while $P_l^{(m)}(\cos\theta)$ is an associated Legendre function.

This function is defined as follows

$$P_l^{(m)}(x) = (-1)^m \left(1 - x^2\right)^{\frac{m}{2}} \frac{d^m P_l(x)}{dx^m}, \tag{2.4.92}$$

where

$$P_l(x) = \frac{1}{2^l l!} \frac{d^l \left(x^2 - 1\right)^l}{dx^l}, \tag{2.4.93}$$

and the notation $x = \cos\theta$ has been adopted.

Spherical harmonics are extensively used for the solution of problems in mathematical physics and electromagnetism by the method of separation of variables in spherical coordinates. The properties of these harmonics are well studied. In particular, it is established that these harmonics have unique properties associated with rotations of spherical coordinates. This is not surprising, because the very definition of the angular momentum operator is connected with such rotations.

Finally, it is clear from formula (2.4.90) that the squared magnitude of the angular momentum and its component along the z-axis do not form a complete set of simultaneously measurable physical quantities. Indeed, the eigenvalues of $\hat{\mathbf{L}}^2$ and $\hat{L}_z$ operators determine only the angular dependence of the corresponding eigenfunction $\psi_{lm}(r, \theta, \varphi)$, while the r-dependence (i.e., function $R(r)$) is not specified. In other words, to completely define the quantum mechanical state with certain values l and m, a value of another physical quantity must be additionally specified. In many applications, this quantity is energy.

2.5 Hamiltonian Operator

The Hamiltonian is the operator of energy. This operator occupies a central place in the structure of quantum mechanics. This is because the Hamiltonian controls the time evolution of a wave function. This is immediately apparent from the time-dependent Schrödinger equation (2.2.9). This equation can be written in a more general form by using Dirac notation as follows:

$$i\hbar\frac{\partial|\psi\rangle}{\partial t} = \hat{H}|\psi\rangle. \tag{2.5.1}$$

It is thus clear that a mathematical expression for the Hamiltonian is needed for the complete formulation of quantum mechanics. However, there is no one universal Hamiltonian that can be used for all physical problems. The mathematical form of Hamiltonians depends on the physical nature of interaction in microscopic systems as well as on the chosen representation of wave functions. Finding the appropriate form of the Hamiltonian is the creative work. Namely, the Hamiltonian has to be derived or properly guessed from our theoretical and experimental knowledge about the physical nature of the system under consideration. In this section, we consider only standard-forms of the Hamiltonian operators for spinless particles and electromagnetic interaction. These Hamiltonians were introduced at the very beginning of the development of quantum mechanics and the remarkable success of quantum theory has been achieved by using these Hamiltonians. The choice of these Hamiltonians was guided by the principle that the relations between quantum mechanical mean (average) values of physical quantities coincide with the relations between the same physical quantities in classical mechanics.

We start with the simplest case of a free particle. In classical mechanics, energy and momentum of free particles are related by the formula

$$\mathcal{E} = \frac{p_x^2 + p_y^2 + p_z^2}{2m}. \tag{2.5.2}$$

In quantum mechanics, the same relation between mean values can be written as follows

$$\overline{\mathcal{E}} = \frac{\overline{p_x^2 + p_y^2 + p_z^2}}{2m}. \tag{2.5.3}$$

By expressing the above mean values in terms of operators, we find

$$\langle \psi | \hat{H} \psi \rangle = \langle \psi | \frac{\hat{p}_x^2 + \hat{p}_y^2 + \hat{p}_z^2}{2m} \psi \rangle. \tag{2.5.4}$$

The last formula suggests the following relation between the Hamiltonian operator and momentum operator:

$$\hat{H} = \frac{\hat{p}_x^2 + \hat{p}_y^2 + \hat{p}_z^2}{2m}, \tag{2.5.5}$$

or

$$\hat{H} = \frac{\hat{\mathbf{p}}^2}{2m}. \tag{2.5.6}$$

The last formula can be viewed as the expression for the Hamiltonian of a free particle in momentum representation. By using formulas from (2.3.33), the expression for the Hamiltonian of a free particle in coordinate representation can be found. This expression is

$$\hat{H} = -\frac{\hbar^2}{2m} \left(\frac{\partial^2}{\partial x^2} + \frac{\partial^2}{\partial y^2} + \frac{\partial^2}{\partial z^2} \right), \tag{2.5.7}$$

or

$$\hat{H} = -\frac{\hbar^2}{2m} \nabla^2, \tag{2.5.8}$$

where ∇^2 is the Laplace operator.

The last formula implies that the time-dependent Schrödinger equation for a free particle has the form:

$$i\hbar \frac{\partial \psi}{\partial t} = -\frac{\hbar^2}{2m} \nabla^2 \psi. \tag{2.5.9}$$

It is interesting to find the expression for a wave function of a free particle. A free particle is the simplest example of a closed system for which space and time are homogeneous. This means that a free particle can be in a state with certain values of momentum and energy. By using formulas (2.2.48) and (2.3.38), we conclude that the wave function for this state is given by the formula

$$\psi_{\mathcal{E},\mathbf{p}}(\mathbf{r},t) = \frac{1}{(2\pi\hbar)^{\frac{3}{2}}} e^{\frac{i}{\hbar}\mathbf{p}\cdot\mathbf{r}} e^{-\frac{i}{\hbar}\mathcal{E}t}. \tag{2.5.10}$$

By substituting the last formula into equation (2.5.9), we find (as expected) that

$$\mathcal{E} = \frac{p^2}{2m}. \tag{2.5.11}$$

From equations (2.5.10) and (2.5.11) follows that

$$\psi_{\mathcal{E},\mathbf{p}}(\mathbf{r},t) = \frac{1}{(2\pi\hbar)^{\frac{3}{2}}} e^{\frac{i}{\hbar}\left(\mathbf{p}\cdot\mathbf{r} - \frac{p^2}{2m}t\right)}. \tag{2.5.12}$$

This is the famous DeBroglie wave function, which played a very important role at the inception of the wave form of quantum mechanics.

Next, we consider the Hamiltonian operator for a charged particle in an external (applied) electric field. According to classical mechanics, the energy of the particle in this case is the sum of kinetic and potential energies:

$$\mathcal{E} = \frac{p_x^2 + p_y^2 + p_z^2}{2m} + U(\mathbf{r},t). \tag{2.5.13}$$

In quantum mechanics, the same relation between mean values can be written as follows

$$\overline{\mathcal{E}} = \frac{\overline{p_x^2 + p_y^2 + p_z^2}}{2m} + \overline{U(\mathbf{r},t)}. \tag{2.5.14}$$

By expressing the above mean values in terms of operators, we find

$$\langle\psi|\hat{H}\psi\rangle = \langle\psi|\left(\frac{\hat{p}_x^2 + \hat{p}_y^2 + \hat{p}_z^2}{2m} + U(\mathbf{r},t)\right)\psi\rangle. \tag{2.5.15}$$

The last formula suggests the following expression for the Hamiltonian:

$$\hat{H} = \frac{\hat{p}_x^2 + \hat{p}_y^2 + \hat{p}_z^2}{2m} + U(\mathbf{r},t), \tag{2.5.16}$$

or

$$\hat{H} = -\frac{\hbar^2}{2m}\nabla^2 + U(\mathbf{r},t). \tag{2.5.17}$$

This means that the time-dependent Schrödinger equation for a charged particle in the external electric field has the form

$$i\hbar\frac{\partial\psi}{\partial t} = -\frac{\hbar^2}{2m}\nabla^2\psi + U(\mathbf{r},t)\psi. \tag{2.5.18}$$

In the case when the applied electric field is constant in time, the potential energy U does not depend on time. According to formula (2.5.17), this implies that the Hamiltonian operator is time-independent. This, in turn, suggests that the energy is conserved. The possible values of energy $\mathcal{E}_n$ (i.e., the energy spectrum) can be found by using the time-independent Schrödinger equation (2.2.49), that is by solving the following eigenvalue problem:

$$-\frac{\hbar^2}{2m}\nabla^2\phi_n(\mathbf{r}) + U(\mathbf{r})\phi_n(\mathbf{r}) = \mathcal{E}_n\phi_n(\mathbf{r}). \qquad (2.5.19)$$

By using the previous line of reasoning, the Hamiltonian in formula (2.5.17) can be extended to the case of many interacting charged particles. Namely, this Hamiltonian is given by the formula:

$$\hat{H} = -\frac{\hbar^2}{2}\sum_{k=1}^{N}\frac{1}{m_k}\nabla_k^2 + U(\mathbf{r}_1, \mathbf{r}_2, \ldots \mathbf{r}_N, t), \qquad (2.5.20)$$

where ∇_k^2 is the Laplacian operator with respect to coordinates of particle number k, and N is the number of particles.

The potential energy $U(\mathbf{r}_1, \mathbf{r}_2, \ldots \mathbf{r}_N, t)$ may have two distinct terms:

$$U(\mathbf{r}_1, \mathbf{r}_2, \ldots \mathbf{r}_N, t) = \frac{1}{8\pi\varepsilon_0}\sum_{j=1}^{N}\sum_{i\neq j}^{N}\frac{q_i q_j}{|\mathbf{r}_i - \mathbf{r}_j|} + \sum_{i=1}^{N}q_i\varphi_i(\mathbf{r}_i, t), \qquad (2.5.21)$$

where the first term in the right-hand side of the last formula is due to the interaction between charged particles, while the second term is due to the presence of an external electric field.

The corresponding time-dependent Schrödinger equation has the form:

$$i\hbar\frac{\partial\psi}{\partial t} = -\frac{\hbar^2}{2}\sum_{k=1}^{N}\frac{1}{m_k}\nabla_k^2\psi + U(\mathbf{r}_1, \mathbf{r}_2, \ldots \mathbf{r}_N, t)\psi. \qquad (2.5.22)$$

If the external electric field does not depend on time, then the energy spectrum of the microsystem of charged particles can be computed by using the time-independent Schrödinger equation

$$-\frac{\hbar^2}{2}\sum_{k=1}^{N}\frac{1}{m_k}\nabla_k^2\phi_n + U(\mathbf{r}_1, \mathbf{r}_2, \ldots \mathbf{r}_N)\phi_n = \mathcal{E}_n\phi_n. \qquad (2.5.23)$$

It is clear from the previous discussion that the Hamiltonian operator $\hat{H}$ has been obtained from the Hamiltonian of the classical mechanics by replacing the momentum $\mathbf{p}$ and coordinate-dependent potential energy U by the momentum operator and by the operator of multiplication by U, respectively. This way of constructing the Hamiltonian operator is consistent with the canonical quantization of classical mechanics.

As discussed above, in the absence of a magnetic field the momentum **p**, which is often called kinetic momentum, is replaced by the momentum operator. The quantization procedure is somewhat different in the presence of a magnetic field. When the magnetic field is present, the classical Hamiltonian function is given by the formula (see the last section of Chapter 1):

$$H = \frac{[\mathbf{p} - q\mathbf{A}(\mathbf{r},t)]^2}{2m} + q\varphi(\mathbf{r},t), \qquad (2.5.24)$$

where **p** is the canonical (not kinetic) momentum. The quantization procedure in this case is performed by replacing the canonical momentum **p** by the momentum operator. This leads to the following expression for the Hamiltonian operator

$$\boxed{\hat{H} = \frac{[\hat{\mathbf{p}} - q\mathbf{A}(\mathbf{r},t)]^2}{2m} + q\varphi(\mathbf{r},t),} \qquad (2.5.25)$$

and the time-dependent Schrödinger equation has the form

$$\boxed{i\hbar\frac{\partial\psi}{\partial t} = \frac{[\hat{\mathbf{p}} - q\mathbf{A}(\mathbf{r},t)]^2}{2m}\psi + q\varphi(\mathbf{r},t)\psi.} \qquad (2.5.26)$$

In some applications, the last equation can be simplified. The simplification is based on the formula:

$$[\hat{\mathbf{p}} - q\mathbf{A}(\mathbf{r},t)]^2 = \hat{\mathbf{p}}^2 - q\left[\mathbf{A}\cdot\hat{\mathbf{p}} + \hat{\mathbf{p}}\cdot\mathbf{A}\right] + q|\mathbf{A}|^2. \qquad (2.5.27)$$

It is clear that

$$\hat{\mathbf{p}}\cdot\mathbf{A} = -i\hbar\mathrm{div}\,\mathbf{A}. \qquad (2.5.28)$$

This term disappears if the Coulomb gauge

$$\mathrm{div}\,\mathbf{A} = 0. \qquad (2.5.29)$$

is adopted.

Furthermore, if the magnetic field is weak, the term $q|\mathbf{A}|^2$ in the right-hand side of formula (2.5.27) can be neglected. Consequently, that formula can be written as

$$[\hat{\mathbf{p}} - q\mathbf{A}(\mathbf{r},t)]^2 \approx \hat{\mathbf{p}}^2 - q\mathbf{A}\cdot\hat{\mathbf{p}}. \qquad (2.5.30)$$

This leads to the following expression for the Hamiltonian operator:

$$\hat{H} = -\frac{\hbar^2}{2m}\nabla^2 + \frac{iq\hbar}{2m}\mathbf{A}(\mathbf{r},t)\cdot\nabla + q\varphi(\mathbf{r},t), \qquad (2.5.31)$$

and the simplified version of the time-dependent Schrödinger equation (2.5.26) has the form

$$i\hbar\frac{\partial\psi}{\partial t} = -\frac{\hbar^2}{2m}\nabla^2\psi + \frac{iq\hbar}{2m}\mathbf{A}(\mathbf{r},t)\cdot\nabla\psi + q\varphi(\mathbf{r},t)\psi. \qquad (2.5.32)$$

Next, we shall consider the important issue of gauge invariance of the time-dependent Schrödinger equation (2.5.26). As discussed before, the state of a microsystem in quantum mechanics is completely described by its wave function ψ. However, it is easy to see that the prediction of the results of measurements will be the same if the wave function ψ is replaced by a wave function ψ' related to ψ by the formula

$$\psi' = e^{i\gamma}\psi, \qquad (2.5.33)$$

where γ is an arbitrary constant. In other words, ψ and ψ' provide equivalent descriptions of the quantum mechanical state. It is easy to verify that all Schrödinger equations discussed above are invariant with respect to the transformation (2.5.33). This means that these equations have the same mathematical forms for ψ and ψ'. The transformation (2.5.33) can be viewed as the global change of phase of wave function, and it reveals the global phase symmetry of the Schrödinger equation.

It can be easily observed that the probability density of the particle localization in space is also invariant with respect to the local change of phase of the wave function described by the following formula

$$\psi'(\mathbf{r},t) = e^{i\gamma(\mathbf{r},t)}\psi(\mathbf{r},t), \qquad (2.5.34)$$

where $\gamma(\mathbf{r},t)$ is not a constant but an arbitrary continuously differentiable function of $\mathbf{r}$ and t. Indeed, from the last formula follows that

$$|\psi'(\mathbf{r},t)|^2 = |\psi(\mathbf{r},t)|^2. \qquad (2.5.35)$$

The transformation described by the formula (2.5.34) leads to the local phase symmetry of the Schrödinger equation (2.5.26). Indeed, it turns out that the time-dependent Schrödinger equation (2.5.26) is invariant with respect to this transformation, and this invariance is acheived by replacing the potentials $\mathbf{A}(\mathbf{r},t)$ and $\varphi(\mathbf{r},t)$ in equation (2.5.26) by the equivalent gauge transformed potentials:

$$\mathbf{A}'(\mathbf{r},t) = \mathbf{A}(\mathbf{r},t) + \frac{\hbar}{q}\nabla\gamma(\mathbf{r},t), \qquad (2.5.36)$$

$$\varphi'(\mathbf{r},t) = \varphi(\mathbf{r},t) - \frac{\hbar}{q}\frac{\partial\gamma(\mathbf{r},t)}{\partial t}. \qquad (2.5.37)$$

The demonstration of this invariance proceeds as follows.

First, from formula (2.5.34), we find

$$\psi(\mathbf{r},t) = \psi'(\mathbf{r},t)e^{-i\gamma(\mathbf{r},t)}, \tag{2.5.38}$$

and

$$\frac{\partial\psi}{\partial t} = \frac{\partial\psi'}{\partial t}e^{-i\gamma(\mathbf{r},t)} - i\psi'\frac{\partial\gamma(\mathbf{r},t)}{\partial t}e^{-i\gamma(\mathbf{r},t)}. \tag{2.5.39}$$

Next, it follows from formula (2.5.38) that

$$[\hat{\mathbf{p}} - q\mathbf{A}]\,\psi = [\hat{\mathbf{p}} - q\mathbf{A}]\,\psi'e^{-i\gamma(\mathbf{r},t)}. \tag{2.5.40}$$

By using formula (2.5.36), the right hand side of the last equation can be transformed as shown below:

$$[\hat{\mathbf{p}} - q\mathbf{A}]\,\psi'e^{-i\gamma(\mathbf{r},t)} = -i\hbar e^{-i\gamma(\mathbf{r},t)}\nabla\psi' - \hbar e^{-i\gamma(\mathbf{r},t)}\psi'\nabla\gamma - q\mathbf{A}\psi'e^{-i\gamma(\mathbf{r},t)}$$
$$= e^{-i\gamma(\mathbf{r},t)}[\hat{\mathbf{p}} - q\mathbf{A}']\,\psi'. \tag{2.5.41}$$

This means that moving the exponent $e^{-i\gamma(\mathbf{r},t)}$ through the bracket results in the replacement of $\mathbf{A}$ by $\mathbf{A}'$ in the bracket. By using this fact, we derive

$$[\hat{\mathbf{p}} - q\mathbf{A}]^2\,\psi'e^{-i\gamma(\mathbf{r},t)} = [\hat{\mathbf{p}} - q\mathbf{A}]\,[\hat{\mathbf{p}} - q\mathbf{A}]\,\psi'e^{-i\gamma(\mathbf{r},t)}$$
$$= [\hat{\mathbf{p}} - q\mathbf{A}]\,e^{-i\gamma(\mathbf{r},t)}[\hat{\mathbf{p}} - q\mathbf{A}']\,\psi' = e^{-i\gamma(\mathbf{r},t)}[\hat{\mathbf{p}} - q\mathbf{A}']^2\,\psi'. \tag{2.5.42}$$

Consequently,

$$[\hat{\mathbf{p}} - q\mathbf{A}]^2\,\psi = e^{-i\gamma(\mathbf{r},t)}[\hat{\mathbf{p}} - q\mathbf{A}']^2\,\psi'. \tag{2.5.43}$$

Finally, by substituting formulas (2.5.38), (2.5.39) and (2.5.43) into equation (2.5.26) and then taking into account formula (2.5.37), we obtain:

$$i\hbar e^{-i\gamma(\mathbf{r},t)}\frac{\partial\psi'}{\partial t} + \hbar e^{-i\gamma(\mathbf{r},t)}\frac{\partial\gamma(\mathbf{r},t)}{\partial t}\psi'$$
$$= \frac{1}{2m}e^{-i\gamma(\mathbf{r},t)}[\hat{\mathbf{p}} - q\mathbf{A}']^2\,\psi' + qe^{-\gamma(\mathbf{r},t)}\varphi\psi', \tag{2.5.44}$$

which leads to

$$\boxed{i\hbar\frac{\partial\psi'}{\partial t} = \frac{[\hat{\mathbf{p}} - q\mathbf{A}'(\mathbf{r},t)]^2}{2m}\psi' + q\varphi'(\mathbf{r},t)\psi'.} \tag{2.5.45}$$

The mathematical form of the last equation for ψ' is identical to the mathematical form of the original equation (2.5.26) for ψ. This means that the time-dependent Schrödinger equation (2.5.26) is invariant with respect to the gauge transformation consisting of simultaneous local change of phase (2.5.34) of the wave function as well as transformations (2.5.36) and (2.5.37) of vector and scalar potentials, respectively. In other words, there exists the class of equivalent wave functions related to one another by the local phase transformation and describing the same quantum mechanical state. Correspondingly, there exist two classes of equivalent vector and scalar potentials defined by the gauge transformation (2.5.36) and (2.5.37) and describing the same electromagnetic field. Any change of ψ within the equivalence class of wave functions is associated with the appropriate changes of vector and scalar potentials.

The described gauge symmetry admits the following interpretation: the invariance of the time-dependent Schrödinger equation with respect to the local change of phase of the wave function requires the **existence** of electromagnetic fields described by the vector and scalar potentials equivalent under the gauge transformation. In other words, the local phase symmetry of the wave functions reveals the existence of electromagnetic interaction. This idea that local symmetries are connected with (and reveal) the physical nature of interactions is dominant in modern physics. Actually, it has been established that all fundamental interactions (electromagnetic, gravitational, weak and strong) are revealed by specific local symmetries. For instance, the existence of gravitation is required by the symmetry among various local non-inertial reference frames. In other words, physical laws have the same mathematical forms in different non-inertial reference frames due to the existence of the gravitational field, and the presence of the gravitational field manifests itself in the appearance of inertial forces. It turns out that the existence of the weak and strong interactions is due to specific internal (not space-time) symmetries between elementary particles. However, the discussion of this matter is beyond the scope of this text. Instead, we conclude this section by illustrating that the quantization of the magnetic flux is related to the local phase symmetry of wave functions.

Consider a doubly connected region (a ring, for instance) free of magnetic field:

$$\mathbf{B} = 0. \tag{2.5.46}$$

We assume that there is a nonzero magnetic field outside this region. It is our intention to demonstrate that the magnetic flux enclosed by this doubly connected region is quantized.

From the last formula follows that inside of the field-free region

$$\mathbf{B} = \nabla \times \mathbf{A}' = 0. \tag{2.5.47}$$

This implies that in the field-free region the vector potential can be represented as

$$\mathbf{A}' = \frac{\hbar}{q}\nabla\gamma. \tag{2.5.48}$$

The last formula can be viewed as the gauge transformation of vector potential $\mathbf{A} = 0$. This gauge transformation corresponds to the following local change of phase of wave function:

$$\psi'(\mathbf{r}, t) = e^{i\gamma(\mathbf{r}, t)}\psi(\mathbf{r}, t). \tag{2.5.49}$$

The flux enclosed by the doubly connected field-free region can be computed as

$$\Phi = \oint_L \mathbf{A} \cdot \mathbf{dl} = \frac{\hbar}{q} \oint_L \nabla\gamma \cdot \mathbf{dl}, \tag{2.5.50}$$

where L is a closed path within the field-free region.

The wave functions ψ' in formula (2.5.49) will be continuous and single valued only if the total change of γ around the path L is proportional to 2π:

$$\oint_L \nabla\gamma \cdot \mathbf{dl} = 2\pi n. \tag{2.5.51}$$

Now, from the last two formulas, we find

$$\Phi = \frac{2\pi\hbar}{q}n = \frac{h}{q}n. \tag{2.5.52}$$

Consider the case when the doubly connected region is occupied by a type-I superconductor. According to the Meissner effect of magnetic flux exclusion in such superconductors, the condition (2.5.46) is realized. In superconductors, electrons are bound in Cooper pairs and, consequently,

$$q = 2e, \tag{2.5.53}$$

where e is the absolute value of electron charge.

From the last two formulas, we find

$$\boxed{\Phi = \frac{h}{2e}n.} \tag{2.5.54}$$

The smallest possible value of the flux, called fluxon, is given by the formula

$$\boxed{\Phi_0 = \frac{h}{2e} = 2.07 \cdot 10^{-15} \text{ Wb,}} \tag{2.5.55}$$

where the symbol Wb stands for the weber, which is the SI unit of flux. The quantization of the magnetic flux enclosed by superconductors and described by the last two formulas has been experimentally observed. Furthermore, this quantization has important implications concerning the structure of the mixed state in type-II superconductors. In type-II superconductors, the transition from the superconducting state to the normal state proceeds through the mixed state. In the mixed state, a periodic pattern (lattice) of cylindrical normal regions surrounded by current vortices appears. This happens because in type-II materials the interface energy between the normal and superconducting regions is negative. As a result, the mixed state is more energetically favorable for some range of magnetic fields than the superconducting state or the normal state. In order to maximize the negative interface energy contribution with respect to the positive energy contribution of the normal regions, these cylindrical normal regions should have exceedingly small cross sections. These cross sections are limited from below by the condition that each current vortex around the normal region should generate the flux equal to one fluxon Φ_0. For this reason, the normal regions in the mixed states are filaments of magnetic flux Φ_0. Type-II superconductors are usually made of metal alloys (niobium-titanium or niobium-tin) or complex oxide ceramics. Type-II superconductors are used in strong electromagnets for medical MRI scanners and particle accelerators.

Problems

(1) Prove Cauchy inequality

$$|\langle \varphi | \psi \rangle| \leq \|\varphi\| \cdot \|\psi\| \tag{P.2.1}$$

and demonstrate that the equality holds only when ψ is a multiple of φ, that is, when $\psi = \alpha \varphi$.

(2) Prove the following identity for commutators

$$\left[\hat{f}\hat{g}, \hat{h}\right] = \hat{f}\left[\hat{g}, \hat{h}\right] + \left[\hat{f}, \hat{h}\right]\hat{g}. \tag{P.2.2}$$

(3) Prove the following identity

$$\left[\left[\hat{f}, \hat{g}\right], \hat{h}\right] + \left[\left[\hat{g}, \hat{h}\right], \hat{f}\right] + \left[\left[\hat{h}, \hat{f}\right], \hat{g}\right] = 0. \tag{P.2.3}$$

(4) Prove formulas (2.2.30) and (2.2.31).

(5) Verify that $\hat{\mathbf{r}}$ is a Hermitian operator.

(6) Verify that $\hat{\mathbf{p}}$ is a Hermitian operator.

(7) Prove that

$$[\varphi(x), \hat{p}_x] = i\hbar \frac{d\varphi}{dx} \tag{P.2.4}$$

for any function $\varphi(x)$.

(8) Prove the generalized uncertainty principle for two noncommuting operators $\hat{f}$ and $\hat{g}$:

$$\sigma_f^2 \sigma_g^2 \geq \left(\frac{1}{2i}\overline{\left[\hat{f}, \hat{g}\right]}\right)^2. \tag{P.2.5}$$

Hint: consider the function

$$F(\alpha) = \langle \left[\alpha\left(\hat{f} - \bar{f}\right) + i\left(\hat{g} - \bar{g}\right)\right]\psi \,|\, \left[\alpha\left(\hat{f} - \bar{f}\right) + i\left(\hat{g} - \bar{g}\right)\right]\psi \rangle \geq 0. \tag{P.2.6}$$

(9) By using the generalized uncertainty principle, derive formula (2.3.68).

(10) Show that the function (2.3.69) is the wave function of minimal uncertainty.

(11) Show that the probability density $|c(p_x)|^2$ of momentum is Gaussian in the state (2.3.69) of minimal uncertainty.

(12) Prove the commutation relations (2.4.24)—(2.4.26).

(13) Prove the commutation relations (2.4.27)—(2.4.29).

(14) Prove the commutation relations (2.4.30)—(2.4.32).

(15) Show that the commutation relations (2.4.37)—(2.4.39) can be written in the following form

$$\left[\hat{L}_k, \hat{L}_n\right] = i\hbar \sum_{m=1}^{3} \varepsilon_{knm} \hat{L}_m, \qquad (P.2.7)$$

where ε_{knm} is the Levi-Civita symbol (tensor) defined by the formula

$$\varepsilon_{knm} = \begin{cases} 1, & \text{if } knm = 123, 231, 312 \\ -1, & \text{if } knm = 132, 213, 321 \\ 0, & \text{otherwise.} \end{cases} \qquad (P.2.8)$$

(16) By using the generalized uncertainty principle (see Problem 8), prove that

$$\sigma_{L_x}^2 \sigma_{L_y}^2 \geq \frac{\hbar^2}{4} \left(\hat{L}_z\right)^2. \qquad (P.2.9)$$

(17) Prove formula (2.4.82).

(18) Prove formula (2.4.83).

(19) Derive the expression (2.4.89) for $\widehat{\mathbf{L}}^2$.

(20) By using quantum mechanical reasoning, explain how spherical harmonics $Y_l^m(\theta, \varphi)$ are transformed as a result of rotation of coordinate axes?

(21) Is the Schrödinger equation (2.5.22) invariant with respect to the local change of phase of the wave function?

(22) Describe the class of Hamiltonians equivalent to the Hamiltonian (2.5.25).

(23) Under what conditions the energy is conserved for the Hamiltonian (2.5.25)?

Chapter 3

Tunneling

3.1 Tunneling Through a Rectangular Barrier

Tunneling is the quantum mechanical phenomenon when microscopic particles may pass through classically prohibited regions. Such regions are usually energy barriers whose energies may exceed the particle energies. Tunneling occurs because particle wave functions extend into barriers and beyond barrier regions. This results in finite measurable probability of finding the particles behind the barrier regions, which is the essence of tunneling. The influx of the particles in the region behind the barrier results in tunneling currents. Tunneling has many technological applications. Some of them are discussed at the end of this section.

First, we consider the simplest problem of tunneling through a rectangular barrier. We shall distinguish two cases when the energy $\mathcal{E}$ of incoming particle is above the potential energy U_0 of the barrier (see Figure 3.1(a)) and when the particle energy is below U_0 (see Figure 3.1(b)). We recall

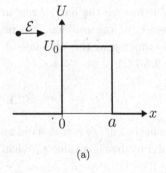

(a)

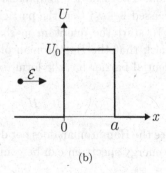

(b)

Fig. 3.1

75

that in classical mechanics the particle energy is given by the formula

$$\mathcal{E} = \frac{mv^2}{2} + U(x).$$ (3.1.1)

This energy is conserved because the potential energy of the external field does not depend on time. In the case a) when

$$\mathcal{E} > U_0,$$ (3.1.2)

the particle moves through the barrier region without any reflection. The particle speed will be reduced in the barrier region. However, its kinetic energy $\frac{1}{2}mv^2$ still will be positive according to the inequality

$$\mathcal{E} = \frac{mv^2}{2} + U_0 > U_0,$$ (3.1.3)

which immediately follows from formulas (3.1.1) and (3.1.2). The situation is different in quantum mechanics which predicts (as discussed below) that the microparticle will experience **over-barrier reflection**.

In the case b) when

$$\mathcal{E} < U_0,$$ (3.1.4)

the classical particle is reflected from the barrier. This follows from the inequality

$$\mathcal{E} = \frac{mv^2}{2} + U_0 < U_0$$ (3.1.5)

which is the consequence of formulas (3.1.1) and (3.1.4) and which suggests that the kinetic energy of the particle cannot be positive within the barrier region. The situation is different in quantum mechanics which predicts (as discussed below) that the particle may tunnel through the barrier region.

To start the quantum mechanical discussion of the problem, we first remark that the Hamiltonian of the particle moving in the presence of a potential barrier has the form (see formula (2.5.17)):

$$\hat{H} = -\frac{\hbar^2}{2m}\frac{d^2}{dx^2} + U(x).$$ (3.1.6)

Since the Hamiltonian does not depend on time, the energy is conserved and the energy spectrum can be computed by solving the eigenvalue problem

$$\hat{H}\Phi(x) = \mathcal{E}\Phi(x).$$ (3.1.7)

By using formula (3.1.6), the last equation can be written for three distinct regions shown in Figure 3.1 as follows:

$$\frac{d^2\Phi(x)}{dx^2} + \frac{2m\mathcal{E}}{\hbar^2}\Phi(x) = 0, \quad (-\infty < x < 0), \tag{3.1.8}$$

$$\frac{d^2\Phi(x)}{dx^2} + \frac{2m\left(\mathcal{E} - U_0\right)}{\hbar^2}\Phi(x) = 0, \quad (0 < x < a), \tag{3.1.9}$$

$$\frac{d^2\Phi(x)}{dx^2} + \frac{2m\mathcal{E}}{\hbar^2}\Phi(x) = 0, \quad (a < x < \infty). \tag{3.1.10}$$

At the boundary of the barrier region, the following boundary conditions are valid

$$\Phi(0_-) = \Phi(0_+), \tag{3.1.11}$$

$$\frac{d\Phi}{dx}(0_-) = \frac{d\Phi}{dx}(0_+), \tag{3.1.12}$$

$$\Phi(a_-) = \Phi(a_+), \tag{3.1.13}$$

$$\frac{d\Phi}{dx}(a_-) = \frac{d\Phi}{dx}(a_+), \tag{3.1.14}$$

where subscripts "$-$" and "$+$" correspond to limits from "below" and "above," respectively.

The above boundary conditions follow from the continuity of the wave function and its derivative. It is apparent that a general solution of equation (3.1.8) can be written as follows

$$\Phi(x) = e^{ikx} + re^{-ikx}, \quad (-\infty < x < 0), \tag{3.1.15}$$

where

$$k = \frac{\sqrt{2m\mathcal{E}}}{\hbar}. \tag{3.1.16}$$

It is apparent that the first term in the right-hand side of equation (3.1.15) represents the wave function of the incoming particle, while the last term represents the wave function of the reflected particle. It is also easy to see that

$$R = |r|^2 = \left|re^{-ikx}\right|^2 \tag{3.1.17}$$

has the physical meaning of the reflection coefficient. Indeed, R can be interpreted as the probability density of spatial position of the reflected particle.

Next, a general solution of differential equation (3.1.9) has the form

$$\Phi(x) = B_1 e^{iqx} + B_2 e^{-iqx}, \quad (0 < x < a), \tag{3.1.18}$$

where

$$q = \frac{\sqrt{2m(\mathcal{E} - U_0)}}{\hbar}. \tag{3.1.19}$$

Finally, a solution of the differential equation (3.1.10) can be written as follows

$$\Phi(x) = te^{ikx}, \quad (a < x < \infty), \tag{3.1.20}$$

where k is given by formula (3.1.16). This solution has only one term, which represents the forward traveling wave corresponding to the particle which passed through the barrier region. It is apparent that

$$T = |t|^2 = |te^{ikx}|^2 \tag{3.1.21}$$

has the physical meaning of the transmission coefficient. Indeed, T can be interpreted as the probability density of spatial position of the particle passed through the barrier.

The derivation of the formula for this transmission coefficient is our immediate goal. To this end, we first remark that the boundary conditions (3.1.1)–(3.1.14) will be satisfied if r, B_1, B_2 and t are the solutions of the following simultaneous algebraic equations

$$1 + r = B_1 + B_2, \tag{3.1.22}$$

$$1 - r = \frac{q}{k}(B_1 - B_2), \tag{3.1.23}$$

$$B_1 e^{iqa} + B_2 e^{-iqa} = te^{ika}, \tag{3.1.24}$$

$$B_1 e^{iqa} - B_2 e^{-iqa} = t\frac{k}{q}e^{ika}. \tag{3.1.25}$$

By adding equations (3.1.22) and (3.1.23), we obtain

$$2 = \frac{k+q}{k}B_1 + \frac{k-q}{k}B_2. \tag{3.1.26}$$

Similarly, by adding equations (3.1.24) and (3.1.25), we obtain

$$2B_1 e^{iqa} = t\frac{k+q}{q}e^{ika}, \tag{3.1.27}$$

which leads to

$$B_1 = t\frac{k+q}{2q}e^{i(k-q)a}. \tag{3.1.28}$$

On the other hand, by subtracting equation (3.1.25) from equation (3.1.24), we arrive at:

$$2B_2 e^{-iqa} = t\frac{q-k}{q}e^{ika},$$
(3.1.29)

which leads to

$$B_2 = -t\frac{k-q}{2q}e^{i(k+q)a}.$$
(3.1.30)

By substituting formulas (3.1.28) and (3.1.30) into equation (3.1.26), we derive

$$2 = t\left[\frac{(k+q)^2}{2kq}e^{i(k-q)a} - \frac{(k-q)^2}{2kq}e^{i(k+q)a}\right],$$
(3.1.31)

from which follows that

$$t = \frac{4kq}{e^{ika}\left[(k+q)^2 e^{-iqa} - (k-q)^2 e^{iqa}\right]}.$$
(3.1.32)

By using Euler's formula in the last equation, we transform this equation as follows

$$t = \frac{4kq}{e^{ika}\left\{\left[(k+q)^2 - (k-q)^2\right]\cos qa - i\left[(k+q)^2 + (k-q)^2\right]\sin qa\right\}},$$
(3.1.33)

which leads to

$$t = \frac{2kq}{e^{ika}\left[2kq\cos qa - i\left(k^2 + q^2\right)\sin qa\right]}.$$
(3.1.34)

Now, consider the case a) when the inequality (3.1.2) is valid. According to formula (3.1.19), q is real in this case. By using this fact and formula (3.1.21), we derive from the last equation that

$$T = \frac{4k^2 q^2}{4k^2 q^2 \cos^2 qa + (k^2 + q^2)^2 \sin^2 qa},$$
(3.1.35)

which can be further transformed as follows

$$T = \frac{4k^2 q^2}{4k^2 q^2 + \left[(k^2 + q^2)^2 - 4k^2 q^2\right]\sin^2 qa}.$$
(3.1.36)

By using simple algebra, we arrive at the final expression for the transmission coefficient:

$$\boxed{T = \frac{1}{1 + \left(\frac{k^2 - q^2}{2kq}\sin qa\right)^2}.}$$
(3.1.37)

It is apparent from the last formula that if $\sin qa \neq 0$, then

$$T < 1. \tag{3.1.38}$$

The last inequality implies that there exists **over-barrier reflection**. This phenomenon is purely of quantum mechanical nature.

It is also clear that

$$T = 1, \tag{3.1.39}$$

if

$$\sin qa = 0. \tag{3.1.40}$$

The last equation is valid if

$$qa = n\pi, \quad (n = 0, 1, 2, \dots). \tag{3.1.41}$$

By recalling formula (3.1.19), the last relation can be transformed as follows:

$$\sqrt{\frac{2m\left(\mathcal{E}_n - U_0\right)}{\hbar}}a = n\pi, \quad (n = 0, 1, 2, \dots), \tag{3.1.42}$$

which leads to

$$\boxed{\mathcal{E}_n = U_0 + \frac{\pi^2 \hbar^2}{2ma^2}n^2,} \quad (n = 0, 1, 2, \dots). \tag{3.1.43}$$

The last formula implies that for the countable set of energies $\mathcal{E}_n$ there is no over-barrier reflection. In other words, only for these special values of energy particles can pass over the energy barrier without any reflection, that is, in the same way as in classical mechanics. It is worthwhile to note that the existence of **infinite** discrete values of energy $\mathcal{E}_n$ for which barriers are transparent is a unique property of rectangular barriers.

Now, we proceed to the discussion of the case b) when the inequality (3.1.4) is valid. This is the case when tunneling of microscopic particles through the energy barrier occurs. It is clear from formula (3.1.19) that in this case q is an imaginary quantity:

$$q = i\gamma, \tag{3.1.44}$$

where

$$\gamma = \frac{\sqrt{2m\left(U_0 - \mathcal{E}\right)}}{\hbar}. \tag{3.1.45}$$

By using the following relations for hyperbolic functions

$$\sin qa = \sin i\gamma a = i\sinh \gamma a, \tag{3.1.46}$$

$$\cos qa = \cos i\gamma a = \cosh \gamma a, \tag{3.1.47}$$

$$\cosh^2 \gamma a - \sinh^2 \gamma a = 1, \tag{3.1.48}$$

formula (3.1.34) is transformed as follows

$$t = \frac{2k\gamma}{e^{ika}\left[2k\gamma \cosh \gamma a - i\left(k^2 - \gamma^2\right)\sinh \gamma a\right]}. \tag{3.1.49}$$

By recalling the definition of the transmission coefficient (see formula (3.1.21)), we derive

$$T = \frac{4k^2\gamma^2}{\left[4k^2\gamma^2 \cosh^2 \gamma a + \left(k^2 - \gamma^2\right)^2 \sinh^2 \gamma a\right]}. \tag{3.1.50}$$

Finally, by using the identity (3.1.48), we find

$$T = \frac{1}{1 + \left[\frac{k^2 + \gamma^2}{2k\gamma}\sinh \gamma a\right]^2}. \tag{3.1.51}$$

It is clear from the last formula that

$$T > 0. \tag{3.1.52}$$

This implies that the physical phenomenon of tunneling is always present. From formulas (3.1.45) and (3.1.16) follows that

$$k^2 + \gamma^2 = \frac{2mU_0}{\hbar^2}, \tag{3.1.53}$$

and

$$k\gamma = \frac{2m\sqrt{\mathcal{E}\left(U_0 - \mathcal{E}\right)}}{\hbar^2}. \tag{3.1.54}$$

By using the last two relations in formula (3.1.51), we derive that

$$T = \frac{1}{1 + \frac{U_0^2}{4\mathcal{E}(U_0 - \mathcal{E})}\sinh^2 \gamma a}. \tag{3.1.55}$$

In the case of wide energy barriers

$$\gamma a \gg 1. \tag{3.1.56}$$

Then,

$$\sinh \gamma a \simeq \frac{e^{\gamma a}}{2} \gg 1. \tag{3.1.57}$$

This leads to the following simplification of formula (3.1.55):

$$T \approx \frac{16\mathcal{E}\left(U_0 - \mathcal{E}\right)}{U_0^2}e^{-2\gamma a}, \tag{3.1.58}$$

which can also be written as

$$T \approx \frac{16\mathcal{E}\,(U_0 - \mathcal{E})}{U_0^2} e^{-\frac{2\sqrt{2m(U_0 - \mathcal{E})}}{\hbar}a}. \qquad (3.1.59)$$

The last formula clearly reveals the **exponential** dependence of the tunneling coefficient (and, consequently, the tunneling current) on barrier width a. It is also clear from formula (3.1.55) that in the case of very narrow barriers, that is, when

$$a \approx 0, \qquad (3.1.60)$$

the transmission coefficient is close to one

$$T \approx 1, \qquad (3.1.61)$$

and tunneling through energy barriers is very efficient.

The presented discussion is illustrated by Figure 3.2, where the graphs of the transmission coefficient T are plotted for the classical (curve 1) and quantum mechanical (curve 2) cases. It is also clear from the presented

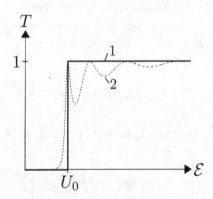

Fig. 3.2

discussion that the energy spectrum (see equation (3.1.7)) is continuous. In the tunneling case (case b)), this spectrum extends from zero to U_0. This is apparent from formula (3.1.16) and equation (3.1.53), which is illustrated by Figure 3.3.

Historically, quantum tunneling was first experimentally observed in the study of radioactive decay of nuclei. The theoretical explanation of the alpha decay of nuclei on the basis of quantum tunneling was first proposed in

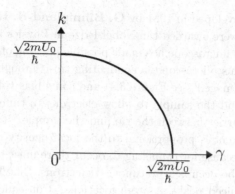

Fig. 3.3

1928 by **G. Gamow**. With the advent of modern semiconductor technology, the physical phenomenon of tunneling found important applications in semiconductor devices. We shall briefly mention two of them. First is tunnel metal-semiconductor contacts which are ubiquitous in transistors. Such contacts have energy barriers at the metal-semiconductor interface. However, semiconductors are highly doped near the interface with a doping density of 10^{19} cm^{-3} or higher. This results in very thin (on the order of 3 nm or less) depletion regions at the metal-semiconductor interface, which manifest themselves as energy barriers. Since these energy barriers are very narrow, carriers can efficiently tunnel across such barriers (see formulas (3.1.60) and (3.1.61)).

Another important example of using quantum tunneling in semiconductor technology is flash memory. This memory uses floating-gate transistors. These transistors resemble standard MOSFETs, except they have two gates instead of one. One of them is the control gate as in standard MOSFETs, while the other is the floating gate completely insulated by very thin silicon oxide layers which represent energy barriers for mobile charge carriers. During writing, carrier tunnel injection is used for charging the floating gate. During erasure, carrier tunnel release is realized to remove charge from the floating gate.

It is also worthwhile to mention that spin-dependent tunneling in magnetic tunnel junctions is a very active research area in spintronics. This tunneling is briefly discussed at the end of Section 7.2.

Finally, we shall point out that quantum tunneling is at the very foundation of operation of the **scanning tunneling microscope**. This microscope was developed in 1981 by **G. Binnig and H. Rohrer** at IBM Zurich, and they were awarded the Nobel Prize in Physics in 1986 for this achievement. This microscope has made possible imaging of surfaces at the atomic level. In this microscope, a conducting tip is brought very close to the surface to be imaged (see Figure 3.4) and some bias voltage is applied between the tip and the sample to allow electrons to tunnel through the vacuum energy barrier between the tip and the sample. It turns out that the tunneling current is proportional to the local density of states of the sample. This is because for tunneling to occur there must be energy levels (states) that can be occupied by tunneled electrons. The local density of states has pronounced peaks at atom locations. Consequently, when the tip scans across the sample surface, peaks of the tunneling current occur at atomic locations, and this eventually results in the image of the atomic structure of the sample surface. The exponential dependence of tunneling current on the barrier width (see formula (3.1.58)) leads to the remarkable sensitivity of scanning tunneling microscopy, which in turn results in high-resolution images of the surface atomic structure.

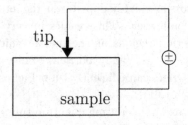

Fig. 3.4

3.2   Resonant Tunneling

In the previous section, tunneling across a single rectangular energy barrier has been discussed. In this section, we extend the previous discussion to the case of tunneling across two rectangular energy barriers (see Figure 3.5 on the next page). At first, it may seem that this problem is of no particular interest because intuition suggests that an additional energy barrier may only result in further reduction of the transmission coefficient, which should

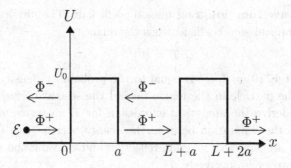

Fig. 3.5

make tunneling through two energy barriers much less probable. However, this is not always the case and, if two energy barriers are **identical**, then the remarkable phenomenon of **resonant tunneling** occurs. The essence of this phenomenon is that for certain energies of incoming microparticles tunneling through two identical energy barriers happens without any reflection, that is, with the transmission coefficient equal to one. In other words, for certain particle energies two identical energy barriers become completely transparent to incident microparticles, which is not possible for a single energy barrier. Remarkably, the phenomenon of resonant tunneling is not sensitive to the width of energy barriers, which is also in contrast with single barrier tunneling.

Next, we proceed to the mathematical analysis of resonant tunneling. As shown in Figure 3.5, in the regions outside of the energy barrier there are forward Φ^+ and backward Φ^- propagating wave functions which are solutions of the Schrödinger equation of the type (3.1.8). These wave functions are given by the formulas

$$\Phi^+(x) = Ae^{ikz}, \tag{3.2.1}$$

$$\Phi^-(x) = Be^{-ikx}, \tag{3.2.2}$$

where, as before, k is as follows

$$k = \frac{\sqrt{2m\mathcal{E}}}{\hbar}. \tag{3.2.3}$$

The transmission coefficient through two energy barriers t' is defined by the formula

$$t' = \frac{\Phi^+(L+2a)}{\Phi^+(0)}. \tag{3.2.4}$$

This is the **wave function** transmission coefficient. The physical meaning of the true transmission coefficient has the quantity

$$T' = |t'|^2. \tag{3.2.5}$$

Indeed, when $\Phi^+(0) = 1$, T' is equal to the probability density of spatial location of the particle in the region behind the second barrier. It is our intention to derive the analytical expression for T' in terms of L, r and t, where L is the separation between the energy barriers (see Figure 3.5) while r and t are the wave function reflection and transmission coefficients for a single barrier, respectively.

First, consider the second energy barrier. According to the definition of t, we have

$$\Phi^+(L + 2a) = t\Phi^+(L + a). \tag{3.2.6}$$

On the other hand, from formula (3.2.1) we find

$$\Phi^+(L + a) = \Phi^+(a)e^{ikL}. \tag{3.2.7}$$

By substituting the last formula into the right-hand side of equation (3.2.6), we obtain

$$\Phi^+(L + 2a) = te^{ikL}\Phi^+(a). \tag{3.2.8}$$

According to the definition of r, we have

$$\Phi^-(L + a) = r\Phi^+(L + a). \tag{3.2.9}$$

From the last equation and formula (3.2.7), we find

$$\Phi^-(L + a) = re^{ikL}\Phi^+(a). \tag{3.2.10}$$

On the other hand, from formula (3.2.2), we conclude that

$$\Phi^-(L + a) = \Phi^-(a)e^{-ikL}. \tag{3.2.11}$$

By substituting the last formula into the left-hand side of equation (3.2.10), we arrive at

$$\Phi^-(a) = re^{2ikL}\Phi^+(a). \tag{3.2.12}$$

Now, consider the first energy barrier. It is clear that $\Phi^+(a)$ is the sum of two distinct terms. The first term $t\Phi^+(0)$ is due to the transmission of the forward traveling wave function through the first energy barrier. The second term $r\Phi^-(a)$ is due to the reflection by the first energy barrier of the wave Φ^- traveling backward between the two barriers. Thus, it can be concluded that

$$\Phi^+(a) = t\Phi^+(0) + r\Phi^-(a). \tag{3.2.13}$$

By using formula (3.2.12) in the last equation, we find

$$\Phi^+(a) = t\Phi^+(0) + r^2 e^{2ikL}\Phi^+(a), \qquad (3.2.14)$$

which leads to

$$\Phi^+(a) = \frac{t}{1 - r^2 e^{2ikL}}\Phi^+(0). \qquad (3.2.15)$$

Next, by substituting the last formula into the right-hand side of equation (3.2.8), we obtain

$$\Phi^+(L + 2a) = \frac{t^2 e^{ikL}}{1 - r^2 e^{2ikL}}\Phi^+(0). \qquad (3.2.16)$$

Now, by recalling the definition of t' (see formula (3.2.4)), we conclude that

$$t' = \frac{t^2 e^{ikL}}{1 - r^2 e^{2ikL}}. \qquad (3.2.17)$$

Next, we shall perform some algebraic transformations of the last formula in order to arrive at the simple expression for T'. First, we shall use the relation

$$r = |r|e^{i\theta} \qquad (3.2.18)$$

to modify formula (3.2.17) as shown below

$$t' = \frac{t^2 e^{ikL}}{1 - |r|^2 e^{2i(kL+\theta)}}. \qquad (3.2.19)$$

The last equation can be next transformed as follows

$$t' = \frac{t^2 e^{ikL}}{1 - |r|^2 \cos 2(kL + \theta) + i|r|^2 \sin 2(kL + \theta)}. \qquad (3.2.20)$$

Now, by using formula (3.2.5), we find

$$T' = \frac{|t|^4}{1 - 2|r|^2 \cos 2(kL + \theta) + |r|^4}. \qquad (3.2.21)$$

As discussed before,

$$|t|^2 = T, \qquad (3.2.22)$$

and

$$|r|^2 = R, \qquad (3.2.23)$$

where T and R are transmission and reflection coefficients for a single energy barrier.

By using the last two formulas, equation (3.2.21) can be written in the form

$$T' = \frac{T^2}{1 - 2R\cos 2(kL + \theta) + R^2}. \tag{3.2.24}$$

By using the trigonometric identity

$$\cos 2(kL + \theta) = 1 - 2\sin^2(kL + \theta), \tag{3.2.25}$$

formula (3.2.24) can be written as follows

$$T' = \frac{T^2}{(1 - R)^2 + 4R\sin^2(kL + \theta)}. \tag{3.2.26}$$

Since

$$(1 - R)^2 = T^2, \tag{3.2.27}$$

formula (3.2.26) can be represented in the following final form

$$\boxed{T' = \frac{1}{1 + \frac{4R}{T^2}\sin^2(kL + \theta)}.} \tag{3.2.28}$$

It is apparent from the last formula that

$$\boxed{T' = 1,} \tag{3.2.29}$$

if

$$\sin(kL + \theta) = 0. \tag{3.2.30}$$

The latter equality occurs when

$$k_n L + \theta_n = n\pi. \tag{3.2.31}$$

According to formula (3.2.3), we have

$$k_n = \frac{\sqrt{2m\mathcal{E}_n}}{\hbar}, \tag{3.2.32}$$

and equation (3.2.31) can be written as

$$\frac{\sqrt{2m\mathcal{E}_n}}{\hbar} = \frac{n\pi - \theta_n}{L}, \tag{3.2.33}$$

which leads to

$$\mathcal{E}_n = \frac{(n\pi - \theta_n)^2 \hbar^2}{2mL^2}. \tag{3.2.34}$$

This is a nonlinear equation for $\mathcal{E}_n$ because the wave function reflection coefficient r and, consequently, θ_n depend on $\mathcal{E}_n$. However, in the case when $\mathcal{E}_n \ll U_0$, we find that

$$r \approx -1, \tag{3.2.35}$$

and

$$\theta_n \approx \pi. \tag{3.2.36}$$

Now, formula (3.2.33) can be written as

$$\boxed{\mathcal{E}_n = \frac{(n-1)^2 \pi^2 \hbar^2}{2mL^2}.} \tag{3.2.37}$$

It is thus established that for microparticle energies specified by formula (3.2.37) the transmission coefficient T' is equal to one and the two identical energy barriers are completely transparent. It must be noted that formula (3.2.37) is valid for sufficiently small n when the condition $\mathcal{E}_n \ll U_0$ is satisfied.

If, however,

$$\mathcal{E} \neq \mathcal{E}_n \text{ and } \mathcal{E} \ll U_0, \tag{3.2.38}$$

then

$$T \ll 1, \tag{3.2.39}$$

while

$$R \approx 1, \tag{3.2.40}$$

and, consequently,

$$\frac{4R}{T^2} \gg 1. \tag{3.2.41}$$

According to the first inequality in formula (3.2.38), $\sin^2(kL + \theta) \neq 0$. This together with inequality (3.2.41) implies (see formula (3.2.28)) that

$$T' \ll 1. \tag{3.2.42}$$

This results in the conclusion that T' as a function of energy $\mathcal{E}$ of an incident microparticle is strongly peaked at $\mathcal{E}_n$. The latter is qualitatively illustrated by Figure 3.6. It is apparent from the last figure that the transmission coefficient and, consequently, tunneling through two identical barriers exhibit resonance nature. That is why this tunneling is called resonant tunneling.

Usually, a tunneling current I is proportional to the transmission coefficient, while the microparticle energy is proportional to the applied voltage V. This suggests that I vs. V characteristics of devices based on resonant tunneling will replicate the T' vs. $\mathcal{E}$ plot shown in Figure 3.6. This means that in such devices differential conductivity g_{dif} can be negative

$$g_{dif} = \frac{dI}{dV} < 0. \tag{3.2.43}$$

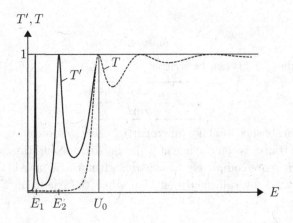

Fig. 3.6

This property of negative differential conductivity (negative differential resistance) can be exploited to design various resonant tunneling semiconductor devices that can be used, for instance, as oscillators. A resonant tunneling diode (RTD) is one and the most prominent example of such devices. The field of resonant tunneling devices has been advanced in the works of **Esaki, Kazarinov and Suris, Capasso** and many other researchers.

We have discussed resonant tunneling for two identical energy barriers. However, the phenomenon of resonant tunneling is present in the case of multiple identical and equally spaced energy barriers. The mathematical analysis in this case is more involved, but it still can be carried out by using the machinery of scattering matrices for individual barriers and their cascading.

It is important to point out that the phenomenon of resonant tunneling is very sensitive to maintaining the identical structure of the energy barriers. Deviations from this identical structure may appreciably degrade resonant tunneling. In some sense, the situation here is similar to the transport of microparticles through a periodic structure of energy barriers discussed in Chapter 5. It is shown there that in the case of an infinite number of identical and periodically placed energy barriers a microparticle may pass through them without any scattering (i.e., reflection) for certain energy values which form energy bands. Scattering appears as a result of deviations from periodic structure of identical energy barriers. In the case when these deviations are due to thermal distortions of periodic structure of identical energy barriers, these deviations manifest themselves as phonon scattering. In the case when there are perturbations of identical structure of energy barriers due to presence of impurities, these perturbations manifest themselves as impurity scattering.

3.3 Josephson Tunneling in Superconductors

Very interesting phenomena of tunneling occur in superconducting junctions. These phenomena were theoretically predicted by **B. Josephson**, and he was awarded the Nobel Prize in Physics in 1973 for this achievement.

Before proceeding to the discussion of **Josephson tunneling** in superconductors, it is worthwhile to briefly review the very basic facts related to superconductivity. Superconductivity was discovered by **H. Kamerlingh Onnes** in 1911 soon after he had succeeded in liquefying helium, making very low temperature experimentation possible. Kamerlingh Onnes observed that the electrical resistivity ρ of solid mercury abruptly and completely disappeared below $T_c = 4.2$ K as shown in Figure 3.7. In other

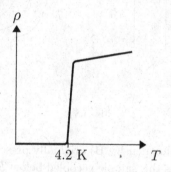

Fig. 3.7

words, below 4.2 K mercury is superconducting. This discovery was followed by observations that other metals (lead and tin) also exhibit zero resistivity below certain critical temperatures T_c. It has been also found that good conductors are bad superconductors and, the other way around, bad conductors can be good superconductors. For instance, gold, silver and copper have not been observed to be superconducting. This suggests that phonon scattering that is responsible for the origin of resistivity in metals above critical temperatures may be instrumental in formation of superconducting states below the critical temperature. Over the years, superconductivity has been discovered in alloys (such as niobium-titanium, niobium-tin) as well as in lanthanum and yttrium-based cuprate ceramics. The development of the latter class of superconductors was initiated by the groundbreaking research of **G. Bednorz and A. Muller in 1986**. Cuprate ceramic superconductors are called high-temperature superconductors and many of them can be cooled down below critical temperatures by using liquid nitrogen (boils at 77 K) instead of liquid helium. Another hallmark property of superconductors is **magnetic flux exclusion** which was discovered by W. Meissner and R. Ochsenfeld and is called the **Meissner effect**. This effect is illustrated by Figure 3.8. This figure demonstrates

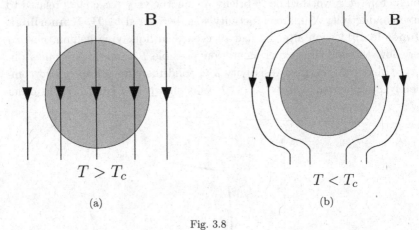

Fig. 3.8

that a magnetic flux (flux density **B**) originally present in a normal sample at $T > T_c$ is expelled as this sample is cooled below T_c. This clearly reveals the difference between superconductor and perfect conductor, in which the magnetic flux would be trapped in instead of being expelled. The Meissner effect implies that superconductors (type I superconductor to be precise) are perfect diamagnetics in which magnetic field **H** is completely canceled by magnetization **M** resulting in zero magnetic flux density ($\mathbf{B} = 0$). In type I materials, the superconducting state can be destroyed by applying magnetic field **H** above some critical value H_c which depends on temperature T. In other words, on the $H - T$ plane, the normal state is separated from the superconducting state by the curve $H_c(T)$ as shown in Figure 3.9. The existence of a critical magnetic field and perfect diamagnetic na-

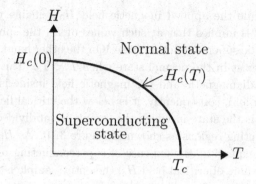

Fig. 3.9

ture of superconductors leads to the existence of the **intermediate state** when normal and superconducting regions coexist within the same sample. The transition from the superconducting state to the normal state proceeds gradually through the intermediate state. This is illustrated by Figure 3.10. Here, a superconducting sphere is subject to a uniform and

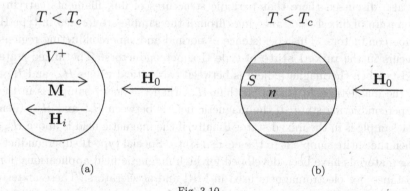

Fig. 3.10

gradually increasing magnetic field H_0. Since the superconductor is the perfect diamagnetic, its magnetization direction is opposite to the direction of $\mathbf{H}_0$ (see Figure 3.10). This results in the demagnetizing field whose direction is the same as the direction of $\mathbf{H}_0$. Consequently, the total field $\mathbf{H}_i$ inside the superconducting sphere is stronger than the applied field. For this reason, the magnetic field inside the spherical sample may exceed the critical field, while the applied magnetic field H_0 remains yet below the critical field. This implies that at such values of H_0 the spherical sample cannot exist in the superconducting state. On the other hand, the spherical sample cannot exist in the normal state when $T < T_c$. This is because it is not anymore diamagnetic and the magnetic field inside is weaker than the applied field and, consequently, it is below the critical field. Then, the only possibility is the state in which the sample is subdivided into normal and superconducting regions as shown in Figure 3.10. As H_0 is increased, normal regions expand at the expense of superconducting regions. When H_0 reaches the value of critical field H_c, the entire sample is in the normal state and the transition from the superconducting state to the normal state is completed.

It is important to stress that the existence of the intermediate state is due to the presence of the demagnetizing field. In this sense, this is the effect of sample shape rather than being the intrinsic property of type I superconducting materials. There are, however, type II superconductors in which coexistence of normal and superconducting regions is the property of these materials. As discussed at the end of section 2.5, this coexistence is energetically favorable because in type II superconductors, the interface energy between the normal and superconducting regions is negative. It is also discussed there that periodic structures of flux filaments carrying quantum of flux Φ_0 and piercing through the sample are formed in type II superconductors. This coexistence of normal and superconducting regions occurs in the **mixed state** of type II superconductors. The mixed state exists when the magnetic field is between two critical values H_{c_1} and H_{c_2}. If the magnetic field is smaller than H_{c_1}, then the entire sample is in the superconducting state. If the magnetic field is between H_{c_1} and H_{c_2}, then the sample is in the mixed state. Finally, if the magnetic field is above H_{c_2}, then the entire sample is in the normal state. Special type II superconducting materials have been developed for high magnetic field applications, for instance, for electromagnets used in MRI and accelerators. These materials have high critical magnetic fields and high current carrying capabilities. These properties are achieved by pinning the motion of flux filaments by defects such as voids, normal inclusions, dislocations, grain boundaries, etc. This pinning results in the multiplicity of meta-stable states, which manifest themselves in hysteresis and some energy losses.

Superconductivity is a macroscopic phenomenon of microscopic origin. For this reason, it requires a quantum mechanical explanation. This explanation was provided by **J. Bardeen, L. Cooper and J. R. Schrieffer** in 1957 and it is called the **BCS theory**. The authors of this theory were awarded the Nobel Prize in Physics in 1972. The central physical idea of this theory is that the presence of (however small) attractive interaction between electrons leads to the instability of the normal state with respect to the formation of electron **Cooper pairs**. Superconductivity is then the result of the condensation of Cooper pairs at the lower (ground) energy level separated from higher energy levels by the energy gap E_g. In conventional superconductors, the Cooper pairing mechanism is due to electron-phonon (i.e., electron–lattice) interaction. The existence of this pairing mechanism is implied by the isotope effect observed in conventional superconductors when the critical temperature is decreased as isotopic mass is increased. The physical nature of the Cooper pairing mechanism in high-T_c supercon-ductors is still debatable. Since the electron pairing mechanism is relatively weak, two attracting electrons should be appreciably separated in order for pairing attraction to overcome Coulomb repulsion. For this reason, Cooper pairs have large spatial extent, about 10^{-4} cm. As a result, a huge number of Cooper pairs overlap and this overlapping extends throughout a sample. As a result, wave functions of Cooper pairs overlap as well, which (as shown in the BCS theory) leads to the formation of a macroscopic wave function and long-range order. This macroscopic wave function is the ensemble av-erage wave function of condensed Cooper pairs. This wave function was introduced by **F. London** before the creation of BCS theory and, for this reason, it is often called the **London wave function**. The square of the magnitude of the macroscopic wave function is interpreted as the local den-sity of Cooper pairs $n(\mathbf{r}, t)$:

$$|\psi(\mathbf{r}, t)|^2 = n(\mathbf{r}, t). \tag{3.3.1}$$

The macroscopic wave function can then be written as

$$\psi(\mathbf{r}, t) = [n(\mathbf{r}, t)]^{\frac{1}{2}} e^{j\theta(\mathbf{r}, t)}. \tag{3.3.2}$$

This formula for the macroscopic wave function will be used below for the analysis of Josephson tunneling. This analysis will follow the ele-gant and very simple approach to superconducting tunneling proposed by **R. Feynman**.

Consider a S-I-S junction when a very thin layer of insulator is sand-wiched between two superconductors (see Figure 3.11). Let ψ_1 and ψ_2 be macroscopic wave functions of the first and second superconductors, respectively. These wave functions are not completely confined to super-conducting regions, but they rather extend (with appreciable attenuation)

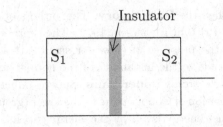

Fig. 3.11

to the insulator region as shown in Figure 3.12. For this reason, the insulator region becomes weakly superconducting, and this makes possible the tunneling of Cooper pairs. Next, we shall discuss the **dc Josephson effect**

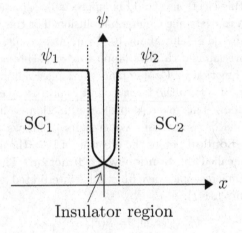

Fig. 3.12

and compute the superconducting current through the junction under the condition that no voltage is applied across the junction. Our analysis will be based on the time-dependent Schrödinger equation for the macroscopic wave function

$$ i\hbar\frac{\partial\psi}{\partial t} = \hat{H}\psi. \tag{3.3.3} $$

The immediate question is how to choose the appropriate Hamiltonian for the problem under consideration. It is apparent that tunneling occurs only if two superconductors are present and very close to one another. This suggests that the time variation of ψ_1 is caused by and proportional to

ψ_2 and, the other way around, the time variation of ψ_2 is caused by and proportional to ψ_1. This can be mathematically expressed as follows

$$i\hbar\frac{\partial\psi_1}{\partial t} = \hbar K\psi_2, \tag{3.3.4}$$

$$i\hbar\frac{\partial\psi_2}{\partial t} = \hbar K\psi_1, \tag{3.3.5}$$

where K is a constant describing the coupling between the two superconducting regions across the barrier.

By invoking formula (3.3.2), we can write:

$$\psi_1 = n_1^{\frac{1}{2}}e^{i\theta_1}, \tag{3.3.6}$$

$$\psi_2 = n_2^{\frac{1}{2}}e^{i\theta_2}. \tag{3.3.7}$$

By substituting these formulas into equations (3.3.4) and (3.3.5), we obtain

$$\frac{1}{2}n_1^{-\frac{1}{2}}\frac{\partial n_1}{\partial t}e^{i\theta_1} + in_1^{\frac{1}{2}}e^{i\theta_1}\frac{\partial\theta_1}{\partial t} = -iKn_2^{\frac{1}{2}}e^{i\theta_2}, \tag{3.3.8}$$

$$\frac{1}{2}n_2^{-\frac{1}{2}}\frac{\partial n_2}{\partial t}e^{i\theta_2} + in_2^{\frac{1}{2}}e^{i\theta_2}\frac{\partial\theta_2}{\partial t} = -iKn_1^{\frac{1}{2}}e^{i\theta_1}. \tag{3.3.9}$$

By introducing the phase shift φ between wave functions ψ_1 and ψ_2

$$\varphi = \theta_2 - \theta_1 \tag{3.3.10}$$

and dividing both sides of equation (3.3.8) by $n_1^{-\frac{1}{2}}e^{i\theta_1}$ and both sides of equation (3.3.9) by $n_2^{-\frac{1}{2}}e^{i\theta_2}$, we derive

$$\frac{1}{2}\frac{\partial n_1}{\partial t} + in_1\frac{\partial\theta_1}{\partial t} = -iK(n_1 n_2)^{\frac{1}{2}}e^{i\varphi}, \tag{3.3.11}$$

$$\frac{1}{2}\frac{\partial n_2}{\partial t} + in_2\frac{\partial\theta_2}{\partial t} = -iK(n_1 n_2)^{\frac{1}{2}}e^{-i\varphi}. \tag{3.3.12}$$

The last two equations can be split in the following four equations, respectively,

$$\frac{\partial n_1}{\partial t} = 2K(n_1 n_2)^{\frac{1}{2}}\sin\varphi, \tag{3.3.13}$$

$$\frac{\partial\theta_1}{\partial t} = -K\left(\frac{n_2}{n_1}\right)^{\frac{1}{2}}\cos\varphi, \tag{3.3.14}$$

$$\frac{\partial n_2}{\partial t} = -2K(n_1 n_2)^{\frac{1}{2}}\sin\varphi, \tag{3.3.15}$$

$$\frac{\partial\theta_2}{\partial t} = -K\left(\frac{n_1}{n_2}\right)^{\frac{1}{2}}\cos\varphi. \tag{3.3.16}$$

In the case when the two superconductors in the junction are of the same material, it can be assumed that

$$n_1 \approx n_2 = n. \tag{3.3.17}$$

By using the last equality in equations (3.3.14) and (3.3.16), we find

$$\frac{\partial \theta_1}{\partial t} = \frac{\partial \theta_2}{\partial t}, \tag{3.3.18}$$

which according to formula (3.3.10) implies that

$$\frac{\partial \varphi}{\partial t} = 0 \tag{3.3.19}$$

and

$$\varphi = const. \tag{3.3.20}$$

On the other hand, from equations (3.3.13) and (3.3.15) we find

$$\frac{\partial n_1}{\partial t} = -\frac{\partial n_2}{\partial t}. \tag{3.3.21}$$

The last equation implies that there may be a flow of Cooper pairs through the junction from one superconductor into another. This flow results in the electric current I through the junction

$$I = 2e\frac{\partial n_1}{\partial t}. \tag{3.3.22}$$

By using equations (3.3.13) and (3.3.17), the last formula can be represented in the form

$$\boxed{I = 4enK \sin \varphi.} \tag{3.3.23}$$

By introducing the notation

$$I_0 = \max I = 4enK, \tag{3.3.24}$$

the previous formula can be written as follows

$$\boxed{I = I_0 \sin \varphi.} \tag{3.3.25}$$

Now, it can be concluded that current I specified by formula (3.3.22) (or by formulas (3.3.23) and (3.3.25)) can flow through the junction without any voltage applied (or appeared) across the junction. It is apparent, according to formula (3.3.20), that this is a dc current. It is also obvious on symmetry grounds that no current can flow through the junction if it is not attached to an external circuit. However, if the junction attached, for instance, to a current source (see Figure 3.13) and

$$I_s < I_0, \tag{3.3.26}$$

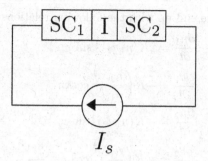

Fig. 3.13

then the phase difference φ adjusts itself to accommodate the flow of I_s through the junction with zero resistance (i.e., with zero voltage across the junction). It is clear from formula (3.3.23) that the maximum value I_0 of the dc superconducting current through the junction depends on coupling coefficient K which, in turn, depends on the thickness of the insulating layer. A typical value of this current is

$$I_0 \approx 1 \text{ mA.} \tag{3.3.27}$$

The smallness of this current is the reason why insulating layers sandwiched between superconductors are called weak links.

Now, we proceed to the discussion of the **ac Josephson effect**. This effect occurs when a dc voltage V is applied across the junction. If the potential of the first superconductor is taken to be zero, then the potential of the second superconductor is equal to V. This means that the potential energy of Cooper pairs in the first superconductor is zero, while the potential energy of the second superconductor is

$$U = -2eV. \tag{3.3.28}$$

This leads to the following modifications of the coupled equations (3.3.4) and (3.3.5):

$$i\hbar \frac{\partial \psi_1}{\partial t} = \hbar K \psi_2, \tag{3.3.29}$$

$$i\hbar \frac{\partial \psi_2}{\partial t} = \hbar K \psi_1 - 2eV \psi_2. \tag{3.3.30}$$

Now, by using formulas (3.3.6) and (3.3.7) for ψ_1 and ψ_2, respectively, and by using the same line of reasoning as in the derivation of equations

(3.3.13)–(3.3.16), we end up with the following relations

$$\frac{\partial n_1}{\partial t} = 2K(n_1 n_2)^{\frac{1}{2}} \sin \varphi, \tag{3.3.31}$$

$$\frac{\partial \theta_1}{\partial t} = -K \left(\frac{n_2}{n_1} \right)^{\frac{1}{2}} \cos \varphi, \tag{3.3.32}$$

$$\frac{\partial n_2}{\partial t} = -K(n_1 n_2)^{\frac{1}{2}} \sin \varphi, \tag{3.3.33}$$

$$\frac{\partial \theta_2}{\partial t} = -K \left(\frac{n_1}{n_2} \right)^{\frac{1}{2}} \cos \varphi + \frac{2eV}{\hbar}. \tag{3.3.34}$$

By using as before the equality (3.3.17), from equations (3.3.32) and (3.3.34) we derive

$$\frac{\partial \varphi}{\partial t} = \frac{\partial \theta_2}{\partial t} - \frac{\partial \theta_1}{\partial t} = \frac{2eV}{\hbar}, \tag{3.3.35}$$

which leads to

$$\varphi(t) = \frac{2eV}{\hbar} t + \varphi_0. \tag{3.3.36}$$

On the other hand, from equations (3.3.31) and (3.3.33) follows that

$$\frac{\partial n_1}{\partial t} = -\frac{\partial n_2}{\partial t}. \tag{3.3.37}$$

The last formula implies that there is a flow of Cooper pairs through the junction, which results in an electric current:

$$i(t) = 2e \frac{\partial n_1}{\partial t}. \tag{3.3.38}$$

By using formulas (3.3.17) and (3.3.31), the last equation can be represented in the form

$$i(t) = 4enK \sin \varphi(t). \tag{3.3.39}$$

Now, by recalling formulas (3.3.24) and (3.3.36), we find

$$\boxed{i(t) = I_0 \sin(\omega t + \varphi_0),} \tag{3.3.40}$$

where

$$\boxed{\omega = \frac{2eV}{\hbar}.} \tag{3.3.41}$$

Thus, it has been established that a dc voltage applied across a S-I-S junction results in an ac current whose frequency is controlled by the voltage. For instance, if a dc voltage bias is $V = 1\mu V$, the frequency of the current through the junction is $\omega = 484$ MHz.

It is interesting to mention that the dual effect when a dc current generates an ac voltage and the frequency of this voltage is controlled by dc current is realized in spin-torque nano-oscillators. The physics of these oscillators is a very active research area in spintronics, but it has nothing to do with superconductivity. For this reason, it will not be further elaborated here. A brief discussion of these oscillators is presented in Section 7.2.

The Josephson effects and Josephson S-I-S junctions have found various applications. In particular, they are used in **SQUIDs**, which is the abbreviation for superconducting quantum interference devices. These devices are utilized in the construction of the most **sensitive magnetometer (SQUID magnetometer)** capable of measuring a single magnetic flux quantum Φ_0. Superconducting tunneling junctions may also find important applications in **quantum computing** as flux qubits (also called Qbits). These qubits are the quantum analogue of classical bits.

In the conclusion of this section, it is worthwhile to mention that there also exists single electron tunneling in superconductor-insulator-normal (S-I-N) junctions. This tunneling has physical properties which are quite different than Josephson tunneling in S-I-S junctions. This single electron tunneling was very instrumental in the experimental demonstration of the existence of the energy gap in superconductors and it has been used for actual measurement of this energy gap. The pioneering work in this area was performed at General Electric (GE) by **I. Giaever** who shared the 1973 Nobel Prize in Physics with **L. Esaki and B. Josephson**.

Problems

(1) Derive the formula for the reflection coefficient R in the case of a rectangular barrier.

(2) Prove that

$$R + T = 1. \qquad (P.3.1)$$

(3) Solve the problem of tunneling in the case of δ-type potential barrier $U(x) = A\delta(x)$.

(4) Derive the expression for θ_n in terms of $\mathcal{E}_n$ (see formula (3.2.34)).

(5) Consider the problem of resonant tunneling through two identical δ-type barriers.

(6) Solve the problem of tunneling for three identical δ-type barriers.

(7) Summarize the main properties of superconductors.

(8) What are the differences between intermediate and mixed states of superconductors?

(9) Derive formulas (3.3.31)—(3.3.34).

(10) Derive the expression for the tunneling current through a S-I-S junction in the case of time-varying applied voltage.

Chapter 4

Problems with Discrete Energy Spectrum

4.1 The Potential Well

We begin this chapter with the discussion of the simplest problem, which is the energy spectrum of a microparticle in a rectangular potential well, shown in Figure 4.1. The potential energy $U(x)$ for this problem is defined

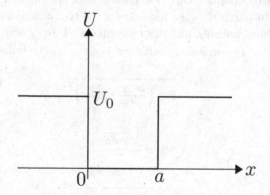

Fig. 4.1

by the following two formulas

$$U(x) = U_0, \text{ if } x < 0 \text{ or } x > a, \tag{4.1.1}$$

$$U(x) = 0, \text{ if } 0 < x < a. \tag{4.1.2}$$

It is apparent that there are two distinct cases: (a) when the particle energy $\mathcal{E}$ is above U_0, i.e.,

$$\mathcal{E} > U_0 \tag{4.1.3}$$

and (b) when the particle energy is below U_0, i.e.,

$$\mathcal{E} < U_0. \tag{4.1.4}$$

In classical mechanics, particles with energies above U_0 move without any reflection (and even with some acceleration in the region $0 < x < a$), while the motion of the particles with energy below U_0 is confined between the walls of the well and the particle energy may assume any value between zero and U_0. In quantum mechanics, the situation is quite different. In the first case specified by the inequality (4.1.3), a microparticle can be scattered from the potential wells. In the second case specified by the inequality (4.1.4), the energy of the microparticle is quantized, i.e., the microparticle energy may assume only certain discrete values. Furthermore, in the latter case there is finite (albeit very small) probability of finding a microparticle outside the potential well. Thus, it is clear that in quantum mechanics the energy spectrum of the microparticle is continuous in the first case and discrete in the second case.

The analysis of the first case is very similar to the analysis of tunneling through a rectangular barrier carried out in the previous chapter (see Section 3.1). In particular, the formula for the transmission coefficient can be obtained from formula (3.1.37) by replacing k by q and q by k. This suggests that the transmission coefficient is given by the following equation

$$T = \frac{1}{1 + \left(\frac{q^2 - k^2}{2kq} \sin ka \right)^2}, \tag{4.1.5}$$

where

$$k = \frac{\sqrt{2m\mathcal{E}}}{\hbar} \tag{4.1.6}$$

and

$$q = \frac{\sqrt{2m(\mathcal{E} - U_0)}}{\hbar}. \tag{4.1.7}$$

It is apparent from the above formula that

$$T < 1, \quad \text{if } \sin ka \neq 0. \tag{4.1.8}$$

This implies that typically a microparticle is scattered (reflected) by the potential well. This scattering does not exist only when

$$\sin ka = 0, \tag{4.1.9}$$

and then

$$T = 1. \tag{4.1.10}$$

From formulas (4.1.6) and (4.1.9) follows that the microparticle can move over the potential well without any scattering for the following values of energy

$$\mathcal{E}_n = \frac{\pi^2 \hbar^2}{2ma^2} n^2. \tag{4.1.11}$$

The scattering of particles by the potential well has been used as a model for the scattering of low-energy electrons from atoms. In this model, the well represents the electrostatic field of the nucleus whose positive charge acts as a scatterer for incident electrons that penetrated the shell structure of atomic electrons. Experiments reveal a low-energy minimum for the reflection coefficient for electrons moving in rare (noble) gases of argon, krypton or xenon. This minimum of the reflection coefficient is consistent with the first (low-energy) maximum of the transmission coefficient T. This transparency of rare gas atoms to low-energy electrons is known as the Ramsauer effect, which is also called the Ramsauer-Townsend effect.

Now, we proceed to the analysis of the discrete energy spectrum which occurs when the inequality (4.1.4) is valid. We shall study the discrete energy spectrum by using the time-independent Schrödinger equation within the well and outside the well which can be written as follows, respectively:

$$-\frac{\hbar^2}{2m}\frac{d^2\Phi(x)}{dx^2} = \mathcal{E}\Phi(x), \quad \text{if } 0 < x < a, \tag{4.1.12}$$

and

$$-\frac{\hbar^2}{2m}\frac{d^2\Phi(x)}{dx^2} + U_0\Phi(x) = \mathcal{E}\Phi(x), \quad \text{if } x < 0 \text{ or } x > a. \tag{4.1.13}$$

A general solution of equation (4.1.12) has the form

$$\Phi(x) = Ae^{ikx} + Be^{-ikx}, \quad (0 < x < a), \tag{4.1.14}$$

where k is given by formula (4.1.6). General solutions of equation (4.1.13) for regions $x < 0$ and $x > a$, which diminish to zero at $-\infty$ and $+\infty$, respectively, are:

$$\Phi_-(x) = Ce^{\gamma x}, \quad \text{if } x < 0, \tag{4.1.15}$$

$$\Phi_+(x) = De^{-\gamma x}, \quad \text{if } x > a, \tag{4.1.16}$$

where

$$\gamma = \frac{\sqrt{2m(U_0 - \mathcal{E})}}{\hbar}. \tag{4.1.17}$$

Coefficients A, B, C and D must be chosen to satisfy the boundary conditions

$$\Phi_-(0) = \Phi(0), \tag{4.1.18}$$

$$\frac{d\Phi_-}{dx}(0) = \frac{d\Phi}{dx}(0), \tag{4.1.19}$$

$$\Phi(a) = \Phi_+(a), \tag{4.1.20}$$

$$\frac{d\Phi}{dx}(a) = \frac{d\Phi_+}{dx}(a). \tag{4.1.21}$$

These four boundary conditions lead to four linear homogeneous simultaneous equations for A, B, C and D. These four homogeneous equations have a nonzero solution if their determinant is equal to zero. The latter condition leads to a nonlinear equation for $\mathcal{E}$ because it appears in formulas (4.1.6) and (4.1.17) for k and γ, respectively. By solving this nonlinear equation, the energy spectrum can be determined.

Below, we shall pursue another approach to the calculation of the energy spectrum, which is based on the "impedance"-type boundary conditions posed on the walls of the well. From the boundary conditions (4.1.18)–(4.1.21) follows that

$$\frac{1}{\Phi(0)}\frac{d\Phi}{dx}(0) = \frac{1}{\Phi_-(0)}\frac{d\Phi_-}{dx}(0) \tag{4.1.22}$$

and

$$\frac{1}{\Phi(a)}\frac{d\Phi}{dx}(a) = \frac{1}{\Phi_+(a)}\frac{d\Phi_+}{dx}(a). \tag{4.1.23}$$

Now, by using formulas (4.1.15) and (4.1.16) in equation (4.1.22) and (4.1.23), respectively, we find

$$\frac{d\Phi}{dx}(0) = \gamma\Phi(0) \tag{4.1.24}$$

and

$$\frac{d\Phi}{dx}(a) = -\gamma\Phi(a). \tag{4.1.25}$$

Boundary conditions (4.1.24) and (4.1.25) are similar to impedance boundary conditions used in electromagnetic field theory. This type of impedance boundary conditions can be used as approximate boundary conditions for potential wells with curved boundaries if $\mathcal{E} \ll U_0$. This is because under the above condition on $\mathcal{E}$, solutions of the time-independent Schrödinger equation in the region exterior to the well vary predominantly along the direction normal to the well boundary and this variation is fairly accurately described by equation (4.1.16).

The impedance boundary conditions (4.1.24) and (4.1.25) can be used to reduce the calculation of energy spectrum to finding the solution of equation (4.1.12) within the well subject to the above boundary conditions. A general solution to equation (4.1.12) can be written in the form

$$\Phi(x) = F\sin(kx + \varphi), \tag{4.1.26}$$

where F and φ are unknowns.

From the last formula and boundary conditions (4.1.24) and (4.1.25), we find, respectively,

$$\cot\varphi = \frac{\gamma}{k} \tag{4.1.27}$$

and

$$\cot(ka + \varphi) = -\frac{\gamma}{k}. \tag{4.1.28}$$

The last two equations imply that

$$\cot(ka + \varphi) = -\cot\varphi. \tag{4.1.29}$$

This equality is only valid if

$$ka + \varphi = n\pi - \varphi, \quad (n = 1, 2, \dots), \tag{4.1.30}$$

which can also be written as

$$\frac{ka}{2} = \frac{n\pi}{2} - \varphi, \quad (n = 1, 2, \dots). \tag{4.1.31}$$

Next, we consider two cases, when n is odd and n is even. In the first case, formula (4.1.31) can be modified as follows

$$\frac{ka}{2} = \frac{(2p + 1)\pi}{2} - \varphi, \quad (p = 0, 1, 2, \dots), \tag{4.1.32}$$

while in the second case formula (4.1.31) can be written in the form

$$\frac{ka}{2} = p\pi - \varphi, \quad (p = 1, 2, \dots). \tag{4.1.33}$$

From the last two formulas we find, respectively,

$$\cos\frac{ka}{2} = \pm\sin\varphi \tag{4.1.34}$$

and

$$\sin\frac{ka}{2} = \pm\sin\varphi. \tag{4.1.35}$$

Next, by using the trigonometric identity

$$\sin \varphi = \frac{1}{\sqrt{1 + \cot^2 \varphi}} \tag{4.1.36}$$

and equation (4.1.27), we derive

$$\sin \varphi = \frac{k}{\sqrt{\gamma^2 + k^2}}. \tag{4.1.37}$$

By invoking formulas (4.1.6) and (4.1.17), we conclude that

$$\gamma^2 + k^2 = \frac{2mU_0}{\hbar^2}. \tag{4.1.38}$$

This means that

$$\sin \varphi = \frac{k\hbar}{\sqrt{2mU_0}}. \tag{4.1.39}$$

By using the last formula in equation (4.1.34) and (4.1.35), we find

$$\cos \frac{ka}{2} = \pm \frac{k\hbar}{\sqrt{2mU_0}} \tag{4.1.40}$$

and

$$\sin \frac{ka}{2} = \pm \frac{k\hbar}{\sqrt{2mU_0}}. \tag{4.1.41}$$

These are nonlinear equations for k and, consequently, for $\mathcal{E}$ that can be used to compute the discrete energy spectrum. It is convenient to introduce the notations

$$\alpha = \frac{ka}{2} \tag{4.1.42}$$

as well as

$$\nu = \frac{2\hbar}{a\sqrt{2mU_0}}, \tag{4.1.43}$$

and to write equations (4.1.40) and (4.1.41) in the following forms, respectively,

$$\boxed{\cos \alpha = \pm \nu \alpha} \tag{4.1.44}$$

$$\boxed{\sin \alpha = \pm \nu \alpha.} \tag{4.1.45}$$

Solution of the last two equations can be found numerically. These solutions can be also found graphically as intersections of straight lines with slopes ν and $-\nu$ with the graphs of cosine and sine functions (see Figures 4.2 and 4.3 on the next page).

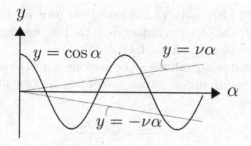

Fig. 4.2

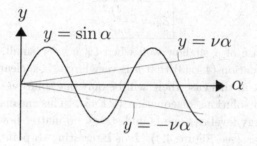

Fig. 4.3

It can be shown that not all solutions of equations (4.1.44) and (4.1.45) correspond to possible energy levels in the well. Actually, only solutions of equation (4.1.44) for which

$$\tan \alpha > 0 \qquad (4.1.46)$$

correspond to possible energy levels. Similarly, only solutions of equation (4.1.45) for which

$$\tan \alpha < 0 \qquad (4.1.47)$$

correspond to possible energy levels. Indeed, it can be derived from formula (4.1.32) that

$$\tan \alpha = \tan \frac{ka}{2} = \cot \varphi > 0, \qquad (4.1.48)$$

while formula (4.1.33) leads to

$$\tan \alpha = \tan \frac{ka}{2} = -\tan \varphi < 0. \qquad (4.1.49)$$

Since formulas (4.1.32) and (4.1.33) have been used in the derivation of equations (4.1.44) and (4.1.45), respectively, the last two inequalities justify the selection rules (4.1.46) and (4.1.47).

Let α_n be appropriate solutions of equations (4.1.44) and (4.1.45) enumerated in increasing order. Then, according to formulas (4.1.6) and (4.1.42) we have

$$\alpha_n = \frac{k_n a}{2} = \frac{a\sqrt{2m\mathcal{E}_n}}{2\hbar}. \tag{4.1.50}$$

Consequently, the possible energy levels $\mathcal{E}_n$ in the well are determined by the formula

$$\boxed{\mathcal{E}_n = \frac{2\alpha_n^2 \hbar^2}{ma^2}.} \tag{4.1.51}$$

Consider the case of a shallow well when U_0 is very small. This implies according to equation (4.1.43) that ν is very large. It is clear from Figures 4.2 and 4.3 that in this case there will be only one solution α_1 of equation (4.1.44) and no solutions of equation (4.1.45). This means that there is always one energy level $\mathcal{E}_1$ (one bound state), no matter how shallow (i.e., weak) the well is (see Figure 4.4). It is interesting to point out that this

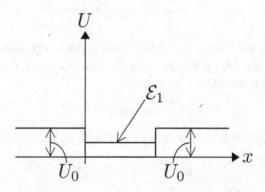

Fig. 4.4

property is not valid for spherical wells. Namely, very shallow spherical wells may not retain microparticles because they may not have allowed energy levels.

It is also clear from formula (4.1.43) and Figures 4.2 and 4.3 that the number of allowed energy levels is increased with the increase of U_0. In the limit of $U_0 \to \infty$, there are infinitely many energy levels $\mathcal{E}_n$. These energy levels can be found from the above figures as well as analytically. Indeed, from formula (4.1.39) we conclude that

$$\varphi \to 0 \text{ as } U_0 \to \infty. \tag{4.1.52}$$

Consequently, according to equation (4.1.31) we have

$$\frac{k_n a}{2} = \frac{n\pi}{2}. \tag{4.1.53}$$

By using formula (4.1.6), the last relation can be written as follows

$$\frac{a\sqrt{2m\mathcal{E}_n}}{\hbar} = n\pi, \tag{4.1.54}$$

which leads to the final expression for $\mathcal{E}_n$:

$$\boxed{\mathcal{E}_n = \frac{\pi^2 \hbar^2}{2ma^2} n^2.} \tag{4.1.55}$$

The corresponding eigenstates (i.e., stationary states) $\Phi_n(x)$ can be found by using formulas (4.1.26), (4.1.52) and (4.1.53). After proper normalization, we end up with the following formula

$$\boxed{\Phi_n(x) = \sqrt{\frac{2}{a}} \sin\frac{n\pi}{a} x.} \tag{4.1.56}$$

The function

$$\Phi_1(x) = \sqrt{\frac{2}{a}} \sin\frac{\pi}{a} x \tag{4.1.57}$$

which corresponds to the lowest energy level

$$\mathcal{E}_1 = \frac{\pi^2 \hbar^2}{2ma^2} \tag{4.1.58}$$

is called the **ground state**. The other $\Phi_n(x)$ corresponding to higher energy levels $\mathcal{E}_n$ are called **excited states**. The graphs of $\Phi_1(x)$, $\Phi_2(x)$ and $\Phi_3(x)$ are shown in Figure 4.5. It is apparent from this figure (as well as from formula (4.1.56)) that the eigenstates $\Phi_n(x)$ are oscillatory in nature and each successive eigenstate has one additional zero-crossing. In general, this oscillatory nature is typical for eigenstates (i.e.; eigenfunctions).

It is clear from the previous discussions that the energy levels $\mathcal{E}_n$ in a one-dimensional well are not degenerate. The latter means that for each energy level there is only one eigenstate. It turns out that this is true for any one-dimensional problem. The proof of this fact proceeds as follows. Suppose that there are two eigenstates $\Phi_n^{(1)}(x)$ and $\Phi_n^{(2)}(x)$ corresponding

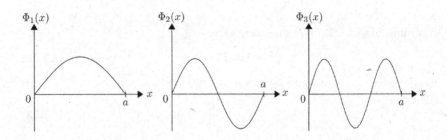

Fig. 4.5

to the same eigenvalue $\mathcal{E}_n$ of the Hamiltonian operator. The corresponding time-independent Schrödinger equations can be written as follows

$$-\frac{\hbar^2}{2m}\frac{d^2\Phi_n^{(1)}(x)}{dx^2} + [U(x) - \mathcal{E}_n]\Phi_n^{(1)}(x) = 0, \qquad (4.1.59)$$

$$-\frac{\hbar^2}{2m}\frac{d^2\Phi_n^{(2)}(x)}{dx^2} + [U(x) - \mathcal{E}_n]\Phi_n^{(2)}(x) = 0. \qquad (4.1.60)$$

By multiplying equation (4.1.59) by $\Phi_n^{(2)}(x)$ and equation (4.1.60) by $\Phi_n^{(1)}(x)$ and then subtracting these equations, we find

$$\Phi_n^{(2)}(x)\frac{d^2\Phi_n^{(1)}(x)}{dx^2} - \Phi_n^{(1)}(x)\frac{d^2\Phi_n^{(2)}(x)}{dx^2} = 0. \qquad (4.1.61)$$

It is clear that the last equation can be also written in the form

$$\frac{d}{dx}\left[\Phi_n^{(2)}(x)\frac{d\Phi_n^{(1)}(x)}{dx} - \Phi_n^{(1)}(x)\frac{d\Phi_n^{(2)}(x)}{dx}\right] = 0. \qquad (4.1.62)$$

By integrating the last equation from $-\infty$ to x and taking into account that $\Phi_n^{(1)}(x)$ and $\Phi_n^{(2)}(x)$ as well as their first derivatives diminish to zero at $-\infty$, we obtain

$$\Phi_n^{(2)}(x)\frac{d\Phi_n^{(1)}(x)}{dx} - \Phi_n^{(1)}(x)\frac{d\Phi_n^{(2)}(x)}{dx} = 0. \qquad (4.1.63)$$

The last equation can be further transformed as follows

$$\frac{d}{dx}\left[\ln\frac{\Phi_n^{(1)}(x)}{\Phi_n^{(2)}(x)}\right] = 0. \qquad (4.1.64)$$

This implies that

$$\frac{\Phi_n^{(1)}(x)}{\Phi_n^{(2)}(x)} = C = const, \qquad (4.1.65)$$

and

$$\Phi_n^{(1)}(x) = C\Phi_n^{(2)}(x). \tag{4.1.66}$$

This means that if $\Phi_n^{(1)}(x)$ and $\Phi_n^{(2)}(x)$ are normalized (i.e., $\|\Phi_n^{(1)}(x)\| = \|\Phi_n^{(2)}(x)\| = 1$), then $\Phi_n^{(1)}(x) = \Phi_n^{(2)}(x)$, and the eigenvalue $\mathcal{E}_n$ is not degenerate.

This nondegeneracy property is not valid for two- or three-dimensional problems. To illustrate this, consider a two-dimensional and infinitely deep square potential well with a cross-section shown in Figure 4.6. By using the method of separation of variables, it can be shown that the eigenvalues

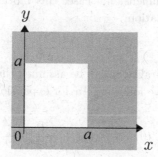

Fig. 4.6

of the Hamiltonian operator for this problem are given by the formula

$$\mathcal{E}_{n_1 n_2} = \frac{\hbar^2 \pi^2}{2ma^2}(n_1^2 + n_2^2). \tag{4.1.67}$$

It is easy to check that there are two linearly independent eigenstates:

$$\Phi_{n_1 n_2}(x, y) = \frac{2}{a} \sin \frac{n_1 \pi}{a} x \sin \frac{n_2 \pi}{a} y, \tag{4.1.68}$$

$$\Phi_{n_2 n_1}(x, y) = \frac{2}{a} \sin \frac{n_2 \pi}{a} x \sin \frac{n_1 \pi}{a} y. \tag{4.1.69}$$

This means that the energy levels $\mathcal{E}_{n_1 n_2}$ are two-fold degenerate. It can be asked what the cause of degeneracy is. Typically, the cause of degeneracy of energy levels is **symmetry**. In the above example, there is symmetry of the problem with respect to the replacement of x by y and y by x. In other words, there is symmetry of the square with respect to its rotations through angle $\frac{\pi}{2}$ around its center. Degeneracy of energy levels caused by symmetry can be studied by using irreducible representations of symmetry groups. However, the discussion of this technique is beyond the scope of this text.

4.2 The Harmonic Oscillator

The harmonic oscillator problem is the model problem for small oscillations around a stable equilibrium. This is also one of the very few quantum mechanical problems for which the exact analytical solution is known. Furthermore, the operator technique employed below for the solution of this problem is used in many different areas of quantum mechanics.

We begin the discussion with a brief review of the harmonic oscillator problem in classical mechanics. Consider the motion of a macroparticle around a stable equilibrium which corresponds to a minimum of potential energy U. In the one-dimensional case, this motion is described by the following differential equation:

$$m\frac{d^2x}{dt^2} = -\frac{dU(x)}{dx}. \tag{4.2.1}$$

Without any loss of generality, it can be assumed that the potential energy minimum is at $x = 0$. The following Taylor expansion can be used for $U(x)$ around $x = 0$:

$$U(x) = U(0) + \frac{dU}{dx}(0)x + \frac{d^2U}{dx^2}(0)\frac{x^2}{2} + \cdots. \tag{4.2.2}$$

It is clear that at the equilibrium point

$$\frac{dU}{dx}(0) = 0. \tag{4.2.3}$$

Since $U(x)$ is defined up to a constant, it can be assumed that

$$U(0) = 0. \tag{4.2.4}$$

By using formulas (4.2.3) and (4.2.4) in the expansion (4.2.2) and taking into account that for small motions the higher-order terms in (4.2.2) can be neglected, we arrive at the following expression for the potential energy

$$U(x) = \frac{k}{2}x^2, \tag{4.2.5}$$

where

$$k = \frac{d^2U}{dx^2}(0). \tag{4.2.6}$$

Now, the differential equation (4.2.7) can be written as

$$m\frac{d^2x}{dt^2} + kx = 0. \tag{4.2.7}$$

A general solution of this equation is

$$x(t) = A\sin(\omega t + \varphi), \tag{4.2.8}$$

where

$$\omega = \sqrt{\frac{k}{m}}. \tag{4.2.9}$$

Thus, small motions of a macroparticle around a stable equilibrium are always harmonic oscillations.

Now, we shall turn to the quantum mechanical treatment of this problem and we shall use the following expression for the potential energy

$$U(x) = \frac{m\omega^2}{2}x^2, \tag{4.2.10}$$

which results from formulas (4.2.5) and (4.2.9).

The equation for possible energy levels $\mathcal{E}_n$ of a microparticle moving in the field described by the potential energy (4.2.10) can be written as

$$\hat{H}\Phi_n(x) = \mathcal{E}_n\Phi_n, \tag{4.2.11}$$

where

$$\hat{H} = \frac{\hat{p}_x^2}{2m} + \frac{m\omega^2}{2}x^2, \tag{4.2.12}$$

or

$$\hat{H} = -\frac{\hbar^2}{2m}\frac{d^2}{dx^2} + \frac{m\omega^2}{2}x^2. \tag{4.2.13}$$

By combining equations (4.2.11) and (4.2.13), we arrive at

$$\boxed{-\frac{\hbar^2}{2m}\frac{d^2\Phi_n(x)}{dx^2} + \frac{m\omega^2}{2}x^2\Phi_n(x) = \mathcal{E}_n\Phi_n(x).} \tag{4.2.14}$$

There are two techniques for computing the energy levels $\mathcal{E}_n$ of harmonic oscillators. The first technique is based on finding nonzero (and square-integrable) solutions of the differential equation (4.2.14). The second technique is operator-algebraic in nature and it is extensively used in many different areas of quantum mechanics. We shall present the second technique. In doing so, we shall first introduce the dimensionless coordinate X and momentum operator $\hat{P}$ by using the following formulas, respectively,

$$X = x\sqrt{\frac{m\omega}{\hbar}}, \tag{4.2.15}$$

$$\hat{P} = \frac{\hat{p}_x}{\sqrt{\hbar m\omega}}. \tag{4.2.16}$$

Now, we shall define two very important operators

$$\hat{a}^+ = \frac{1}{\sqrt{2}}(X - i\hat{P}) \tag{4.2.17}$$

and

$$\hat{a} = \frac{1}{\sqrt{2}}(X + i\hat{P}).$$ (4.2.18)

These operators are called raising and lowering operators, respectively. They are also called ladder operators as well as creation and annihilation operators. The latter terminology is used in quantum field theory.

It can be shown that $\hat{a}^+$ and $\hat{a}$ are mutually adjoint operators in the sense that the following equality is valid

$$\langle \hat{a}^+ \psi_1 | \psi_2 \rangle = \langle \psi_1 | \hat{a} \psi_2 \rangle.$$ (4.2.19)

The proof is left as an exercise for the reader.

From the definitions of operators $\hat{a}^+$ and $\hat{a}$ (see formulas (4.2.17) and (4.2.18), respectively), we find

$$\hat{a}^+ \hat{a} = \frac{1}{2}(X - i\hat{P})(X + i\hat{P}) = \frac{1}{2}(X^2 + iX\hat{P} - i\hat{P}X + \hat{P}^2).$$ (4.2.20)

In terms of x and $\hat{p}_x$, the last equality can be written as follows

$$\hat{a}^+ \hat{a} = \frac{m\omega^2}{2\hbar}x^2 + \frac{\hat{p}_x^2}{2\hbar m\omega} + \frac{i}{2\hbar}(x\hat{p}_x - \hat{p}_x x).$$ (4.2.21)

By recalling the commutation relation (see Chapter 2)

$$x\hat{p}_x - \hat{p}_x x = i\hbar,$$ (4.2.22)

we arrive at

$$\hat{a}^+ \hat{a} = \frac{\hat{p}_x^2}{2\hbar m\omega} + \frac{m\omega^2}{2\hbar}x^2 - \frac{1}{2},$$ (4.2.23)

which leads to

$$\hbar\omega\left(\hat{a}^+ \hat{a} + \frac{1}{2}\right) = \frac{\hat{p}_x^2}{2m} + \frac{m\omega^2}{2}x^2.$$ (4.2.24)

By comparing formulas (4.2.12) and (4.2.24), we conclude that

$$\boxed{\hat{H} = \hbar\omega\left[\hat{a}^+ \hat{a} + \frac{1}{2}\right].}$$ (4.2.25)

By using the same line of reasoning as before, another representation of the Hamiltonian operator $\hat{H}$ in terms of operators $\hat{a}$ and $\hat{a}^+$ can be established. This representation is

$$\boxed{\hat{H} = \hbar\omega\left[\hat{a}\hat{a}^+ - \frac{1}{2}\right].}$$ (4.2.26)

It is apparent from the last two formulas that

$$\hat{a}^+\hat{a} + \frac{1}{2} = \hat{a}\hat{a}^+ - \frac{1}{2}, \tag{4.2.27}$$

and

$$\boxed{\hat{a}\hat{a}^+ - \hat{a}^+\hat{a} = 1.} \tag{4.2.28}$$

Next, we shall apply operator $\hat{a}^+$ to both sides of equation (4.2.11):

$$\hat{a}^+\hat{H}\Phi_n = \mathcal{E}_n\hat{a}^+\Phi_n. \tag{4.2.29}$$

By using formula (4.2.26), the following transformations can be performed:

$$\hat{a}^+\hat{H}\Phi_n = \hbar\omega\left(\hat{a}^+\hat{a}\hat{a}^+ - \frac{1}{2}\hat{a}^+\right)\Phi_n = \hbar\omega\left[\left(\hat{a}^+\hat{a} + \frac{1}{2}\right)\hat{a}^+ - \hat{a}^+\right]\Phi_n. \tag{4.2.30}$$

By recalling formula (4.2.25), we obtain

$$\hat{a}^+\hat{H}\Phi_n = \hat{H}\hat{a}^+\Phi_n - \hbar\omega\hat{a}^+\Phi_n. \tag{4.2.31}$$

From equations (4.2.29) and (4.2.31), we derive

$$\boxed{\hat{H}(\hat{a}^+\Phi_n) = (\mathcal{E}_n + \hbar\omega)\hat{a}^+\Phi_n.} \tag{4.2.32}$$

The last equation implies that if Φ_n is the eigenstate corresponding to the energy level $\mathcal{E}_n$, then $\hat{a}^+\Phi_n$ is the eigenstate corresponding to the energy level $\mathcal{E}_n + \hbar\omega$.

By using the same line of reasoning as before, it can be established that

$$\boxed{\hat{H}(\hat{a}\Phi_n) = (\mathcal{E}_n - \hbar\omega)\hat{a}\Phi_n.} \tag{4.2.33}$$

The last equation implies that if Φ_n is the eigenstate corresponding to the energy level $\mathcal{E}_n$, then $\hat{a}\Phi_n$ is the eigenstate corresponding to the energy level $\mathcal{E}_n - \hbar\omega$.

The last two formulas reveal the reason why $\hat{a}^+$ and $\hat{a}$ are called ladder operators. This is because these operators allow climbing up and down in energy levels. It is also clear why $\hat{a}^+$ and $\hat{a}$ are called **raising** and **lowering** operators, respectively.

The last two formulas are instrumental in derivation of formulas for the energy levels and the corresponding eigenstates. Indeed, if somehow we find the ground energy level $\mathcal{E}_0$ and the corresponding eigenstate Φ_0 then the higher energy levels can be found by repeatedly incrementing $\mathcal{E}_0$ by $\hbar\omega$, and the corresponding eigenstates can be derived from Φ_0 by repeatedly applying operator $\hat{a}^+$.

To derive the expressions for $\mathcal{E}_0$ and Φ_0, we shall first prove that all energy levels are strictly positive, i.e.,

$$\mathcal{E}_n > 0, \quad (n = 0, 1, 2, \ldots). \tag{4.2.34}$$

The proof proceeds as follows. By using formula (4.2.12), we write equation (4.2.11) in the form

$$\frac{1}{2m}\hat{p}_x^2\Phi_n + \frac{m\omega^2}{2}x^2\Phi_n = \mathcal{E}_n\Phi_n. \tag{4.2.35}$$

Now, we multiply both sides by Φ_n^* and integrate with respect to x from $-\infty$ to $+\infty$. This leads to the following equality

$$\frac{1}{2m}\langle\Phi_n|\hat{p}_x^2\Phi_n\rangle + \frac{m\omega^2}{2}\langle\Phi_n|x^2\Phi_n\rangle = \mathcal{E}_n\langle\Phi_n|\Phi_n\rangle, \tag{4.2.36}$$

which can be also written as

$$\frac{1}{2m}\langle\hat{p}_x\Phi_n|\hat{p}_x\Phi_n\rangle + \frac{m\omega^2}{2}\langle x\Phi_n|x\Phi_n\rangle = \mathcal{E}_n\|\Phi_n\|^2. \tag{4.2.37}$$

From the last formula, we find

$$\mathcal{E}_n = \frac{\|\hat{p}_x\Phi_n\|^2}{2m\|\Phi_n\|^2} + \frac{m\omega^2\|x\Phi_n\|^2}{2\|\Phi_n\|^2} > 0. \tag{4.2.38}$$

Thus, the inequality (4.2.34) is established. This inequality suggests that the ground energy level $\mathcal{E}_0$ defined as

$$\mathcal{E}_0 = \min \mathcal{E}_n \tag{4.2.39}$$

is positive and there is no energy below it. In accordance with equation (4.2.33) this implies that

$$\hat{a}\Phi_0 = 0. \tag{4.2.40}$$

Indeed, only under this condition equation (4.2.33) can be satisfied, because there is no energy level below $\mathcal{E}_0$. Now, by writing equation (4.2.11) for $\mathcal{E}_0$ and Φ_0

$$\hat{H}\Phi_0 = \mathcal{E}_0\Phi_0 \tag{4.2.41}$$

and by using the formula (4.2.25) for the Hamiltonian, we obtain

$$\hbar\omega\hat{a}^+\hat{a}\Phi_0 + \frac{\hbar\omega}{2}\Phi_0 = \mathcal{E}_0\Phi_0. \tag{4.2.42}$$

According to equation (4.2.40), from (4.2.42) we find

$$\frac{\hbar\omega}{2}\Phi_0 = \mathcal{E}_0\Phi_0, \tag{4.2.43}$$

which means that

$$\mathcal{E}_0 = \frac{\hbar\omega}{2}. \tag{4.2.44}$$

By applying the raising operator $\hat{a}^+$ to Φ_0, we find according to formula (4.2.32) that $\hat{a}^+\Phi_0$ is the eigenstate corresponding to the energy level $\mathcal{E}_1$ given below

$$\mathcal{E}_1 = \mathcal{E}_0 + \hbar\omega = \frac{3}{2}\hbar\omega. \tag{4.2.45}$$

It must be remarked that there is no energy level between $\mathcal{E}_0$ and $\mathcal{E}_1$. Indeed, if it is assumed that there exists an energy level $\mathcal{E}_1' < \mathcal{E}_1$ and the corresponding eigenstate $\Phi_1'(x)$, then according to formula (4.2.33) the eigenstate $\hat{a}\Phi_1'(x)$ must exist with energy level $\mathcal{E}_1' - \hbar\omega < \mathcal{E}_0$, which is a contradiction (see equation (4.2.39)).

The eigenstate $\Phi_1(x)$ corresponding to the energy level $\mathcal{E}_1$ is defined by the formula

$$\hat{a}^+\Phi_0 = \alpha_1\Phi_1, \tag{4.2.46}$$

where α_1 is a normalization coefficient, i.e., a coefficient that guarantees that $\|\Phi_1\| = 1$.

By repeatedly applying operator $\hat{a}^+$, we can find that energy levels and the corresponding eigenstates are related by formulas

$$\mathcal{E}_n = \mathcal{E}_{n-1} + \hbar\omega, \tag{4.2.47}$$

$$\hat{a}^+\Phi_{n-1} = \alpha_n\Phi_n. \tag{4.2.48}$$

From (4.2.44) and (4.2.47) we obtain the final expression for energy levels $\mathcal{E}_n$ of the harmonic oscillator

$$\mathcal{E}_n = \hbar\omega(n + \frac{1}{2}). \tag{4.2.49}$$

These energy levels are graphically represented in Figure 4.7.

Now, we turn to the derivation of formulas for eigenstates $\Phi_n(x)$. According to equation (4.2.40) and the definition of the lowering operator $\hat{a}$, we have

$$\hat{a}\Phi_0 = \frac{1}{\sqrt{2}}\left(\sqrt{\frac{m\omega}{\hbar}}x + \sqrt{\frac{\hbar}{m\omega}}\frac{d}{dx}\right)\Phi_0 = 0. \tag{4.2.50}$$

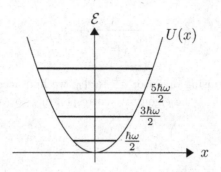

Fig. 4.7

This differential equation for Φ_0 can be easily integrated. Indeed, from the last formula we find that

$$\frac{1}{\Phi_0}\frac{d\Phi_0}{dx} = -\frac{m\omega}{\hbar}x, \qquad (4.2.51)$$

which leads to

$$\Phi_0(x) = Ce^{-\frac{m\omega}{2\hbar}x^2}. \qquad (4.2.52)$$

The constant C is determined from the condition that $\|\Phi_0\| = 1$. The latter is the case when

$$C = \left(\frac{m\omega}{\hbar\pi}\right)^{\frac{1}{4}}. \qquad (4.2.53)$$

By combining formulas (4.2.52) and (4.2.53), we obtain

$$\boxed{\Phi_0(x) = \left(\frac{m\omega}{\pi\hbar}\right)^{\frac{1}{4}} e^{-\frac{m\omega}{2\hbar}x^2}.} \qquad (4.2.54)$$

Having found $\Phi_0(x)$, the eigenstates for higher energy levels can be derived by repeatedly using formula (4.2.48). It can be shown that the normalization constant α_n in the latter equation is given by the formula

$$\alpha_n = \sqrt{n}. \qquad (4.2.55)$$

The proof of formula (4.2.55) proceeds as follows. It is clear according to formula (4.2.48) that

$$|\alpha_n|^2\|\Phi_n\|^2 = \langle \hat{a}^+\Phi_{n-1}|\hat{a}^+\Phi_{n-1}\rangle. \qquad (4.2.56)$$

Since operators $\hat{a}^+$ and $\hat{a}$ are mutually adjoint, we derive

$$|\alpha_n|^2 \|\Phi_n\|^2 = \langle \hat{a}\hat{a}^+ \Phi_{n-1} | \Phi_{n-1} \rangle. \tag{4.2.57}$$

Furthermore,

$$\hat{H}\Phi_{n-1} = \mathcal{E}_{n-1}\Phi_{n-1}. \tag{4.2.58}$$

By using formulas (4.2.26) and (4.2.49) in the last equation, we find

$$\hat{a}\hat{a}^+ \Phi_{n-1} = n\Phi_{n-1}. \tag{4.2.59}$$

By using the last formula in equation (4.2.57), we obtain

$$|\alpha_n|^2 \|\Phi_n\|^2 = n\|\Phi_{n-1}\|^2. \tag{4.2.60}$$

Since Φ_n and Φ_{n-1} are assumed to be normalized, the last equality leads to formula (4.2.55). Now, it can be easily shown that

$$\boxed{\Phi_n(x) = \frac{1}{\sqrt{n!}}(\hat{a}^+)^n \Phi_0(x).} \tag{4.2.61}$$

According to formulas (4.2.15)–(4.2.17), we have

$$\hat{a}^+ = \frac{1}{\sqrt{2}}\left(\sqrt{\frac{m\omega}{\hbar}}x - \sqrt{\frac{\hbar}{m\omega}}\frac{d}{dx}\right). \tag{4.2.62}$$

By using the last two equations, the formulas for $\Phi_n(x)$ can be derived. These formulas are summarized in the following equation where X is used instead of x for the sake of simplicity

$$\boxed{\Phi_n(X) = \frac{1}{\sqrt{\sqrt{\pi}2^n n!}}\mathcal{H}_n(X)e^{-\frac{X^2}{2}}.} \tag{4.2.63}$$

Here $\mathcal{H}_n(X)$ are **Hermite polynomials**. These polynomials are orthogonal with respect to a weight function e^{-X^2}

$$\int_{-\infty}^{\infty} \mathcal{H}_n(X)\mathcal{H}_m(X)e^{-X^2}dX = \sqrt{\pi}2^n n!\delta_{nm}. \tag{4.2.64}$$

The last formula implies that eigenstates $\Phi_n(X)$ are orthonormal. The first five Hermite polynomials are given by the formulas

$$\mathcal{H}_1(X) = 2X, \ \mathcal{H}_2(X) = -2 + 4X^2, \tag{4.2.65}$$

$$\mathcal{H}_3(X) = -12X + 8X^3, \ \mathcal{H}_4(X) = 16X^4 - 48X^2 + 12, \tag{4.2.66}$$

$$\mathcal{H}_5(X) = 32X^5 - 160X^3 + 120X. \tag{4.2.67}$$

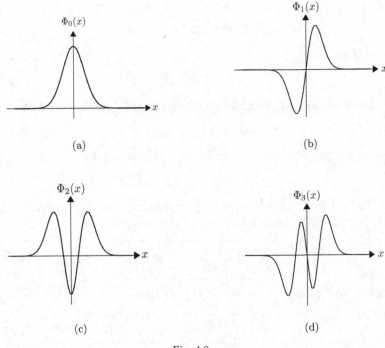

Fig. 4.8

These polynomials have been extensively studied in mathematics and many important facts related to these polynomials have been established. Just two of such facts are mentioned below. First, there exists the following explicit formula, called the **Rodrigues formula**, for Hermite polynomials

$$\mathcal{H}_n(X) = (-1)^n e^{X^2} \frac{d^n}{dX^n} \left(e^{-X^2} \right). \tag{4.2.68}$$

Second, the Hermite polynomials can be also computed by using the following recursion relation

$$\mathcal{H}_{n+1}(X) = 2X\mathcal{H}_n(X) - 2n\mathcal{H}_{n-1}(X). \tag{4.2.69}$$

The plots of the first four eigenstates $\Phi_n(X)$ given by formula (4.2.63) are shown in Figure 4.8 above.

It is apparent from this figure that the eigenstates are oscillatory in nature and each successive eigenstate has one additional zero-crossing. This oscillatory nature is a general property of eigenstates.

The presented analysis of the one-dimensional harmonic oscillator problem can be extended to two and three dimensions. Consider a two-dimensional case when the Hamiltonian operator is given by the formula

$$\hat{H} = \frac{\hat{p}_x^2}{2m} + \frac{\hat{p}_y^2}{2m} + \frac{m\omega^2}{2}(x^2 + y^2). \qquad (4.2.70)$$

The analysis of the energy spectrum requires the solution of the eigenvalue problem

$$\hat{H}\Phi_n(x, y) = \mathcal{E}_n\Phi_n(x, y). \qquad (4.2.71)$$

It is apparent from formula (4.2.70) that the Hamiltonian operator is the sum of two operators $\hat{H}_1(x)$ and $\hat{H}_2(y)$ depending only on x and y, respectively,

$$\hat{H}(x, y) = \hat{H}_1(x) + \hat{H}_2(y), \qquad (4.2.72)$$

where

$$\hat{H}_1(x) = \frac{\hat{p}_x^2}{2m} + \frac{m\omega^2}{2}x^2, \qquad (4.2.73)$$

$$\hat{H}_2(y) = \frac{\hat{p}_y^2}{2m} + \frac{m\omega^2}{2}y^2. \qquad (4.2.74)$$

For this reason, the method of separation of variables can be used for the solution of the eigenvalue problem (4.2.71). According to this method, the eigenstates $\Phi_n(x, y)$ are represented in the form

$$\Phi_n(x, y) = \Phi_{n_1}(x)\Phi_{n_2}(y). \qquad (4.2.75)$$

By using formulas (4.2.72) and (4.2.75), the eigenvalue equation (4.2.71) can be written as follows:

$$\Phi_{n_2}(y)\hat{H}_1(x)\Phi_{n_1}(x) + \Phi_{n_1}(x)\hat{H}_2\Phi_{n_2}(y) = \mathcal{E}_n\Phi_{n_1}(x)\Phi_{n_2}(y). \qquad (4.2.76)$$

By dividing both sides of the last equation by $\Phi_{n_1}(x)\Phi_{n_2}(y)$, we find

$$\frac{1}{\Phi_{n_1}(x)}\hat{H}_1(x)\Phi_{n_1}(x) + \frac{1}{\Phi_{n_2}(y)}\hat{H}_2\Phi_{n_2}(y) = \mathcal{E}_n. \qquad (4.2.77)$$

The left-hand side of the last formula is the sum of two terms which depend only on x or y, respectively. This sum can be constant only if each of the two terms in the left-hand side of (4.2.77) is constant. Consequently,

$$\frac{1}{\Phi_{n_1}(x)}\hat{H}_1(x)\Phi_{n_1}(x) = \mathcal{E}_{n_1}, \qquad (4.2.78)$$

$$\frac{1}{\Phi_{n_2}(y)}\hat{H}_2(y)\Phi_{n_2}(y) = \mathcal{E}_{n_2}, \qquad (4.2.79)$$

and

$$\mathcal{E}_{n_1} + \mathcal{E}_{n_2} = \mathcal{E}_n. \tag{4.2.80}$$

By using formulas (4.2.73) and (4.2.74), equations (4.2.78) and (4.2.79) can be written, respectively, as follows:

$$-\frac{\hbar^2}{2m}\frac{d^2\Phi_{n_1}(x)}{dx^2} + \frac{m\omega^2}{2}x^2\Phi_{n_1}(x) = \mathcal{E}_{n_1}\Phi_{n_1}(x), \tag{4.2.81}$$

$$-\frac{\hbar^2}{2m}\frac{d^2\Phi_{n_2}(y)}{dy^2} + \frac{m\omega^2}{2}y^2\Phi_{n_2}(y) = \mathcal{E}_{n_2}\Phi_{n_2}(y). \tag{4.2.82}$$

Now, it can be easily recognized that the last two equations are mathematically identical to equation (4.2.14) for the one-dimensional harmonic oscillator. Consequently, the solutions of equations (4.2.81) and (4.2.82) are, respectively, as follows

$$\mathcal{E}_{n_1} = \hbar\omega\left(n_1 + \frac{1}{2}\right), \tag{4.2.83}$$

$$\Phi_{n_1}(X) = \frac{1}{\sqrt{\sqrt{\pi}2^{n_1}n_1!}}\mathcal{H}_{n_1}(X)e^{-\frac{X^2}{2}}, \tag{4.2.84}$$

$$\mathcal{E}_{n_2} = \hbar\omega\left(n_2 + \frac{1}{2}\right), \tag{4.2.85}$$

$$\Phi_{n_2}(Y) = \frac{1}{\sqrt{\sqrt{\pi}2^{n_2}n_2!}}\mathcal{H}_{n_2}(Y)e^{-\frac{Y^2}{2}}, \tag{4.2.86}$$

where Y is related to y by the formula similar to (4.2.15).

By substituting formulas (4.2.83) and (4.2.85) into equation (4.2.80), we obtain the following expression of energy levels of the two-dimensional harmonic oscillator

$$\boxed{\mathcal{E}_n = \hbar\omega(n+1)}, \quad (n = 0, 1, 2, \ldots), \tag{4.2.87}$$

where

$$\boxed{n = n_1 + n_2.} \tag{4.2.88}$$

It can be easily observed that in contrast with the one-dimensional harmonic oscillator, the energy levels $\mathcal{E}_n$ of the two-dimensional harmonic oscillator are **degenerate** (except the ground level $\mathcal{E}_0$). It is easy to see that the degeneracy $\mathcal{D}_n$ of energy level $\mathcal{E}_n$ is equal to the number of ways an integer n can be represented as the **ordered** sum of two integers n_1 and n_2 (see formula (4.2.88)). Indeed, for different representations of n we have different eigenfunctions $\Phi_{n_1}(x)$ and $\Phi_{n_2}(y)$ and, consequently, different eigenstates $\Phi_n(x, y)$ (see formula (4.2.75)) corresponding to the same energy level $\mathcal{E}_n$. It is easy to see that the following formula is valid for the degeneracy of energy levels

$$\mathcal{D}_n = n + 1. \tag{4.2.89}$$

The presented analysis of the two-dimensional harmonic oscillator can be extended to three dimensions. It can be shown that in the three-dimensional case the energy levels $\mathcal{E}_n$ of the harmonic oscillator are given by the formula

$$\boxed{\mathcal{E}_n = \hbar\omega\left(n + \frac{3}{2}\right),} \quad (n = 0, 1, 2, \ldots) \tag{4.2.90}$$

and the degeneracy $\mathcal{D}_n$ of energy level $\mathcal{E}_n$ is

$$\mathcal{D}_n = \frac{(n+1)(n+2)}{2}. \tag{4.2.91}$$

It must be remarked that the above results are valid for the isotropic harmonic oscillator when the potential energy is given by the formula

$$U(x, y, z) = \frac{m\omega^2}{2}(x^2 + y^2 + z^2), \tag{4.2.92}$$

which can also be written as

$$U(\mathbf{r}) = \frac{m\omega^2}{2}r^2. \tag{4.2.93}$$

It is apparent that this potential energy has rotational symmetry, i.e., it is invariant with respect to rotations of coordinate systems. As discussed in the previous section, the symmetry is the cause of degeneracy of energy levels $\mathcal{E}_n$. It turns out that indeed the rotational symmetry is **partially** the cause of degeneracy of energy levels $\mathcal{E}_n$ given by formula (4.2.90). The word "partially" is used in the last sentence because the total degeneracy $\mathcal{D}_n$ cannot be explained only in terms of rotational symmetry. There is additional degeneracy of energy levels $\mathcal{E}_n$, which is called **accidental** degeneracy, which is related to the additional invariance (i.e., additional symmetry) properties of the Hamiltonian operator $\hat{H}$. This situation is very similar to the degeneracy of energy levels in the hydrogen atom, which is discussed in the next section.

It is clear from the previous discussion that the minimum energy value of the harmonic oscillator in the ground state $\mathcal{E}_0 = \frac{\hbar\omega}{2}$ is not equal to zero. This is in contrast with classical mechanics where the smallest energy value of the harmonic oscillator is zero, and this zero energy value corresponds to the state of the static equilibrium when the particle momentum and its position are simultaneously equal to zero. This essential difference between minimum energy values in quantum and classical mechanics occurs because of the uncertainty principle which precludes the possibility of ascertaining simultaneously the particle location and its momentum. This results in dynamic (rather than static) equilibrium in quantum mechanics. This equilibrium is described by the wave function $\Phi_0(x)$ which is centered around (and narrowly peaked at) the location of classical static equilibrium (see Figure 4.8(a)). It can be shown that $\Phi_0(x)$ is a **coherent state**, i.e., the state that most resembles the state of classical equilibrium. This is because the product of variances $\overline{x^2}$ and $\overline{p_x^2}$ achieves its minimum value (see Section 2.3 of Chapter 2)

$$\overline{x^2} \cdot \overline{p_x^2} = \frac{\hbar^2}{4}. \qquad (4.2.94)$$

In this sense, Φ_0 is a state of minimal uncertainty. In the state Φ_0, expected values $\overline{x}$ and $\overline{p_x}$ are equal to zero, and this is reflected in the notations for variances in formula (4.2.94). There are, however, coherent states of harmonic oscillators for which expected values $\overline{x}$ and $\overline{p_x}$ are not equal to zero and for which the product of variances $\overline{(x - \overline{x})^2}$ and $\overline{(p_x - \overline{p_x})^2}$ achieves its minimum value as well

$$\overline{(x - \overline{x})^2} \cdot \overline{(p_x - \overline{p_x})^2} = \frac{\hbar^2}{4}. \qquad (4.2.95)$$

These coherent states are eigenfunctions of the lowering (annihilation) operator $\hat{a}$:

$$\hat{a}\psi = \lambda\psi. \qquad (4.2.96)$$

It can be shown that the coherent state wave functions ψ are congruent to (i.e., have the same shape as) the wave function Φ_0 of the ground state, however they are shifted in position and momentum by $\overline{x}$ and $\overline{p_x}$, respectively. For this reason, these coherent states are sometimes called "displaced ground states." The coherent state wave functions ψ evolve in time in such a way that their shape remains unchanged, while the x and p_x coordinates of their centers move along the classical trajectories. This means that coherent states can be viewed as compact wave packets of minimal uncertainty traveling along classical trajectories. In this sense, these time-varying coherent states most resemble classical dynamics.

The theory of coherent states has been extensively developed and many important results have been accumulated in this research area. These states have important applications in quantum optics and laser physics, and there they are usually called **Glauber states**. R. J. Glauber introduced and used these states for a quantum description of electromagnetic field coherence observed in laser radiation, and he was awarded the Nobel Prize in Physics in 2005 for his achievements.

4.3 The Hydrogen Atom

The hydrogen atom problem is another, albeit rare, example of a quantum mechanical problem for which exact analytical solution can be obtained. This problem has played a unique role in the historical development of quantum mechanics by serving as a testing ground for advanced theories.

The hydrogen-like atom consists of a heavy (and practically motionless) nucleus of positive charge Ze and a single electron of negative charge $-e$, bound to the nucleus by attractive Coulomb force between opposite charges. It can be assumed that the nucleus is at the origin of the coordinate system (see Figure 4.9). The potential energy of the bound electron in the electric

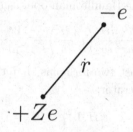

Fig. 4.9

field of the nucleus is given by the formula

$$U(r) = -\frac{Ze^2}{r}. \tag{4.3.1}$$

This leads to the following expression for the Hamiltonian operator

$$\hat{H} = -\frac{\hbar^2}{2m}\nabla^2 - \frac{Ze^2}{r}, \tag{4.3.2}$$

where ∇^2 is the Laplacian operator.

In order to find discrete energy levels of hydrogen-like atoms, **square-integrable solutions** of the time-independent Schrödinger equation

$$\hat{H}\Phi_{\mathcal{E}} = \mathcal{E}\Phi_{\mathcal{E}} \tag{4.3.3}$$

must be found.

By combining formulas (4.3.2) and (4.3.3) we end up with the following differential equation

$$-\frac{\hbar^2}{2m}\nabla^2\Phi_{\mathcal{E}} - \frac{Ze^2}{r}\Phi_{\mathcal{E}} = \mathcal{E}\Phi_{\mathcal{E}}. \qquad (4.3.4)$$

In the analysis of the hydrogen-like atom spectrum, we shall use the spherical coordinates (r, θ, ψ). In these coordinates, the Laplacian operator can be written as follows

$$-\frac{\hbar^2}{2m}\nabla^2 = -\frac{\hbar^2}{2mr^2}\frac{\partial}{\partial r}\left(r^2\frac{\partial}{\partial r}\right) - \frac{\hbar^2}{2mr^2}\left[\frac{1}{\sin\theta}\frac{\partial}{\partial\theta}\left(\sin\theta\frac{\partial}{\partial\theta}\right) + \frac{1}{\sin^2\theta}\frac{\partial^2}{\partial\varphi^2}\right].$$
$$(4.3.5)$$

Now, we shall recall (see Section 2.4) that the squared magnitude of angular momentum operator $\widehat{\mathbf{L}}^2$ is given by the formula

$$\widehat{\mathbf{L}}^2 = -\hbar^2\left[\frac{1}{\sin\theta}\frac{\partial}{\partial\theta}\left(\sin\theta\frac{\partial}{\partial\theta}\right) + \frac{1}{\sin^2\theta}\frac{\partial^2}{\partial\varphi^2}\right]. \qquad (4.3.6)$$

By using formulas (4.3.5) and (4.3.6) in equation (4.3.2), we arrive at the following expression for the Hamiltonian operator

$$\hat{H} = -\frac{\hbar^2}{2mr^2}\frac{\partial}{\partial r}\left(r^2\frac{\partial}{\partial r}\right) + \frac{\widehat{\mathbf{L}}^2}{2mr^2} - \frac{Ze^2}{r}. \qquad (4.3.7)$$

It is apparent from the last two formulas that the Hamiltonian operator commutes with the $\widehat{\mathbf{L}}^2$-operator:

$$\left[\hat{H}, \widehat{\mathbf{L}}^2\right] = 0. \qquad (4.3.8)$$

Furthermore, since

$$\hat{L}_z = -i\hbar\frac{\partial}{\partial\varphi} \qquad (4.3.9)$$

and since $\hat{L}_z$ commutes with $\widehat{\mathbf{L}}^2$ (see Section 2.4)

$$\left[\widehat{\mathbf{L}}^2, \hat{L}_z\right] = 0, \qquad (4.3.10)$$

it is easy to conclude that the Hamiltonian operator $\hat{H}$ commutes with $\hat{L}_z$

$$\left[\hat{H}, \hat{L}_z\right] = 0. \qquad (4.3.11)$$

Commutation relations (4.3.8), (4.3.10) and (4.3.11) imply that energy $\mathcal{E}$, squared magnitude of angular momentum L^2 and angular momentum projection L_z on the z-axis are simultaneously measurable. This means that we can consider states $\Phi_{\mathcal{E},l,m}(r, \theta, \varphi)$ of hydrogen-like atoms in which the above three quantities have definite (certain) values which are used as subscripts for Φ. For this reason, we shall look for the solution of equation (4.3.3) in the form

$$\Phi_\mathcal{E}(r, \theta, \varphi) = \Phi_{\mathcal{E},l,m}(r, \theta, \varphi). \qquad (4.3.12)$$

It is also apparent that $\Phi_{\mathcal{E},l,m}(r, \theta, \varphi)$ is the eigenfunction of $\widehat{\mathbf{L}}^2$ and $\hat{L}_z$ operators:

$$\widehat{\mathbf{L}}^2 \Phi_{\mathcal{E},l,m}(r, \theta, \varphi) = \hbar^2 l(l+1) \Phi_{\mathcal{E},l,m}(r, \theta, \varphi), \qquad (4.3.13)$$

$$\hat{L}_z \Phi_{\mathcal{E},l,m}(r, \theta, \varphi) = \hbar m \Phi_{\mathcal{E},l,m}(r, \theta, \varphi), \quad (m = 0, \pm 1, \pm 2, \cdots \pm l). \qquad (4.3.14)$$

As discussed in the second chapter (see Section 2.4), the solution of simultaneous equations (4.3.13) and (4.3.14) for given l and m is the spherical harmonic $Y_l^m(\theta, \varphi)$. This implies that the state $\Phi_{\mathcal{E},l,m}(r, \theta, \varphi)$ can be represented as follows

$$\Phi_{\mathcal{E},l,m}(r, \theta, \varphi) = T(r) Y_l^m(\theta, \varphi). \qquad (4.3.15)$$

By substituting the last formula into equation (4.3.3) and taking into account formulas (4.3.7) and (4.3.13), we end up with

$$-\frac{\hbar^2}{2mr^2} \frac{d}{dr}\left(r^2 \frac{dT(r)}{dr}\right) Y_l^m(\theta, \varphi) + \frac{\hbar^2 l(l+1)}{2mr^2} T(r) Y_l^m(\theta, \varphi)$$
$$-\frac{Ze^2}{r} T(r) Y_l^m(\theta, \varphi) = \mathcal{E} T(r) Y_l^m(\theta, \varphi), \qquad (4.3.16)$$

which leads to

$$\frac{1}{r^2} \frac{d}{dr}\left(r^2 \frac{dT(r)}{dr}\right) - \frac{l(l+1)}{r^2} T(r) + \frac{2Ze^2 m}{\hbar^2 r} T(r) = -\frac{2m\mathcal{E}}{\hbar^2} T(r). \qquad (4.3.17)$$

To avoid any misunderstanding, it must be noted here that "m" in the last equation is the electron (rest) mass, and it must not be confused with "m" which defines the value of angular momentum projection L_z on the z-axis (see formula (4.3.14)). Actually, the last equation does not contain the latter "m" and, consequently, the energy levels of the hydrogen-like atom do not depend on the angular momentum orientation. This implies that the energy levels of the hydrogen-like atom are degenerate with respect to the value of this "m". In other words, for a given value of l in equation (4.3.17), there are $2l+1$ different values of L_z (see formula (4.3.14)) and there are $2l+1$ different spherical harmonics $Y_l^m(\theta, \varphi)$ and, consequently, there are $2l+1$ distinct eigenstates defined by formula (4.3.15). This degeneration of energy levels with respect to the orientation of angular momentum is due to the rotational symmetry (i.e., rotational invariance) of the Hamiltonian operator $\hat{H}$ given by formula (4.3.2). The latter means that the mathematical form of the Hamiltonian (4.3.2) does not depend on the orientation of the chosen coordinate system. It is shown below that there exists additional degeneracy of energy levels in hydrogen-like atoms which cannot be attributed to the rotational symmetry of the Hamiltonian.

Now, we return to the solution of equation (4.3.17). First, it can be easily shown that

$$\frac{1}{r^2}\frac{d}{dr}\left(r^2\frac{dT(r)}{dr}\right) = \frac{1}{r}\frac{d^2}{dr^2}(rT(r)). \tag{4.3.18}$$

By using this fact, we introduce the function

$$N(r) = rT(r) \tag{4.3.19}$$

and write equation (4.3.17) in terms of $N(r)$ as follows

$$\frac{d^2N(r)}{dr^2} - \frac{l(l+1)}{r^2}N(r) + \frac{2Ze^2m}{\hbar^2r}N(r) = -\frac{2m\mathcal{E}}{\hbar^2}N(r). \tag{4.3.20}$$

Next, we introduce dimensionless quantities

$$\beta = \frac{e^2m}{\hbar^2}r, \tag{4.3.21}$$

$$\varepsilon = -\frac{\hbar^2\mathcal{E}}{me^4}, \tag{4.3.22}$$

and represent equation (4.3.20) in the form

$$\frac{d^2N(\beta)}{d\beta^2} - \frac{l(l+1)}{\beta^2}N(\beta) + \frac{2Z}{\beta}N(\beta) = 2\varepsilon N(\beta). \tag{4.3.23}$$

To find nonsingular solutions of the last equation which decay to zero at infinity, we shall analyze the asymptotics of solutions of the last equation for very small and very large values of β. For very small values of β, equation (4.3.23) can be written as

$$\frac{d^2N(\beta)}{d\beta^2} - \frac{l(l+1)}{\beta^2}N(\beta) = 0. \tag{4.3.24}$$

By looking for the solution of the last equation in the form

$$N(\beta) = C\beta^\alpha \tag{4.3.25}$$

and by substituting formula (4.3.25) into equation (4.3.24), we find

$$\alpha(\alpha - 1) - l(l+1) = 0. \tag{4.3.26}$$

The last equality is valid if

$$\alpha_1 = l+1 \text{ or } \alpha_2 = -l. \tag{4.3.27}$$

It is apparent now that nonsingular and integrable solutions of equation (4.3.23) should have the following asymptotics for very small values of β

$$N(\beta) \sim C\beta^{l+1}. \tag{4.3.28}$$

For very large values of β, equation (4.3.23) can be written as

$$\frac{d^2 N(\beta)}{d\beta^2} = 2\varepsilon N(\beta). \tag{4.3.29}$$

It is clear that the solution of the last equation decays to zero at infinity if

$$\varepsilon > 0. \tag{4.3.30}$$

Indeed, if $\varepsilon < 0$, then the solution of equation (4.3.29) is oscillatory in nature and does not decay to zero at infinity.

The last inequality implies (according to formula (4.3.22)) that

$$\mathcal{E} < 0. \tag{4.3.31}$$

Thus, **energy levels corresponding to the discrete spectrum are negative**.

From formulas (4.3.29) and (4.3.30), we find that for very large values of β, the solution of equation (4.3.23) has the following asymptotics

$$N(\beta) \sim Ce^{-\sqrt{2\varepsilon}\beta}. \tag{4.3.32}$$

By using asymptotics (4.3.28) and (4.3.32), we shall look for the solution of equation (4.3.23) in the form

$$N(\beta) = \beta^{l+1} e^{-\sqrt{2\varepsilon}\beta} \sum_{k=0}^{\infty} a_k \beta^k, \tag{4.3.33}$$

which can be also written as follows:

$$N(\beta) = e^{-\sqrt{2\varepsilon}\beta} \sum_{k=0}^{\infty} a_k \beta^{k+l+1}. \tag{4.3.34}$$

From the last formula, we find

$$\frac{d^2 N(\beta)}{d\beta^2} = 2\varepsilon e^{-\sqrt{2\varepsilon}\beta} \sum_{k=0}^{\infty} a_k \beta^{k+l+1} - 2\sqrt{2\varepsilon} e^{-\sqrt{2\varepsilon}\beta} \sum_{k=0}^{\infty} (k+l+1) a_k \beta^{k+l}$$

$$+ e^{-\sqrt{2\varepsilon}\beta} \sum_{k=0}^{\infty} (k+l+1)(k+l) a_k \beta^{k+l-1}.$$

$$\tag{4.3.35}$$

By substituting formulas (4.3.34) and (4.3.35) into equation (4.3.23), after simple transformations we obtain

$$-2\sqrt{2\varepsilon}\sum_{k=0}^{\infty}(k+l+1)a_k\beta^{k+l} + \sum_{k=0}^{\infty}(k+l+1)(k+l)a_k\beta^{k+l-1}$$

$$-l(l+1)\sum_{k=0}^{\infty}a_k\beta^{k+l-1} + 2Z\sum_{k=0}^{\infty}a_k\beta^{k+l} = 0. \qquad (4.3.36)$$

By combining the similar terms in the last formula, we arrive at

$$2\sum_{k=0}^{\infty}[Z - \sqrt{2\varepsilon}(k+l+1)]a_k\beta^{k+l}$$

$$= \sum_{k=0}^{\infty}[l(l+1) - (k+l+1)(k+l)]a_k\beta^{k+l-1}. \qquad (4.3.37)$$

By equating terms of the same power of β in the last equation, we get

$$2[Z - \sqrt{2\varepsilon}(k+l+1)]a_k = [l(l+1) - (k+l+2)(k+l+1)]a_{k+1}, \qquad (4.3.38)$$

which leads to

$$a_{k+1} = \frac{2Z - 2\sqrt{2\varepsilon}(k+l+1)}{l(l+1) - (k+l+2)(k+l+1)}a_k. \qquad (4.3.39)$$

From the last formula we find that in the case of very large k

$$a_{k+1} \simeq \frac{2\sqrt{2\varepsilon}}{k+1}a_k. \qquad (4.3.40)$$

Consider the function

$$e^{2\sqrt{2\varepsilon}\beta} = \sum_{k=0}^{\infty}c_k\beta^k. \qquad (4.3.41)$$

It is clear that

$$c_{k+1} = \frac{(2\sqrt{2\varepsilon})^{k+1}}{(k+1)!} = \frac{2\sqrt{2\varepsilon}}{k+1}c_k. \qquad (4.3.42)$$

By comparing formulas (4.3.40) and (4.3.42), it can be concluded that the infinite series in formula (4.3.33) grows at infinity in the same way as the function in (4.3.41), i.e.,

$$\sum_{k=0}^{\infty}a_k\beta^k \sim e^{2\sqrt{2\varepsilon}\beta}. \qquad (4.3.43)$$

This implies according to formula (4.3.33) that at very large values of β

$$N(\beta) \sim \beta^{l+1}e^{\sqrt{\varepsilon}\beta}. \qquad (4.3.44)$$

This means that $N(\beta)$ can decay to zero at infinity only if **the sum in formula (4.3.33) contains a finite number of terms**. This is the case if

$$a_{k+1} = 0. \tag{4.3.45}$$

According to formula (4.3.39), the last equality is valid only if

$$Z - \sqrt{2\varepsilon}(k + l + 1) = 0. \tag{4.3.46}$$

This brings us to the conclusion that solutions to equation (4.3.23) and, consequently, the eigenstates $\Phi_{\varepsilon,l,m}(r, \theta, \varphi)$ decay to zero at infinity only for special values of ε for which equality (4.3.46) holds. These values are

$$\varepsilon_n = \frac{Z^2}{2n^2}, \tag{4.3.47}$$

where

$$\boxed{n = k + l + 1.} \tag{4.3.48}$$

Now, by recalling formula (4.3.22), we find the energy levels of the hydrogen-like atom

$$\boxed{\mathcal{E}_n = -\frac{me^4 Z^2}{2\hbar^2 n^2}.} \tag{4.3.49}$$

For a hydrogen atom when $Z = 1$, we have

$$\boxed{\mathcal{E}_n = -\frac{me^4}{2\hbar^2 n^2}.} \tag{4.3.50}$$

These energy levels are degenerate for $n > 1$ and their degeneracy is two-fold. First, according to formula (4.3.48), there is degeneracy with respect to l. Namely, for any given n, l may assume values from 0 to $n - 1$. For each of these values of l, there is a different value of k in formula (4.3.48) and, consequently, a different finite sum in formula (4.3.33). This results in a different function $T(r)$ in formula (4.3.15). Actually, this formula now can be written as

$$\Phi_{nlm}(r, \theta, \varphi) = T_{nl}(r) Y_l^m(\theta, \varphi). \tag{4.3.51}$$

This is because the energy level is defined according to formula (4.3.49) (or (4.3.50)) by n, while function $T(r)$ depends on n and l as explained above. The second form of degeneracy is $2l + 1$ degeneracy with respect to m, because m may assume the values $m = 0, \pm 1, \pm 2, \cdots \pm l$ for any chosen l. We have already discussed that this degeneracy is with respect to the orientation of angular momentum and it results in different spherical harmonics Y_l^m in formula (4.3.51). It is clear from the above discussion that the degeneracy $\mathcal{D}_n$ can be computed as

$$\mathcal{D}_n = \sum_{l=0}^{n-1} (2l+1) = n^2. \tag{4.3.52}$$

Degeneracy with respect to l is called accidental degeneracy. It is due to the "hidden" symmetry of the hydrogen-like atom, which can be attributed to the $\frac{1}{r}$-dependence of potential energy. This symmetry manifests itself in the existence of the Runge–Lenz operator

$$\widehat{\mathbf{R}} = \frac{1}{me^2} \left(\widehat{\mathbf{L}} \times \hat{\mathbf{p}} - \hat{\mathbf{p}} \times \widehat{\mathbf{L}} \right) + \frac{\mathbf{r}}{r} \tag{4.3.53}$$

that commutes with the Hamiltonian operator

$$\left[\hat{H}, \widehat{\mathbf{R}} \right] = 0, \tag{4.3.54}$$

therefore resulting in the specific conservation law corresponding to the conservation of the Runge–Lenz vector in classical mechanics. We shall not discuss this matter further. However, it is worthwhile to mention that the Runge–Lenz operator can be used for the derivation of formula (4.3.49) by the pure operator-algebraic method. Actually, this is how the hydrogen atom energy spectrum (4.3.50) was first derived by W. Pauli within the framework of the matrix formulation of quantum mechanics.

Returning to formula (4.3.50), we can see that the ground energy level $\mathcal{E}_1$ of the hydrogen atom is given by the formula

$$\mathcal{E}_1 = -\frac{me^4}{2\hbar^2}. \tag{4.3.55}$$

The numerical value of this energy level is

$$\mathcal{E}_1 = -13.55 \text{ eV}. \tag{4.3.56}$$

Formula (4.3.50) is sometimes written in terms of the Bohr radius a_0 as

$$\mathcal{E}_n = -\frac{e^2}{2a_0 n^2} \tag{4.3.57}$$

and

$$\mathcal{E}_1 = -\frac{e^2}{2a_0}, \tag{4.3.58}$$

where

$$a_0 = \frac{\hbar^2}{me^2} = 5.29 \times 10^{-9} \text{ cm}. \tag{4.3.59}$$

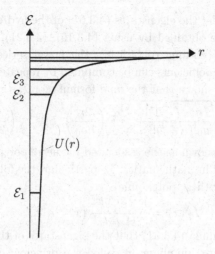

Fig. 4.10

The Bohr radius is approximately equal to the most probable distance between the nucleus and electron and can be considered as an approximate geometrical measure of the hydrogen atom size. The variable β used in our derivation is the distance normalized by a_0, while ε is the energy normalized by $2\mathcal{E}_1$.

The discrete energy levels $\mathcal{E}_n$ correspond to bound states of the electron in the hydrogen atom. They are graphically illustrated by Figure 4.10. There is also the continuous energy spectrum of the hydrogen atom which consists of the positive value of energy $\mathcal{E} > 0$ corresponding to the free (unbound) electron.

The energy level structure of the hydrogen atom (as well as other atoms and molecules) is studied by spectroscopic methods by measuring frequencies of emitted radiation by excited atoms or frequencies of resonance absorption by atoms. Such frequencies are called spectral lines or spectral series. In the case of hydrogen atoms two of such series are most prominent: the Lyman series $\nu^{(1,n)}$ and the Balmer series $\nu^{(2,n)}$ which are defined by the following formulas, respectively,

$$\mathcal{E}_n - \mathcal{E}_1 = h\nu^{(1,n)}, \tag{4.3.60}$$

$$\mathcal{E}_n - \mathcal{E}_2 = h\nu^{(2,n)}. \tag{4.3.61}$$

The spectral series predicted by formula (4.3.50) are in a remarkable agreement with experimental observations.

Next, we consider the eigenstates (4.3.51) of the hydrogen atom. Functions $T_{nl}(r)$ can be obtained by using (4.3.19), (4.3.21), and (4.3.33). In formula (4.3.33), according to (4.3.45), the power series is reduced to a polynomial whose coefficients can be computed by repeatedly using formula (4.3.39). It can be shown that the final formula for $T_{nl}(r)$ is as follows

$$T_{nl}(r) = \frac{2}{n^2 a_0} \sqrt{\frac{(n-l-1)!}{a_0[(n+l)!]^3}} e^{-\frac{r}{na_0}} \left(\frac{2r}{na_0}\right)^l L_{n-l-1}^{2l+1}\left(\frac{r}{2na_0}\right), \quad (4.3.62)$$

where L_{n-l-1}^{2l+1} are known as the associated Laguerre polynomials which are extensively studied in mathematics. In particular, the following Rodrigues formula is valid for these polynomials

$$L_k^\lambda(x) = \frac{x^{-\lambda} e^x}{k!} \frac{d^k}{dx^k}\left(x^{k+\lambda} e^{-x}\right). \quad (4.3.63)$$

It is clear from formula (4.3.51) that the eigenstates of the hydrogen atom are specified by three numbers: n called the principal quantum number, l called the orbital quantum number and m called the magnetic quantum number. Spectroscopic notations are often used in literature for different values of the orbital quantum number. These notations are given below for the first four values of l:

$$
\begin{array}{cccc}
l = 0 & l = 1 & l = 2 & l = 3 \\
\downarrow & \downarrow & \downarrow & \downarrow \\
\text{s} & \text{p} & \text{d} & \text{f}
\end{array}
\quad (4.3.64)
$$

Here: abbreviations "s" stands for "sharp", "p" stands for "principal", "d" stands for "diffuse" and "f" stands for "fundamental". Letters for subsequent values of l are assigned in alphabetical order omitting only the letter "j". In hydrogen-like atoms those letters are preceded by the value of the principal quantum number. For instance, the first six lowest energy states have the notations: 1s, 2s, 2p, 3s, 3p, 3d.

The presented analysis of the hydrogen spectrum is based on the Hamiltonian operator given by the formula (4.3.2). This Hamiltonian does not account for electron spin and, associated with it, spin-orbit coupling, which occurs because the electric field of the nucleus manifests itself as the magnetic field in the reference frame of the moving electron and, consequently, interacts with the spin magnetic moment. This Hamiltonian does not account for other relativistic corrections as well as for the quantized nature of electromagnetic field. The neglected effects mentioned above are very small and they are controlled by different powers of the fine-structure constant α which is given by the formula (in cgs units)

$$\alpha = \frac{e^2}{\hbar c} = \frac{1}{137.036}. \quad (4.3.65)$$

These neglected effects result in the splitting of energy levels (splitting of spectral lines) which is called the fine structure and Lamb shift depending on the physical origin of the splitting. These small effects are accounted for with remarkable accuracy within the framework of relativistic quantum electrodynamics.

4.4 Landau Levels. Quantum Hall Effect

In this section, we present the quantum mechanical treatment of electron motion in a uniform magnetic field $\mathbf{B}_0$ directed along the z-axis. We begin by briefly reviewing the classical results related to this problem. We first assume that there is no electron velocity component along the field $\mathbf{B}_0$. This implies that the electron dynamics is on the xy plane. Due to the

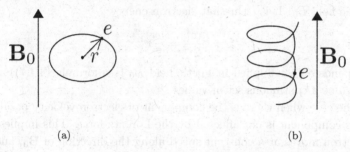

(a)   (b)

Fig. 4.11

rotational symmetry of the problem with respect to the z-axis, an electron trajectory is circular (see Figure 4.11(a)). At each point of this trajectory magnetic (Lorentz) force $\mathbf{F}_m$ is equal to the centripetal force $\mathbf{F}_c$:

$$\mathbf{F}_m = \mathbf{F}_c. \tag{4.4.1}$$

By equating the magnitudes of these two forces, we find

$$evB_0 = \frac{mv^2}{r}, \tag{4.4.2}$$

which leads to the following expression for the radius r of the circular trajectory

$$r = \frac{mv}{eB_0}. \tag{4.4.3}$$

The period T of electron motion along the circular trajectory is given by the formula

$$T = \frac{2\pi r}{v}. \tag{4.4.4}$$

By combining the last two equations, we find

$$T = \frac{2\pi m}{eB_0}. \qquad (4.4.5)$$

The angular (cyclotron) frequency can be computed as

$$\omega_c = \frac{2\pi}{T}, \qquad (4.4.6)$$

which leads to

$$\boxed{\omega_c = \frac{eB_0}{m}.} \qquad (4.4.7)$$

It is apparent from the last formula that the cyclotron frequency does not depend on electron speed v or radius r of the circular trajectory, but it rather depends on electron parameters e and m as well as on the applied magnetic field B_0. It is clear that electron energy

$$\mathcal{E} = \frac{mv^2}{2} = \frac{m}{2}r^2\omega_c^2 \qquad (4.4.8)$$

also depends on the applied magnetic field B_0 (see formula (4.4.7)) and it may assume a continuous set of values.

In the case when there is the component of electron velocity parallel to $\mathbf{B}_0$, this component is not affected by the Lorentz force. This implies that the electron moves at a constant speed along the direction of $\mathbf{B}_0$, and the electron path is a helix (see Figure 4.11(b)).

Now, we turn to the quantum mechanical treatment of the problem and shall demonstrate that in contrast with the classical mechanics and electrodynamics the cyclotron energy $\mathcal{E}$ is quantized. The discrete levels of the cyclotron energy are called **Landau levels**, and they play a central role in the origin of the **quantum Hall effect**.

In our analysis, we shall use the Hamiltonian operator (see Section 2.5 of Chapter 2)

$$\hat{H} = \frac{(\hat{\mathbf{p}} + e\mathbf{A})^2}{2m} - e\varphi. \qquad (4.4.9)$$

In the absence of the electric field

$$\varphi = 0, \qquad (4.4.10)$$

and, consequently,

$$\hat{H} = \frac{(\hat{\mathbf{p}} + e\mathbf{A})^2}{2m}. \qquad (4.4.11)$$

Next, we have to choose the expression for the vector potential $\mathbf{A}$. The problem under consideration has translational symmetry in the xy plane and rotational symmetry around the z-axis. However, it is not possible to find the expression (the gauge) for $\mathbf{A}$ that preserves these two symmetries simultaneously. For this reason, different expressions for $\mathbf{A}$ that preserve one of the above symmetries can be used. These expressions are solutions of equations

$$\text{curl } \mathbf{A} = \mathbf{B}_0, \tag{4.4.12}$$

$$\text{div } \mathbf{A} = 0, \tag{4.4.13}$$

where

$$\mathbf{B}_0 = B_0 \mathbf{e}_z. \tag{4.4.14}$$

One of these expressions is

$$\mathbf{A} = -B_0 y \mathbf{e}_x \tag{4.4.15}$$

and it is usually called the **Landau gauge**. This expression preserves the translation invariance along the x-axis.

Another similar gauge is

$$\mathbf{A} = B_0 x \mathbf{e}_y, \tag{4.4.16}$$

Finally, the so called **symmetric gauge** is

$$\mathbf{A} = -\frac{B_0}{2} y \mathbf{e}_x + \frac{B_0}{2} x \mathbf{e}_y, \tag{4.4.17}$$

which preserves the rotational invariance around the z-axis.

In our subsequent discussion, we shall adopt the Landau gauge (4.4.15) and shall use the following Hamiltonian operator

$$\hat{H} = \frac{[\hat{\mathbf{p}} - eB_0 y \mathbf{e}_x]^2}{2m}. \tag{4.4.18}$$

To find the energy spectrum, we have to solve the time-independent Schrödinger equation

$$\frac{[\hat{\mathbf{p}} - eB_0 y \mathbf{e}_x]^2}{2m} \Phi = \mathcal{E}\Phi. \tag{4.4.19}$$

Since

$$\hat{\mathbf{p}} = \mathbf{e}_x \hat{p}_x + \mathbf{e}_y \hat{p}_y + \mathbf{e}_z \hat{p}_z, \tag{4.4.20}$$

equation (4.4.19) can be written as follows

$$\frac{1}{2m}[(\hat{p}_x - eB_0 y)^2 + \hat{p}_y^2 + \hat{p}_z^2]\Phi = \mathcal{E}\Phi. \tag{4.4.21}$$

It is clear that the Hamiltonian operator in equation (4.4.21) commutes with $\hat{p}_x$ and $\hat{p}_z$ operators:

$$\hat{H}\hat{p}_x - \hat{p}_x\hat{H} = 0, \tag{4.4.22}$$

$$\hat{H}\hat{p}_z - \hat{p}_z\hat{H} = 0. \tag{4.4.23}$$

This implies that the energy and momentum components along the x- and z-axes are simultaneously measurable. This suggests that we can look for the solutions of equation (4.4.21) which are also eigenfunctions of $\hat{p}_x$ and $\hat{p}_z$ operators. The latter is the case when $\Phi(x, y, z)$ is represented in the form

$$\Phi(x, y, z) = \zeta(y)e^{\frac{i}{\hbar}(p_x x + p_z z)}. \tag{4.4.24}$$

Next, by substituting the last formula into equation (4.4.19), we shall derive the equation for $\zeta(y)$. The derivation proceed as follows. First, since p_x and $e^{\frac{i}{\hbar}p_x x}$ are an eigenvalue and eigenfunction of operator $\hat{p}_x$, respectively, we find

$$(\hat{p}_x - eB_0 y)^2 \zeta(y)e^{\frac{i}{\hbar}(p_x x + p_z z)} = e^{\frac{i}{\hbar}(p_x x + p_z z)}(p_x - eB_0 y)^2 \zeta(y). \tag{4.4.25}$$

Similarly,

$$\hat{p}_z^2[\zeta(y)e^{\frac{i}{\hbar}(p_x x + p_z z)}] = e^{\frac{i}{\hbar}(p_x x + p_z z)}p_z^2 \zeta(y). \tag{4.4.26}$$

Furthermore,

$$\hat{p}_y^2[\zeta(y)e^{\frac{i}{\hbar}(p_x x + p_z z)}] = e^{\frac{i}{\hbar}(p_x x + p_z z)}\hat{p}_y^2 \zeta(y). \tag{4.4.27}$$

By substituting formulas (4.4.24)–(4.4.27) into equation (4.4.21) and by canceling the exponential factor $e^{\frac{i}{\hbar}(p_x x + p_z z)}$, we arrive at the following equation for $\zeta(y)$

$$\frac{\hat{p}_y^2}{2m}\zeta(y) + \frac{1}{2m}(eB_0 y - p_x)^2 \zeta(y) = \left(\mathcal{E} - \frac{p_z^2}{2m}\right)\zeta(y), \tag{4.4.28}$$

which by using formula (4.4.7) can be also written as

$$-\frac{\hbar^2}{2m}\frac{d^2\zeta(y)}{dy^2} + \frac{m\omega_c^2}{2}\left(y - \frac{p_x}{eB_0}\right)^2 \zeta(y) = \left(\mathcal{E} - \frac{p_z^2}{2m}\right)\zeta(y). \tag{4.4.29}$$

By introducing the notations

$$\tilde{y} = y - \frac{p_x}{eB_0}, \tag{4.4.30}$$

$$\tilde{\mathcal{E}} = \mathcal{E} - \frac{p_z^2}{2m}, \tag{4.4.31}$$

the last equation can be written in the form

$$-\frac{\hbar^2}{2m}\frac{d^2\zeta(\tilde{y})}{d\tilde{y}^2} + \frac{m\omega_c^2}{2}\tilde{y}^2\zeta(\tilde{y}) = \tilde{\mathcal{E}}\zeta(\tilde{y}). \qquad (4.4.32)$$

This equation is mathematically identical to equation (4.2.14) for the harmonic oscillator. For this reason, we conclude that the eigenvalues $\tilde{\mathcal{E}}$ in equation (4.4.32) are given by the formula

$$\tilde{\mathcal{E}}_k = \hbar\omega_c\left(k + \frac{1}{2}\right), \quad (k = 0, 1, \ldots). \qquad (4.4.33)$$

Now, by recalling formula (4.4.31), we arrive at the following expression for the energy spectrum

$$\boxed{\mathcal{E}_k = \hbar\omega_c\left(k + \frac{1}{2}\right) + \frac{p_z^2}{2m}, \quad (k = 0, 1, 2, \ldots).} \qquad (4.4.34)$$

Thus, the energy spectrum is parameterized by two quantum numbers: k and p_z. The first term in the right-hand side of formula (4.4.34) depends on k and describes discrete energy levels of cyclotron oscillations on the xy plane. These energy levels are called the **Landau levels** in honor of L. Landau who derived formula (4.4.34) in 1930. The second term in formula (4.4.34) describes the motion along the direction of the field $\mathbf{B}_0$ with constant momentum p_z which may assume a continuous set of values. In this sense, the motion along the z-direction is not quantized. It is also important to stress that energy levels $\mathcal{E}_k$ given by formula (4.4.34) are infinitely degenerate. This is because these energy levels do not depend on p_x which assumes a continuous set of values. As a result, for each k there are infinitely many eigenfunctions which, according to formula (4.4.30), are shifted along the y-axis by $\frac{p_x}{eB_0}$.

The continuous energy spectrum with respect to p_z and the infinite degeneracy of $\mathcal{E}_k$ with respect to p_x occur because the electron motion is considered in infinite free space. In the case when electrons are confined inside a box with sides L_x, L_y, and L_z, then the momentum spectrum is discrete. In other words, the momentum is quantized in the latter case. We demonstrate this for p_x in the case of periodic boundary conditions. As discussed before (see Chapter 2), the eigenfunctions of operator $\hat{p}_x$ are $\lambda(y, z)e^{\frac{i}{\hbar}p_x x}$. These eigenfunctions assume the same values at $x = 0$ and $x = L_x$ if

$$\frac{i}{\hbar}p_x^{(\nu)}L_x = i\nu 2\pi. \qquad (4.4.35)$$

Consequently, the quantized values of p_x are

$$p_x^{(\nu)} = \nu \frac{2\pi\hbar}{L_x} = \nu \frac{h}{L_x}. \tag{4.4.36}$$

These quantized values $p_x^{(\nu)}$ result in discrete shifts (translations) of eigen-functions along the y-axis. According to formula (4.4.30), these shifts $y^{(\nu)}$ are

$$y^{(\nu)} = \frac{p_x^{(\nu)}}{eB_0} = \nu \frac{h}{eB_0 L_x}. \tag{4.4.37}$$

These shifts cannot exceed L_y. Consequently, the maximum value of ν can be found from the equation

$$y^{(\nu_{max})} = L_y = \nu_{max} \frac{h}{eB_0 L_x}. \tag{4.4.38}$$

This leads to the following expression for the degeneracy $\mathcal{D} = \nu_{max}$ of Landau levels

$$\boxed{\mathcal{D} = \frac{eB_0 L_x L_y}{h}.} \tag{4.4.39}$$

This degeneracy is extremely large for appreciable B_0. This can be easily seen by representing the last formula in the form

$$\mathcal{D} = \frac{\Phi}{2\Phi_0}, \tag{4.4.40}$$

where $\Phi = B_0 L_x L_y$, while $\Phi_0 = \frac{h}{2e} = 2.07 \times 10^{-15}$ Wb (see formula (2.5.55)).

It is also convenient to introduce the degeneracy d per unit area. From formula (4.4.39) follows that

$$\boxed{d = \frac{eB_0}{h}.} \tag{4.4.41}$$

Next, we consider the case of two-dimensional electron confinement. In this case, the electron motion along the z-direction can be neglected ($p_z = 0$) and, by using equation (4.4.7), formula (4.4.34) can be written as follows

$$\boxed{\mathcal{E}_k = \frac{\hbar e B_0}{m} \left(k + \frac{1}{2} \right), \quad (k = 1, 2, \ldots).} \tag{4.4.42}$$

It is clear from the last two formulas that an increase in magnetic field B_0 results in two effects: the first is the increase in the degeneracy of Landau levels, while the second is the increase in the energy separations (energy gaps) between Landau levels. At very low temperatures and strong applied magnetic fields, when thermal excitations between Landau levels are negligible, the above two effects result in unique physical phenomena. These physical phenomena stem from successive emptying of high-energy Landau levels and filling in lower-energy Landau levels with increased degeneracy. Indeed, when the magnetic field is continuously increased, the capacity of each Landau level is increased as well, and electrons from higher energy levels drop to lower Landau levels until the point is reached when all lower energy levels are completely filled while higher energy levels are empty. With the further increase in magnetic field, this process of completely filling of lower energy levels and emptying of higher energy levels is repeated. This results in oscillations of electrical resistance which are the essence of the **Shubnikov–de Haas effect**. There is also another related effect, called the **de Haas–van Alphen effect**, of magnetization oscillations with the increase in the applied magnetic field.

One of the most remarkable effects related to the Landau levels is the **integer quantum Hall effect** discovered in 1980 by **Klaus von Klitzing**. For this discovery, von Klitzing was awarded the Nobel Prize in Physics in 1985. The first experiments on the integer quantum Hall effect were performed by using MOSFET (metal-oxide-semiconductor-field-effect transistor) devices. In these devices, the electron flow is confined within a very thin two-dimensional inversion layer near the dioxide-semiconductor interface (see Figure 4.12). The measurements of the Hall resistance R_H

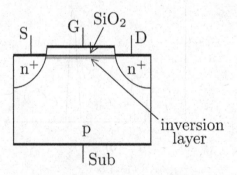

Fig. 4.12

were performed at temperatures about 4 K and at magnetic fields above 10 Tesla applied normal to the inversion layer. It was found that the Hall resistance R_H as a function of applied magnetic field exhibits a staircase sequence of plateaus shown in Figure 4.13(a). The value of $R_H^{(k)}$ at these plateaus is described with unanticipated precision by the formula

$$kR_h^{(k)} = \frac{e^2}{h},\qquad(4.4.43)$$

where k is an integer that can be viewed as a plateau number. It was also observed that, within the range of the plateaus, the resistance R measured along the direction of current flow drops practically to zero (see Figure 4.13(b)). Because the value of resistance R_H at various plateau is controlled by the integer k, the described phenomenon is called the integer quantum Hall effect. For comparison purposes, we shall briefly discuss the

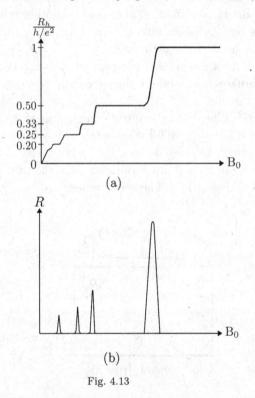

Fig. 4.13

classical Hall effect by considering an electron flow along the x-direction in the xy plane of a very thin film in the presence of the normal magnetic field $\mathbf{B}_0 = B_0\mathbf{e}_z$. The Lorentz force created by the magnetic field results in the accumulation of electrons and excess positive charges on opposite sides of the film in the transverse (with respect to the current flow) direction. These charges produce electric field E_y which grows until the balance between the electrostatic force and the Lorentz force is achieved:

$$eE_y = evB_0. \tag{4.4.44}$$

From the last formula, the following expression for the Hall voltage V_y follows

$$V_y = wE_y = wvB_0, \tag{4.4.45}$$

where w is the width of the film in the y-direction.

On the other hand, the electric current I_x through the film can be computed as follows

$$I_x = w\Delta j_x, \tag{4.4.46}$$

where Δ is the film thickness, while j_x is the current density given by the formula

$$j_x = Nev, \tag{4.4.47}$$

with N being the number of electrons per unit volume.

From the last two equations, we find

$$I_x = w\Delta Nev, \tag{4.4.48}$$

which leads to the following expression for the Hall resistance

$$R_H = \frac{V_y}{I_x} = \frac{B_0}{N\Delta e}. \tag{4.4.49}$$

By introducing the surface electron density $n = N\Delta$, the last formula can be written as

$$R_H = \frac{B_0}{ne}. \tag{4.4.50}$$

It is apparent from the last formula that the classical Hall resistance is a linear function of B_0 in contrast with the experimental results shown in Figure 4.13(a). Nevertheless, the last formula combined with the expression (4.4.41) for the degeneracy of Landau levels can be used to derive the expression for the Hall resistance at the onsets of plateaus in Figure 4.13(a). The derivation proceeds as follows. Consider the situation when a certain number of low-energy Landau levels are completely filled, while all higher-energy Landau levels are completely empty. If this situation is realized at very low temperatures and high magnetic fields when the Landau levels are separated by sufficiently high energy (see formula (4.4.42)), then the thermal scattering is not possible and the resistance R must drop to zero as shown in Figure 4.13(b). This means that the onsets of plateaus coincide with the value of magnetic fields $B_0^{(k)}$ for which a certain number k of low-energy Landau levels are completely filled. This value of magnetic field can be computed from the condition

$$n = kd, \tag{4.4.51}$$

which according to formula (4.4.41) means that

$$n = k \frac{e B_0^{(k)}}{h}. \tag{4.4.52}$$

Consequently,

$$B_0^{(k)} = \frac{nh}{ke}. \tag{4.4.53}$$

According to formula (4.4.50), we find

$$\boxed{R_H^{(k)} = \frac{B_0^{(k)}}{ne} = \frac{e^2}{kh},} \tag{4.4.54}$$

which coincides with experimentally measured values of the Hall resistance at the onsets of plateaus (see formula (4.4.43)). The presented simple derivation of formula (4.4.54) is valid only at specific points when various integer numbers of Landau levels are completely filled. This derivation cannot be extended to predict the existence of plateaus and their widths. The latter can be accomplished on the basis of more sophisticated theory based on the notion of extended and localized states. The discussion of this theory is beyond the scope of this text.

It is important to point out the remarkable precision of the integer quantum Hall effect. This precision led to the definition of the new resistance standard R_k often called the **von Klitzing constant**

$$\boxed{R_k = \frac{h}{e^2} = 25,812.8075 \, \Omega.} \tag{4.4.55}$$

The remarkable precision of the integer quantum Hall effect is also very attractive for the very accurate measurement of the fine-structure constant α given by formula (4.3.65).

The question can be naturally asked about the origin of this remarkable precision of the integer quantum Hall effect, the precision that has been observed for samples with different impurities, different geometries and different electron concentrations. This precision suggests that the integer k in formula (4.4.43) is somehow immune to variations in experimental conditions. The modern point of view for this immunity is that the integers appearing in the quantum Hall effect are topological quantum numbers (also known as Chern numbers) which are invariant with respect to some continuous variations of the Hamiltonian.

It is worthwhile to mention in the conclusion of this section that there also exists the **fractional quantum Hall effect** when plateaus in the Hall resistance appear at fractional values of k in formula (4.4.43). Examples of such values of k are $\frac{1}{3}, \frac{2}{5}, \frac{3}{7}, \frac{2}{3}, \frac{3}{5}$ and so on. **R. Laughlin, H. Störmer and D. Tsui** were awarded the 1998 Nobel Prize in Physics for the discovery and explanation of the fractional Hall effect. The microscopic origin of this effect is a very active research topic in condensed matter physics.

4.5 Model Problem for Energy Gap Formation in Superconductors

It was mentioned in Section 3.3 of Chapter 3 that the presence of the small attractive interaction between electrons in superconductors may lead to the instability of the normal state with respect to the formation of electron Cooper pairs. Furthermore, this instability may lead to the formation of lower (ground) energy level separated from higher energy levels by some energy gap. In this section, we consider a model problem for this energy gap formation.

Consider the Hamiltonian $\hat{H}_0$ with very large number N of energy levels which are very close to one another. Then, these energy levels can be approximated by one N-times degenerate energy level $\mathcal{E}_0$. Mathematically, this means that there are N eigenfunctions ψ_k, $(k = 1, 2, \ldots N)$ corresponding to the eigenvalue $\mathcal{E}_0$, i.e.,

$$\hat{H}_0 \psi_k = \mathcal{E}_0 \psi_k, \quad (k = 1, 2, \ldots N), \tag{4.5.1}$$

and

$$\langle \psi_k | \psi_n \rangle = \delta_{kn}, \tag{4.5.2}$$

where δ_{kn} is the Kronecker symbol.

Now, we introduce a small perturbation $\hat{V}$ of the Hamiltonian $\hat{H}_0$ which models weak attractive interaction between electrons. Thus,

$$\hat{H} = \hat{H}_0 + \hat{V}. \tag{4.5.3}$$

We are interested in understanding how this perturbation may affect the energy spectrum. This requires the calculation of eigenvalues of operator $\hat{H}_0 + \hat{V}$:

$$\left(\hat{H}_0 + \hat{V} \right) \tilde{\psi} = \tilde{\mathcal{E}} \tilde{\psi}. \tag{4.5.4}$$

Since the perturbation $\hat{V}$ is small, $\tilde{\mathcal{E}}$ and $\tilde{\psi}$ can be represented as follows

$$\tilde{\mathcal{E}} = \mathcal{E}_0 + \varepsilon \qquad (4.5.5)$$

and

$$\tilde{\psi} = \psi_0 + \varphi, \qquad (4.5.6)$$

where ε and φ are small, while

$$\psi_0 = \sum_{n=1}^{N} a_n \psi_n. \qquad (4.5.7)$$

This means, in accordance with formula (4.5.1), that

$$\hat{H}_0 \psi_0 = \mathcal{E}_0 \psi_0. \qquad (4.5.8)$$

By substituting formulas (4.5.5) and (4.5.6) into equation (4.5.4), we find

$$\left(\hat{H}_0 + \hat{V} \right) (\psi_0 + \varphi) = (\mathcal{E}_0 + \varepsilon)(\psi_0 + \varphi). \qquad (4.5.9)$$

By taking into account the relation (4.5.8), we obtain

$$\hat{H}_0 \varphi + \hat{V} \psi_0 + \hat{V} \varphi = \mathcal{E}_0 \varphi + \varepsilon \psi_0 + \varepsilon \varphi. \qquad (4.5.10)$$

It is apparent that the terms $\hat{V} \varphi$ and $\varepsilon \varphi$ are of higher order of smallness in comparison with all other terms in the last formula. For this reason, these terms can be neglected, and this leads to the equation

$$\hat{H} \varphi + \hat{V} \psi_0 = \mathcal{E}_0 \varphi + \varepsilon \psi_0, \qquad (4.5.11)$$

which can also be written as

$$\left(\hat{H} - \mathcal{E}_0 \hat{I} \right) \varphi = - \left(\hat{V} - \varepsilon \hat{I} \right) \psi_0, \qquad (4.5.12)$$

where $\hat{I}$ is the identity operator.

The last equation results in the following equalities for the inner products:

$$\langle \psi_k | \left(\hat{H}_0 - \mathcal{E}_0 \hat{I} \right) \varphi \rangle = -\langle \psi_k | \left(\hat{V} - \varepsilon \hat{I} \right) \psi_0 \rangle, \quad (k = 1, 2, \dots N). \qquad (4.5.13)$$

By taking into account that $\hat{H}_0 - \mathcal{E}_0 \hat{I}$ is a Hermitian operator and by using formula (4.5.1), we find that

$$\langle \psi_k | \left(\hat{H}_0 - \mathcal{E}_0 \hat{I} \right) \varphi \rangle = \langle \left(\hat{H}_0 - \mathcal{E}_0 \hat{I} \right) \psi_k | \varphi \rangle = 0. \qquad (4.5.14)$$

By using the last equation in formula (4.5.13), we obtain

$$\langle \psi_k | \left(\hat{V} - \varepsilon \hat{I} \right) \psi_0 \rangle = 0, \quad (k = 1, 2, \dots N). \qquad (4.5.15)$$

Next, by substituting formula (4.5.7) into the last equations, we get

$$\langle \psi_k | \left(\hat{V} - \varepsilon \hat{I} \right) \sum_{n=1}^{N} a_n \psi_n \rangle = 0, \quad (k = 1, 2, \dots N), \tag{4.5.16}$$

which can be further transformed as follows

$$\sum_{n=1}^{N} \langle \psi_k | \hat{V} \psi_n \rangle a_n = \varepsilon \sum_{n=1}^{N} a_n \langle \psi_k | \psi_n \rangle, \quad (k = 1, 2, \dots N). \tag{4.5.17}$$

By introducing the notation

$$V_{kn} = \langle \psi_k | \hat{V} \psi_n \rangle = \langle \hat{V} \psi_k | \psi_n \rangle \tag{4.5.18}$$

and taking into account formula (4.5.2), we obtain

$$\sum_{n=1}^{N} V_{kn} a_n = \varepsilon a_k, \quad (k = 1, 2, \dots N). \tag{4.5.19}$$

We shall write equations (4.5.19) in the matrix form

$$\hat{T} \mathbf{a} = \varepsilon \mathbf{a}, \tag{4.5.20}$$

where the matrix $\hat{T}$ and vector $\mathbf{a}$ are defined as follows

$$\hat{T} = \begin{pmatrix} V_{11} & V_{12} & \cdots & V_{1N} \\ V_{21} & V_{22} & \cdots & V_{2N} \\ \vdots & \vdots & \ddots & \vdots \\ V_{N1} & V_{N2} & \cdots & V_{NN} \end{pmatrix}, \tag{4.5.21}$$

$$\mathbf{a} = \begin{pmatrix} a_1 \\ a_2 \\ \vdots \\ a_N \end{pmatrix}. \tag{4.5.22}$$

Thus, the analysis of the energy spectrum is reduced to the eigenvalue problem (4.5.20) for the matrix $\hat{T}$. Explicit formulas for the eigenvalues can be found by assuming that

$$V_{kn} = -V, \quad (V > 0). \tag{4.5.23}$$

The negativity of matrix elements V_{kn} reflects the fact that the perturbation operator describes weak attractive interaction between electrons.

The eigenvalues of matrix $\hat{T}$ with the elements defined by formula (4.5.23) can be computed by using the following mathematical fact.

Let $\lambda_n, (n = 1, 2, \ldots N)$ be the eigenvalues of symmetric matrix $\hat{C}$ with real elements, i.e.,

$$\hat{C}\mathbf{x}_n = \lambda_n x_n. \tag{4.5.24}$$

Then,

$$\sum_{k=1}^{N}\sum_{n=1}^{N} c_{kn}^2 = \sum_{n=1}^{N} \lambda_n^2. \tag{4.5.25}$$

The proof of the last formula proceeds as follows. Consider matrix

$$\hat{B} = \hat{C}^2 \tag{4.5.26}$$

and the eigenvalue problem

$$\hat{B}\mathbf{x}_n = \nu_n \mathbf{x}_n. \tag{4.5.27}$$

Then,

$$\nu_n = \lambda_n^2. \tag{4.5.28}$$

Indeed, from formulas (4.5.24) and (4.5.26), we derive

$$\hat{C}^2 \mathbf{x}_n = \hat{C}(\hat{C}\mathbf{x}_n) = \lambda_n \hat{C}\mathbf{x}_n = \lambda_n^2 \mathbf{x}_n, \tag{4.5.29}$$

which proves the relation (4.5.28).

It is known from linear algebra that the sum of matrix diagonal elements (called the matrix trace) is equal to the sum of matrix eigenvalues

$$\sum_{n=1}^{N} b_{nn} = \sum_{n=1}^{N} \nu_n. \tag{4.5.30}$$

From formula (4.5.26) follows that

$$\sum_{n=1}^{N} b_{nn} = \sum_{n=1}^{N}\sum_{k=1}^{N} c_{nk}c_{kn} = \sum_{n=1}^{N}\sum_{k=1}^{N} c_{kn}^2. \tag{4.5.31}$$

By substituting formulas (4.5.28) and (4.5.31) into equation (4.5.30), we arrive at the relation (4.5.25).

According to this relation, the following formula is valid for eigenvalues ε_n of matrix $\hat{T}$ with matrix elements defined by equation (4.5.23):

$$\sum_{n=1}^{N} \varepsilon_n^2 = N^2 V^2. \tag{4.5.32}$$

On the other hand, it is easy to check that the vector

$$\mathbf{a}_1 = \begin{pmatrix} 1 \\ 1 \\ \vdots \\ 1 \end{pmatrix} \tag{4.5.33}$$

is the eigenvector of matrix $\hat{T}$. Namely, the following formula is valid:

$$\hat{T}\mathbf{a}_1 = -NV\mathbf{a}_1. \tag{4.5.34}$$

From the last formula we conclude that the eigenvalue ε_1 of matrix $\hat{T}$ corresponding to eigenvector $\mathbf{a}_1$ is

$$\varepsilon_1 = -NV. \tag{4.5.35}$$

By substituting the last formula into equation (4.5.32), we find that

$$\sum_{n=2}^{N} \varepsilon_n^2 = 0, \tag{4.5.36}$$

which implies that

$$\varepsilon_n = 0 \text{ for } n = 2, 3, \ldots N. \tag{4.5.37}$$

Now, by using equations (4.5.5), (4.5.35) and (4.5.37), we conclude that the energy levels of the perturbed Hamiltonian (4.5.3) are given by the formulas

$$\tilde{\mathcal{E}}_1 = \mathcal{E}_0 - NV, \tag{4.5.38}$$

$$\tilde{\mathcal{E}}_n = \mathcal{E}_0 \text{ for } n = 2, 3, \ldots N. \tag{4.5.39}$$

This modification of energy levels due to the attractive perturbation of the Hamiltonian is illustrated below in Figure 4.14. Thus, if N is very large, then small attractive perturbation $\hat{V}$ may result in appreciable lowering of

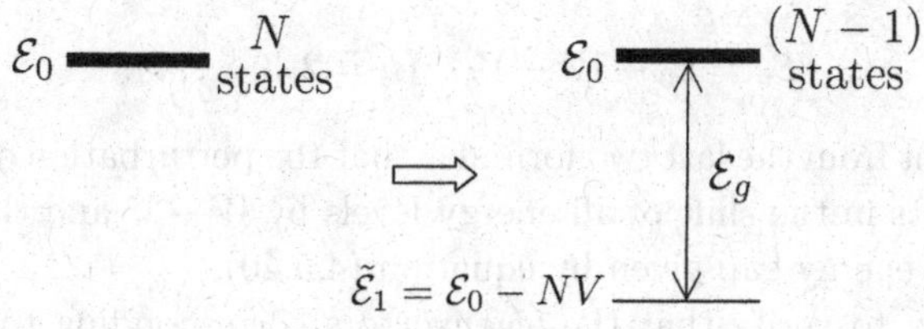

Fig. 4.14

the energy and in the formation of energy gap

$$\boxed{\mathcal{E}_g = NV} \tag{4.5.40}$$

between the ground energy level and excited states. This fact has been established by assuming the validity of equality (4.5.23). It turns out, and it is demonstrated in the BCS theory, that this assumption may be fairly accurate for states where electrons with opposite momenta and spins are bound in Cooper pairs. It also can be demonstrated that the equality (4.5.23) can be somewhat relaxed without affecting the formation of the energy gap given by formula (4.5.40). Indeed, consider the case when all off-diagonal elements of matrix $\hat{T}$ are the same and given by the formula (4.5.23) and all diagonal elements of $\hat{T}$ are the same and given by the formula

$$V_{kk} = -W. \tag{4.5.41}$$

It can be demonstrated (and suggested as an exercise for the reader) that the eigenvalues ε'_n of this matrix are related to the eigenvalues ε_n by the formula

$$\varepsilon'_n = \varepsilon_n + (W - V), \quad (n = 1, 2, \ldots N). \tag{4.5.42}$$

Consequently, in accordance with equations (4.5.35) and (4.5.37), we have

$$\varepsilon'_1 = W - (N + 1)V, \tag{4.5.43}$$

and

$$\varepsilon'_n = W - V, \quad (n = 2, 3, \ldots N). \tag{4.5.44}$$

Consequently, the energy levels of the perturbed Hamiltonian are

$$\tilde{\mathcal{E}}_1 = \mathcal{E}_0 + W - (N + 1)V, \tag{4.5.45}$$

and

$$\tilde{\mathcal{E}}_n = \mathcal{E}_0 + W - V, \quad (n = 2, 3, \ldots N). \tag{4.5.46}$$

It is apparent from the last two formulas that the perturbation of the Hamiltonian results in the shift of all energy levels by $W - V$ and the formation of the same energy gap given by equation (4.5.40).

It is easy to verify that the eigenvectors corresponding to eigenvalues ε_n and ε'_n for the two versions of matrix $\hat{T}$ are the same. Namely, the first eigenvector $\mathbf{a}_1$ corresponding to eigenvalues ε_1 and ε'_1 is given by the equation (4.5.33). It can be demonstrated that all eigenvectors $\mathbf{a}_k$ can be described by the formula

$$
\mathbf{a}_k =
\begin{pmatrix}
1 \\
\alpha^{k-1} \\
\alpha^{2(k-1)} \\
\vdots \\
\alpha^{(N-1)(k-1)}
\end{pmatrix},
\quad (k = 1, 2, \ldots N),
\tag{4.5.47}
$$

where

$$
\alpha = e^{i\frac{2\pi}{N}}.
\tag{4.5.48}
$$

Next, we discuss the importance of the energy gap for the existence of persisting superconducting current by using the phenomenological reasoning belonging to **L. Landau.**

A superconducting electric current is a collective motion of electron Cooper pairs with respect to the lattice of superconductors. Consider this motion in the reference frame moving with the same velocity as the paired electrons. In this reference frame, the lattice moves with some velocity v. For scattering to occur, certain energy $\mathcal{E}_p$ and momentum $\mathbf{p}$ must be transferred from the lattice to the superconducting electrons. The conservation of energy and momentum yields

$$
\frac{Mv^2}{2} - \mathcal{E}_p = \frac{M\tilde{v}^2}{2},
\tag{4.5.49}
$$

and

$$
M\boldsymbol{v} - \mathbf{p} = M\tilde{\boldsymbol{v}},
\tag{4.5.50}
$$

where M is the mass of the lattice, while v and $\tilde{v}$ are its velocity before and after scattering.

From the last equation follows that

$$
M^2 v^2 - 2M\boldsymbol{v} \cdot \mathbf{p} + p^2 = M^2 \tilde{v}^2,
\tag{4.5.51}
$$

which can also be written as

$$
M^2 v^2 - M^2 \tilde{v}^2 - 2M\boldsymbol{v} \cdot \mathbf{p} + p^2 = 0.
\tag{4.5.52}
$$

From formula (4.5.49), we find

$$
M^2 v^2 - M^2 \tilde{v}^2 = 2M\mathcal{E}_p.
\tag{4.5.53}
$$

By substituting the last formula into equation (4.5.52), we end up with

$$
2M\mathcal{E}_p - 2M\boldsymbol{v} \cdot \mathbf{p} + p^2 = 0,
\tag{4.5.54}
$$

which leads to

$$\mathcal{E}_p = \boldsymbol{v} \cdot \mathbf{p} - \frac{p^2}{2M}. \tag{4.5.55}$$

From the last equation follows that

$$\mathcal{E}_p < |\boldsymbol{v} \cdot \mathbf{p}| \leq vp, \tag{4.5.56}$$

and

$$v > \frac{\mathcal{E}_p}{p}. \tag{4.5.57}$$

For scattering to occur, the following inequality should hold

$$\mathcal{E}_p > \mathcal{E}_g. \tag{4.5.58}$$

From the last two inequalities, we find

$$v > \frac{\mathcal{E}_g}{p}. \tag{4.5.59}$$

Suppose that p_0 is the maximum value of magnitude of momentum that can be transferred as a result of scattering. Then, from inequality (4.5.59), we find

$$v > \frac{\mathcal{E}_g}{p_0} = v_c. \tag{4.5.60}$$

This means that the magnitude of velocity v of superconducting electrons should exceed some critical value v_c for scattering to occur. If

$$v < v_c, \tag{4.5.61}$$

then the superconducting electron moves through the lattice without scattering. It is also clear from formula (4.5.60) that the larger the energy gap, the larger the critical value v_c and the larger the superconducting current which can flow without any scattering, that is, with zero resistance. The latter is true because the superconducting current density is given by the formula

$$\mathbf{j} = 2en_c\boldsymbol{v}, \tag{4.5.62}$$

where n_c is the volume density of Cooper pairs.

It may seem from the presented discussion that the energy gap is necessary for superconductivity to occur. However, this is not always the case, and there are so-called "gapless" superconductors. The discussion of these superconductors is beyond the scope of this text.

In the conclusion of this section, it is worthwhile to mention some connections of the presented discussion to electric power engineering. First, matrices with the same diagonal and the same off-diagonal elements (see

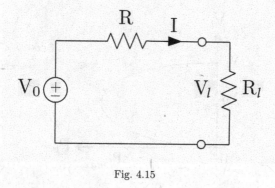

Fig. 4.15

formulas (4.5.41) and (4.5.23), respectively) are encountered in the analysis of polyphase power systems, and eigenvectors $\mathbf{a}_k$ are symmetrical components which are extensively used in the analysis of faults in power systems. Second, the use of superconductivity may have far-reaching applications in electric power engineering. We shall illustrate this by one example of high-voltage transmission lines. High voltages in such lines are used for two reasons: 1) to reduce power losses in transmission line conductors and 2) to increase the power transmission capacity of the transmission lines. We shall briefly discuss the second issue in the case of dc transmission lines. A simplified model of dc transmission line is shown in Figure 4.15, where R stands for the overall resistance of connecting wires and R_l is the resistance of the load to which dc power P_l is delivered at voltage V_l. It is apparent that

$$P_l = V_l I, \tag{4.5.63}$$

and

$$I = \frac{V_0 - V_l}{R}. \tag{4.5.64}$$

From the last two formulas, we find

$$P_l = \frac{1}{R}(V_l V_0 - V_l^2) = \frac{V_0^2}{4R} - \frac{1}{R}\left(V_l - \frac{V_0}{2}\right)^2. \tag{4.5.65}$$

The last equation is graphically illustrated in Figure 4.16. It is clear from the last equation that the maximum transmitted power (i.e., transmission capacity) is given by the formula

$$P_{l,max} = \frac{V_0^2}{4R}. \tag{4.5.66}$$

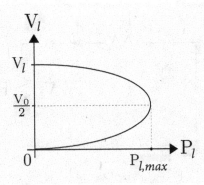

Fig. 4.16

No power above $P_{l,max}$ can be transmitted. The last formula implies that the higher the voltage of a dc transmission line, the higher its transmission capacity. The last formula also implies that the smaller the resistance of the dc line conductors, the higher the transmission capacity. The latter fact suggests that the use of superconducting wires may dramatically increase the transmission capacity. This capacity will be limited by the value of critical current of a superconductor, i.e., by the current at which superconductivity is destroyed. This discussion clearly suggests that the progress in high-temperature superconductor physics and technology may have a dramatic effect on future structures of power systems.

Problems

(1) Derive formula (4.1.5).

(2) By using the method of separation of variables, derive formulas (4.1.67), (4.1.68) and (4.1.69).

(3) Show that in the case of three-dimensional infinitely deep ($U_0 = \infty$) potential well ($0 < x < a, 0 < y < b, 0 < z < c$) the energy levels and the corresponding eigenstates are given by the following formulas, respectively,

$$\mathcal{E}_{n_1,n_2,n_3} = \frac{\pi^2\hbar^2}{2m}\left(\frac{n_1^2}{a^2} + \frac{n_2^2}{b^2} + \frac{n_3^2}{c^2}\right), \qquad (\text{P.4.1})$$

$$\Phi_{n_1,n_2,n_3}(x,y,z) = \sqrt{\frac{8}{abc}}\sin\frac{\pi n_1}{a}x\sin\frac{\pi n_2}{b}y\sin\frac{\pi n_3}{c}z, \quad (\text{P.4.2})$$

$$n_1, n_2, n_3 = 1, 2, 3, \dots$$

(4) Determine the degeneracy $\mathcal{D}_n$ of energy levels in the case when $a = b = c$.

(5) Solve the problem for asymmetric potential well shown below.

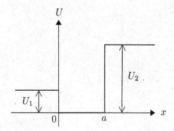

Fig. P.1

(6) Prove formula (4.2.19).

(7) Prove formula (4.2.26).

(8) Prove the validity of formula (4.2.33).

(9) Prove formula (4.2.61).

(10) Prove formula (4.2.90).

(11) Establish the formula (4.2.91).

(12) Verify the commutation relation (4.3.8).

(13) Verify the validity of formula (4.3.11).

(14) Derive formula (4.3.36).

(15) Verify the validity of the commutation relation (4.3.54).

(16) By using formulas (4.3.19), (4.3.21), (4.3.33) and (4.3.45) derive the expressions for $T_{nl}(r)$ for 1s, 2s, 2p, 3s, 3p and 3d states. Compare these results with those obtained by using formulas (4.3.62) and (4.3.63).

(17) Prove formulas (4.4.22) and (4.4.23).

(18) Prove formula (4.4.25).

(19) Verify formula (4.5.35).

(20) Prove formulas (4.5.42), (4.5.43) and (4.5.44).

(21) Prove that eigenvectors of matrix $\hat{T}$ (see formula (4.5.20)) with matrix elements specified by equations (4.5.23) and (4.5.41) are given by formulas (4.5.47) and (4.5.48).

Chapter 5

Periodic Potentials

5.1 Bloch Function and Energy Bands

Many solids have crystalline structures with atoms being arranged in regular, periodically repeated patterns which extend throughout the samples. These crystalline structures can be studied by using the concept of the crystal lattice. In the case of a crystal lattice, one deals with three- or two-dimensional arrays of periodically located points (sites) in space (see

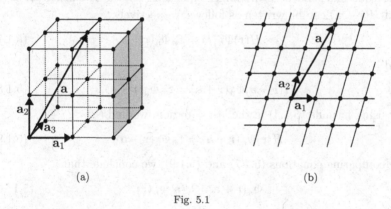

(a)   (b)

Fig. 5.1

Figure 5.1). Such arrays can be fully defined by the primitive basis vectors $\mathbf{a}_1$, $\mathbf{a}_2$ and $\mathbf{a}_3$ for the three-dimensional case and $\mathbf{a}_1$ and $\mathbf{a}_2$ for the two-dimensional case. These basis vectors are the shortest in length and independent vectors connecting lattice point sites. All vectors $\mathbf{a}$ connecting lattice sites can be represented as linear (integer) combinations of primitive vectors:

$$\mathbf{a} = n_1\mathbf{a}_1 + n_2\mathbf{a}_2 + n_3\mathbf{a}_3 \tag{5.1.1}$$

in the three-dimensional case, and

$$\mathbf{a} = n_1\mathbf{a}_1 + n_2\mathbf{a}_2 \tag{5.1.2}$$

in the two-dimensional case, with n_1, n_2 and n_3 being integers.

Next, we assume that **identical** (positive) charges (ions) exist at lattice points. Then, the electric field and its potential have **discrete translation symmetry**, which implies that for any lattice vector $\mathbf{a}$ we have

$$U(\mathbf{r} + \mathbf{a}) = U(\mathbf{r}). \tag{5.1.3}$$

This leads to the translational symmetry of the Hamiltonian operator

$$\hat{H}(\mathbf{r}) = \hat{H}(\mathbf{r} + \mathbf{a}), \tag{5.1.4}$$

where

$$\hat{H}(\mathbf{r}) = -\frac{\hbar^2}{2m}\nabla^2 + U(\mathbf{r}), \tag{5.1.5}$$

and

$$\hat{H}(\mathbf{r} + \mathbf{a}) = -\frac{\hbar^2}{2m}\nabla^2 + U(\mathbf{r} + \mathbf{a}). \tag{5.1.6}$$

The time-independent Schrödinger equations for the Hamiltonians $\hat{H}(\mathbf{r})$ and $\hat{H}(\mathbf{r} + \mathbf{a})$ can be written as follows, respectively,

$$\hat{H}(\mathbf{r})\Phi_n(\mathbf{r}) = \mathcal{E}_n\Phi_n(\mathbf{r}), \tag{5.1.7}$$

and

$$\hat{H}(\mathbf{r} + \mathbf{a})\Phi_n(\mathbf{r} + \mathbf{a}) = \mathcal{E}_n\Phi_n(\mathbf{r} + \mathbf{a}). \tag{5.1.8}$$

By using formula (5.1.4) in the last equation, we find

$$\hat{H}(\mathbf{r})\Phi_n(\mathbf{r} + \mathbf{a}) = \mathcal{E}_n\Phi_n(\mathbf{r} + \mathbf{a}). \tag{5.1.9}$$

By comparing equations (5.1.7) and (5.1.9), we conclude that

$$\Phi_n(\mathbf{r} + \mathbf{a}) = c(\mathbf{a})\Phi_n(\mathbf{r}), \tag{5.1.10}$$

which, in turn, leads to

$$|\Phi_n(\mathbf{r} + \mathbf{a})| = |c(\mathbf{a})||\Phi_n(\mathbf{r})|. \tag{5.1.11}$$

The last equation implies that

$$|\Phi_n(\mathbf{r} + 2\mathbf{a})| = |c(\mathbf{a})||\Phi_n(\mathbf{r} + \mathbf{a})| = |c(\mathbf{a})|^2|\Phi_n(\mathbf{r})|. \tag{5.1.12}$$

Similarly, we find that

$$|\Phi_n(\mathbf{r} + m\mathbf{a})| = |c(\mathbf{a})|^m|\Phi_n(\mathbf{r})|. \tag{5.1.13}$$

It can be likewise proven that

$$|\Phi_n(\mathbf{r} - m\mathbf{a})| = \frac{1}{|c(\mathbf{a})|^m}|\Phi_n(\mathbf{r})|. \tag{5.1.14}$$

By using the last two formulas, it is easy to establish that

$$|c(\mathbf{a})| = 1. \tag{5.1.15}$$

Indeed, if

$$|c(\mathbf{a})| > 1, \tag{5.1.16}$$

then according to formula (5.1.13), we have

$$\lim_{m \to \infty} |\Phi_n(\mathbf{r} + m\mathbf{a})| = \infty, \tag{5.1.17}$$

which is not possible because $\Phi_n(\mathbf{r})$ must be finite.

On the other hand, if

$$|c(\mathbf{a})| < 1, \tag{5.1.18}$$

then, according to formula (5.1.14), it follows that

$$\lim_{m \to \infty} |\Phi_n(\mathbf{r} - m\mathbf{a})| = \infty, \tag{5.1.19}$$

which is not possible as well.

Thus, equality (5.1.15) is established. This equality implies that

$$c(\mathbf{a}) = e^{i\nu(\mathbf{a})}, \tag{5.1.20}$$

where $\nu(\mathbf{a})$ is some function of $\mathbf{a}$.

By combining formulas (5.1.10) and (5.1.20), we end up with

$$\Phi_n(\mathbf{r} + \mathbf{a}) = e^{i\nu(\mathbf{a})}\Phi_n(\mathbf{r}). \tag{5.1.21}$$

The last equation suggests that

$$\Phi_n(\mathbf{r} + \mathbf{a} + \mathbf{b}) = e^{i\nu(\mathbf{a}+\mathbf{b})}\Phi_n(\mathbf{r}). \tag{5.1.22}$$

On the other hand, we find

$$\Phi_n(\mathbf{r} + \mathbf{a} + \mathbf{b}) = e^{i[\nu(\mathbf{a})+\nu(\mathbf{b})]}\Phi_n(\mathbf{r}). \tag{5.1.23}$$

From the last two formulas, it follows that

$$\nu(\mathbf{a} + \mathbf{b}) = \nu(\mathbf{a}) + \nu(\mathbf{b}). \tag{5.1.24}$$

Thus, $\nu(\mathbf{a})$ is an additive function, and it can be represented as

$$\nu(\mathbf{a}) = \mathbf{k} \cdot \mathbf{a}, \tag{5.1.25}$$

where $\mathbf{k}$ is an arbitrary vector.

By substituting the last formula into equation (5.1.21), we obtain

$$\boxed{\Phi_n(\mathbf{r} + \mathbf{a}) = e^{i\mathbf{k}\cdot\mathbf{a}}\Phi_n(\mathbf{r}).} \qquad (5.1.26)$$

The last formula admits the following interpretation. Consider the operator $\hat{T}_{\mathbf{a}}$ of translation by $\mathbf{a}$ defined as

$$\hat{T}_{\mathbf{a}}\Phi(\mathbf{r}) = \Phi(\mathbf{r} + \mathbf{a}). \qquad (5.1.27)$$

It is clear from the last two formulas that

$$\hat{T}_{\mathbf{a}}\Phi_n(\mathbf{r}) = e^{i\mathbf{k}\cdot\mathbf{a}}\Phi_n(\mathbf{r}). \qquad (5.1.28)$$

This means that the eigenvalues $\lambda(\mathbf{a})$ of the translation operator $\hat{T}_{\mathbf{a}}$ are

$$\lambda(\mathbf{a}) = e^{i\mathbf{k}\cdot\mathbf{a}}. \qquad (5.1.29)$$

Furthermore, the operators $\hat{H}(\mathbf{r})$ and $\hat{T}_{\mathbf{a}}$ have the same eigenfunctions. This implies that these two operators commute

$$\hat{H}\hat{T}_{\mathbf{a}} - \hat{T}_{\mathbf{a}}\hat{H} = 0. \qquad (5.1.30)$$

Now, by using formula (5.1.26), we shall establish the Bloch theorem, which is also known in mathematics as the Floquet theorem. This theorem states that the eigenfunctions $\Phi_n(\mathbf{r})$ are of the form

$$\boxed{\Phi_n(\mathbf{r}) = e^{i\mathbf{k}\cdot\mathbf{r}}u_{n\mathbf{k}}(\mathbf{r}),} \qquad (5.1.31)$$

where functions $u_{n\mathbf{k}}(\mathbf{r})$ have the periodicity of the lattice

$$\boxed{u_{n\mathbf{k}}(\mathbf{r} + \mathbf{a}) = u_{n\mathbf{k}}(\mathbf{r}).} \qquad (5.1.32)$$

Since any bounded function can be represented in the form (5.1.31), the only thing that must be proven is the validity of formula (5.1.32). The proof proceeds as follows. By using formulas (5.1.26) and (5.1.31), we find

$$\Phi_n(\mathbf{r} + \mathbf{a}) = e^{i\mathbf{k}\cdot\mathbf{a}}\Phi_n(\mathbf{r}) = e^{i\mathbf{k}\cdot\mathbf{a}}e^{i\mathbf{k}\cdot\mathbf{r}}u_{n\mathbf{k}}(\mathbf{r}), \qquad (5.1.33)$$

or

$$\Phi_n(\mathbf{r} + \mathbf{a}) = e^{i\mathbf{k}\cdot(\mathbf{r}+\mathbf{a})}u_{n\mathbf{k}}(\mathbf{r}). \qquad (5.1.34)$$

On the other hand, it follows from formula (5.1.31) that

$$\Phi_n(\mathbf{r} + \mathbf{a}) = e^{i\mathbf{k}\cdot(\mathbf{r}+\mathbf{a})}u_{n\mathbf{k}}(\mathbf{r} + \mathbf{a}). \qquad (5.1.35)$$

The validity of equality (5.1.32) is the immediate consequence of the last two formulas.

Functions $\Phi_n(\mathbf{r})$ defined by formulas (5.1.31) and (5.1.32) are called Bloch functions or Bloch waves. They can be viewed as periodically modulated plane waves. It is remarkable that electrons described by Bloch functions move through periodically varying electric fields without any scattering. The scattering appears only when the ideal periodicity of the lattice structure is perturbed by thermal vibrations of lattice sites or by introduction of impurities, that is, when not all lattice sites are occupied by identical ions.

It is interesting to point out some similarity between the Bloch functions and the wave functions for free electrons with certain momentum $\mathbf{p}$. The latter functions can be written in the form (see formula (2.3.38) from Chapter 2):

$$\psi_{\mathbf{p}}(\mathbf{r}) = \frac{1}{(2\pi\hbar)^{3/2}}e^{i\mathbf{k}\cdot\mathbf{r}}, \tag{5.1.36}$$

where

$$\mathbf{p} = \hbar\mathbf{k}. \tag{5.1.37}$$

It is because of this similarity that the momentum $\mathbf{p}$ associated with the wave vector $\mathbf{k}$ in the Bloch function (5.1.31) is called **crystal momentum** or **quasimomentum**.

Next, we shall discuss how the Bloch theorem can be used for the calculation of electron energy bands in crystalline solids. For this purpose, we shall return to equation (5.1.7) and shall write it in the form

$$-\frac{\hbar^2}{2m}\nabla^2\Phi_n(\mathbf{r}) + U(\mathbf{r})\Phi_n(\mathbf{r}) = \mathcal{E}_n\Phi_n(\mathbf{r}). \tag{5.1.38}$$

Next, by using formula (5.1.31), we find that

$$\nabla^2\Phi_n(\mathbf{r}) = e^{i\mathbf{k}\cdot\mathbf{r}}\left[\nabla^2 u_{n\mathbf{k}} + 2i\mathbf{k}\cdot\nabla u_{n\mathbf{k}} - |\mathbf{k}|^2 u_{n\mathbf{k}}(\mathbf{r})\right]. \tag{5.1.39}$$

By using the last formula and formula (5.1.31), equation (5.1.38) can be transformed as follows

$$-\frac{\hbar^2}{2m}\left[\nabla^2 u_{n\mathbf{k}} + 2i\mathbf{k}\cdot\nabla u_{n\mathbf{k}}\right] + \left[\frac{\hbar^2}{2m}|\mathbf{k}|^2 + U(\mathbf{r})\right]u_{n\mathbf{k}} = \mathcal{E}_n(\mathbf{k})u_{n\mathbf{k}}(\mathbf{r}). \tag{5.1.40}$$

The last equation can be also written in another, more concise form:

$$\frac{(\hat{\mathbf{p}} + \hbar\mathbf{k})^2}{2m}u_{n\mathbf{k}} + U(\mathbf{r})u_{n\mathbf{k}} = \mathcal{E}_n(\mathbf{k})u_{n\mathbf{k}}. \tag{5.1.41}$$

The last two equations can be used for the calculations of energy bands. This is done as follows. By fixing the wave vector $\mathbf{k}$ in equation (5.1.41) and by solving this equation, we find the eigenvalues $\mathcal{E}_1, \mathcal{E}_2, \ldots \mathcal{E}_n, \ldots$ corresponding to the fixed wave vector. Now, by continuously changing $\mathbf{k}$ and solving equation (5.1.41) for those $\mathbf{k}$, we find eigenvalues $\mathcal{E}_n(\mathbf{k})$ as functions of $\mathbf{k}$. These functions $\mathcal{E}_n(\mathbf{k})$ are called energy bands. For a fixed n, $\mathcal{E}_n(\mathbf{k})$ represents the possible values of energy for the band number n for various values of $\mathbf{k}$. When these bands do not overlap in energy, then there are energy gaps between "adjacent" bands.

The described computations are very laborious because they have to be performed for many different wave vectors $\mathbf{k}$. Fortunately, it turns out that there is no need to solve equation (5.1.41) for all possible values of $\mathbf{k}$. This is because functions $\mathcal{E}_n(\mathbf{k})$ are **periodic** in k-space. This can be demonstrated by using the concepts of **reciprocal** lattice and **Brillouin zone**.

The primitive vectors $\mathbf{K}_1$, $\mathbf{K}_2$ and $\mathbf{K}_3$ of the reciprocal lattice are solutions of the following linear equations

$$\mathbf{K}_i \cdot \mathbf{a}_j = 2\pi \delta_{ij}, \tag{5.1.42}$$

where, as before, $\mathbf{a}_j$ stands for the primitive vectors of the crystal lattice, while δ_{ij} is the Kronecker delta.

It is easy to demonstrate that the solutions of equations (5.1.42) are given by the formulas

$$\mathbf{K}_1 = 2\pi \frac{\mathbf{a}_2 \times \mathbf{a}_3}{\mathbf{a}_1 \cdot (\mathbf{a}_2 \times \mathbf{a}_3)}, \tag{5.1.43}$$

$$\mathbf{K}_2 = 2\pi \frac{\mathbf{a}_3 \times \mathbf{a}_1}{\mathbf{a}_1 \cdot (\mathbf{a}_2 \times \mathbf{a}_3)}, \tag{5.1.44}$$

$$\mathbf{K}_3 = 2\pi \frac{\mathbf{a}_1 \times \mathbf{a}_2}{\mathbf{a}_1 \cdot (\mathbf{a}_2 \times \mathbf{a}_3)}. \tag{5.1.45}$$

Vectors $\mathbf{K}$ connecting the sites of the reciprocal lattice are defined by the formula

$$\mathbf{K} = m_1 \mathbf{K}_1 + m_2 \mathbf{K}_2 + m_3 \mathbf{K}_3, \tag{5.1.46}$$

where m_1, m_2 and m_3 are some integers.

By using formulas (5.1.42) and (5.1.46), it can be demonstrated that

$$\boxed{e^{i\mathbf{K} \cdot \mathbf{a}} = 1} \tag{5.1.47}$$

for any crystal lattice vector $\mathbf{a}$.

Next, we are going to prove that for any $\mathbf{K}$ the following relations are valid:

$$\boxed{\mathcal{E}_n(\mathbf{k} + \mathbf{K}) = \mathcal{E}_n(\mathbf{k})} \tag{5.1.48}$$

and

$$\boxed{\Phi_{n,\mathbf{k}+\mathbf{K}}(\mathbf{r}) = \Phi_{n\mathbf{k}}(\mathbf{r}),} \tag{5.1.49}$$

where the following notation is adopted

$$\Phi_{n\mathbf{k}}(\mathbf{r}) = e^{i\mathbf{k}\cdot\mathbf{r}} u_{n\mathbf{k}}(\mathbf{r}). \tag{5.1.50}$$

It is clear from the last formula that

$$\Phi_{n,\mathbf{k}+\mathbf{K}}(\mathbf{r}) = e^{i(\mathbf{k}+\mathbf{K})\cdot\mathbf{r}} u_{n,\mathbf{k}+\mathbf{K}}(\mathbf{r}). \tag{5.1.51}$$

According to the Bloch theorem, we have

$$u_{n,\mathbf{k}+\mathbf{K}}(\mathbf{r} + \mathbf{a}) = u_{n,\mathbf{k}+\mathbf{K}}(\mathbf{r}). \tag{5.1.52}$$

Now, we introduce the function

$$v_{n,\mathbf{k}}(\mathbf{r}) = e^{i\mathbf{K}\cdot\mathbf{r}} u_{n,\mathbf{k}+\mathbf{K}}(\mathbf{r}). \tag{5.1.53}$$

It is clear from the last formula that

$$v_{n,\mathbf{k}}(\mathbf{r} + \mathbf{a}) = e^{i\mathbf{K}\cdot(\mathbf{r}+\mathbf{a})} u_{n,\mathbf{k}+\mathbf{K}}(\mathbf{r} + \mathbf{a}). \tag{5.1.54}$$

According to formula (5.1.47), we find

$$e^{i\mathbf{K}\cdot(\mathbf{r}+\mathbf{a})} = e^{i\mathbf{K}\cdot\mathbf{r}}. \tag{5.1.55}$$

By using formulas (5.1.52) and (5.1.55) in equation (5.1.54), we obtain

$$v_{n,\mathbf{k}}(\mathbf{r} + \mathbf{a}) = e^{i\mathbf{K}\cdot\mathbf{r}} u_{n,\mathbf{k}+\mathbf{K}}(\mathbf{r}). \tag{5.1.56}$$

By comparing formulas (5.1.53) and (5.1.56), we find

$$v_{n,\mathbf{k}}(\mathbf{r} + \mathbf{a}) = v_{n,\mathbf{k}}(\mathbf{r}). \tag{5.1.57}$$

From formulas (5.1.51) and (5.1.53), it follows that

$$\Phi_{n,\mathbf{k}+\mathbf{K}}(\mathbf{r}) = e^{i\mathbf{k}\cdot\mathbf{r}} v_{n,\mathbf{k}}(\mathbf{r}). \tag{5.1.58}$$

It is apparent that function $\Phi_{n,\mathbf{k}+\mathbf{K}}(\mathbf{r})$ satisfies the equation

$$-\frac{\hbar^2}{2m}\nabla^2\Phi_{n,\mathbf{k}+\mathbf{K}}(\mathbf{r}) + U(\mathbf{r})\Phi_{n,\mathbf{k}+\mathbf{K}}(\mathbf{r}) = \mathcal{E}_n(\mathbf{k}+\mathbf{K})\Phi_{n,\mathbf{k}+\mathbf{K}}(\mathbf{r}). \tag{5.1.59}$$

By substituting function $\Phi_{n,\mathbf{k}+\mathbf{K}}(\mathbf{r})$ given by formula (5.1.58) into the last equation and by using the same line of reasoning as in the derivation of equation (5.1.41), we obtain

$$\frac{(\hat{\mathbf{p}} + \hbar\mathbf{k})^2}{2m}v_{n,\mathbf{k}}(\mathbf{r}) + U(\mathbf{r})v_{n,\mathbf{k}}(\mathbf{r}) = \mathcal{E}_n(\mathbf{k}+\mathbf{K})v_{n,\mathbf{k}}(\mathbf{r}). \tag{5.1.60}$$

Now, we observe that the operators acting on functions $u_{n,\mathbf{k}}(\mathbf{r})$ and $v_{n,\mathbf{k}}(\mathbf{r})$ in the left-hand sides of equations (5.1.41) and (5.1.60), respectively, are identical. This implies that for the same **k** these operators must have the same eigenvalues and eigenfunctions. This means that

$$\mathcal{E}_n(\mathbf{k}) = \mathcal{E}_n(\mathbf{k} + \mathbf{K}) \tag{5.1.61}$$

and

$$u_{n,\mathbf{k}}(\mathbf{r}) = v_{n,\mathbf{k}}(\mathbf{r}). \tag{5.1.62}$$

By substituting the last formula into equation (5.1.58) and taking into account formula (5.1.50), we derive

$$\Phi_{n,\mathbf{k}+\mathbf{K}}(\mathbf{r}) = e^{i\mathbf{k}\cdot\mathbf{r}} u_{n,\mathbf{k}}(\mathbf{r}) = \Phi_{n,\mathbf{k}}(\mathbf{r}). \tag{5.1.63}$$

Thus, the periodic relations (5.1.48) and (5.1.49) have been established. These relations can be utilized in band structure computations by using the concept of a Brillouin zone. The Brillouin zone is constructed for any node (any site) of the reciprocal lattice. This is done by normally bisecting by planes all lattice vectors emanating from a lattice site to adjacent (nearest) sites. In particular, the first Brillouin zone is formed by using planes that normally bisect the shortest in length lattice vectors emanating from the origin of **k**-space. This is illustrated in Figure 5.2 for a two-dimensional lattice. It is apparent that the entire **k**-space can be covered by the Brillouin zones because such zones can be constructed for all lattice sites. Any

Fig. 5.2

Brillouin zone can be obtained from the first Brillouin zone by translation through an appropriate lattice vector **K** connecting the centers of these zones. This means that all **k**-vectors can be obtained from **k**-vectors in the first Brillouin zone as a result of translation by some **K**. Thus, if $\mathcal{E}_n(\mathbf{k})$ and $\Phi_{n,\mathbf{k}}(\mathbf{r})$ are computed in the first Brillouin zone, then $\mathcal{E}_n(\mathbf{k})$ and $\Phi_{n,\mathbf{k}}(\mathbf{r})$ can be found for any Brillouin zone (i.e., for any **k**) by using formulas (5.1.48) and (5.1.49). This suggests that it is sufficient to perform band structure computations for **k**-vectors inside the first Brillouin zone.

It is worthwhile to mention that cells similar to Brillouin zones can be constructed for crystal lattices, and they are called Wigner-Seitz cells. Similar cells are used in geometry and numerical analysis, and they are called Voronoi cells.

5.2 Kronig-Penney Model

It is discussed in the previous section that the energy band structure calculations require numerous solutions of the time-independent Schrödinger equations (5.1.41) for various wave vectors **k** within the first Brillouin zone. This is usually done numerically, and special techniques have been developed for this purpose. Most notable among them are the pseudo-potential method and the density functional method. It is apparent that band structure calculations are laborious and computer-intensive. It this section, we illustrate the energy band calculations by considering the simplest version of one-dimensional Kronig-Penney model in which the periodic potential $U(x)$ is modeled by periodically spaced Dirac delta functions (see Figure 5.3). For such a potential, the analytical treatment is possible, and the

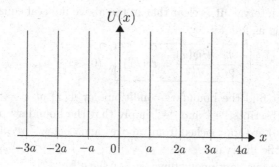

Fig. 5.3

main features of energy band structures discussed in the previous section can be vividly illustrated.

In our analysis, we shall use the following Hamiltonian

$$\hat{H} = -\frac{\hbar^2}{2m}\frac{d^2}{dx^2} + U(x), \qquad (5.2.1)$$

where

$$U(x) = \beta \sum_{\nu=-\infty}^{\infty} \delta(x - \nu a), \qquad (5.2.2)$$

and β is some constant.

It is apparent that

$$U(x + a) = U(x). \qquad (5.2.3)$$

The time-independent Schrödinger equation for the above potential can be written as follows

$$-\frac{\hbar^2}{2m}\frac{d^2\Phi(x)}{dx^2} + \beta \left(\sum_{\nu=-\infty}^{\infty} \delta(x - \nu a) \right) \Phi(x) = \mathcal{E}\Phi(x). \qquad (5.2.4)$$

According to formula (5.1.26), any solution of this equation has the following property

$$\Phi(x + a) = e^{ika}\Phi(x), \qquad (5.2.5)$$

which implies that if $\Phi(x)$ is known on the interval

$$0 < x < a, \qquad (5.2.6)$$

then by using the last formula it can be found for any x, i.e., it can be extended to the entire x-axis. For this reason, it is sufficient to find $\Phi(x)$ on the above interval. It is clear that in the above interval equation (5.2.4) can be written as follows

$$\boxed{-\frac{\hbar^2}{2m}\frac{d^2\Phi(x)}{dx^2} = \mathcal{E}\Phi(x),} \quad (0 < x < a). \qquad (5.2.7)$$

We shall next find the boundary conditions for $\Phi(x)$ at $x = 0_+$ and $x = a_-$, where subscripts "+" and "−" imply that the boundary points of the interval (5.2.6) are approached from "above" and "below," respectively, i.e., from within the interval.

Since $\Phi(x)$ is a continuous function, we have

$$\Phi(a_+) = \Phi(a_-). \qquad (5.2.8)$$

According to formula (5.2.5), we have

$$\Phi(a_+) = e^{ika}\Phi(0_+). \tag{5.2.9}$$

By substituting the last formula into equation (5.2.8), we end up with the following boundary condition

$$e^{ika}\Phi(0_+) = \Phi(a_-). \tag{5.2.10}$$

These boundary conditions relate the values of $\Phi(x)$ at the ends of the interval (5.2.6).

Next, we shall derive the boundary condition that relates the values of $\frac{d\Phi}{dx}(x)$ at the ends of the same interval. The derivation proceeds as follows.

Consider the interval

$$0 < x < 2a. \tag{5.2.11}$$

On this interval, equation (5.2.4) has the form

$$-\frac{\hbar^2}{2m}\frac{d^2\Phi(x)}{dx^2} + \beta\delta(x-a)\Phi(x) = \mathcal{E}\Phi(x). \tag{5.2.12}$$

By integrating both sides of the last equation over the very small interval $a - \alpha < x < a + \alpha$, we obtain

$$-\frac{\hbar^2}{2m}\left[\frac{d\Phi}{dx}(a+\alpha) - \frac{d\Phi}{dx}(a-\alpha)\right] + \beta\Phi(a_+) = \mathcal{E}\int_{a-\alpha}^{a+\alpha}\Phi(x)dx. \tag{5.2.13}$$

In the limit when $\alpha \to 0$, from the last equation we find

$$\frac{d\Phi}{dx}(a_+) - \frac{d\Phi}{dx}(a_-) = \frac{2m\beta}{\hbar^2}\Phi(a_+). \tag{5.2.14}$$

From formula (5.2.5), it follows that

$$\frac{d\Phi}{dx}(a_+) = e^{ika}\frac{d\Phi}{dx}(0_+). \tag{5.2.15}$$

By substituting formulas (5.2.9) and (5.2.15) into equation (5.2.14), we end up with the boundary condition

$$e^{ika}\frac{d\Phi}{dx}(0_+) - \frac{d\Phi}{dx}(a_-) = \frac{2m\beta}{\hbar^2}e^{ika}\Phi(0_+). \tag{5.2.16}$$

Thus, in order to find $\Phi(x)$ in the interval (5.2.6) we have to find a solution of equation (5.2.7) subject to the boundary conditions (5.2.10) and (5.2.16).

A general solution of equation (5.2.7) can be written in the form

$$\Phi(x) = A\cos qx + B\sin qx, \quad (0 < x < a), \tag{5.2.17}$$

where

$$q = \frac{\sqrt{2m\mathcal{E}}}{\hbar}. \tag{5.2.18}$$

It is easy to derive the following equations for A and B from the boundary conditions (5.2.10) and (5.2.16), respectively,

$$e^{ika} A = A \cos qa + B \sin qa, \tag{5.2.19}$$

$$qa[e^{ika} B + A \sin qa - B \cos qa] = 2\gamma e^{ika} A, \tag{5.2.20}$$

where

$$\gamma = \frac{m\beta a}{\hbar^2}. \tag{5.2.21}$$

From equation (5.2.19), we find

$$\frac{A}{B} = \frac{\sin qa}{e^{ika} - \cos qa}. \tag{5.2.22}$$

Similarly, from equation (5.2.20) it follows that

$$\frac{A}{B} = \frac{e^{ika} - \cos qa}{\frac{2\gamma}{qa}e^{ika} - \sin qa}. \tag{5.2.23}$$

From the last two equations, we conclude that

$$\frac{\sin qa}{e^{ika} - \cos qa} = \frac{e^{ika} - \cos qa}{\frac{2\gamma}{qa}e^{ika} - \sin qa}, \tag{5.2.24}$$

which yields

$$(e^{ika} - \cos qa)^2 - \sin qa \left(\frac{2\gamma}{qa}e^{ika} - \sin qa \right) = 0. \tag{5.2.25}$$

Further algebraic transformations lead to

$$e^{i2ka} - 2e^{ika} \left(\cos qa + \gamma \frac{\sin qa}{qa} \right) + 1 = 0. \tag{5.2.26}$$

Dividing the last equation by $2e^{ika}$, we find

$$\cos qa + \gamma \frac{\sin qa}{qa} = \frac{e^{ika} + e^{-ika}}{2}, \tag{5.2.27}$$

which leads to the final equation

$$\boxed{\cos qa + \gamma \frac{\sin qa}{qa} = \cos ka.} \tag{5.2.28}$$

For any given k, the last equation can be solved numerically to find the corresponding values of q and, according to formula (5.2.18), the corresponding values of $\mathcal{E}$. In this way, the band structures $\mathcal{E}_n(k)$ can be computed. The described solution process can be illustrated graphically and the emergence of energy bands becomes clearly visible. To this end, we introduce the function

$$F(q) = \cos qa + \gamma \frac{\sin qa}{qa} \tag{5.2.29}$$

and write the equation (5.2.28) in the form

$$F(q) = \cos ka. \tag{5.2.30}$$

We next plot function $F(q)$ and identify graphically disjoint intervals on the horizontal q-axis for which (see Figure 5.4)

$$|F(q)| \leq 1. \tag{5.2.31}$$

These intervals are marked in bold, and they represent the values of q and,

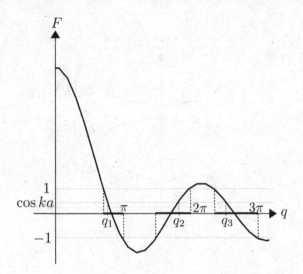

Fig. 5.4

consequently, the values of energy

$$\mathcal{E} = \frac{q\hbar^2}{2m} \tag{5.2.32}$$

corresponding to different energy bands. Indeed, for any chosen value of k we find the corresponding value of $\cos ka$. We plot this value of $\cos ka$

along the vertical F-axis and then draw the horizontal line and identify the values of q within the bold intervals for which the equality (5.2.30) holds. These values of q are marked on Figure 5.4 as q_1, q_2 and q_3. Having found these values of q, we can immediately compute the values of energy corresponding to the chosen value of k as follows:

$$\mathcal{E}_1(k) = \frac{q_1\hbar^2}{2m}, \tag{5.2.33}$$

$$\mathcal{E}_2(k) = \frac{q_2\hbar^2}{2m}, \tag{5.2.34}$$

$$\mathcal{E}_3(k) = \frac{q_3\hbar^2}{2m}. \tag{5.2.35}$$

By repeating the described process for different values of k, we find band structure $\mathcal{E}_1(k)$, $\mathcal{E}_2(k)$, $\mathcal{E}_3(k)$ and so on. These band structures are illustrated graphically on Figure 5.5. It is clear from Figure 5.4 and reflected

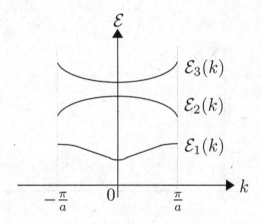

Fig. 5.5

in Figure 5.5 that, for the first and third bands, energy achieves minimum values at $k = 0$, while for the second band energy achieves its maximum value at $k = 0$. It is also apparent that the interval $\left(-\frac{\pi}{a}, \frac{\pi}{a}\right)$ in k-space serves as the first Brillouin zone. Furthermore, $K = \frac{2\pi}{a}$ can be viewed as the length of the reciprocal lattice and

$$Ka = 2\pi, \tag{5.2.36}$$

which is consistent with formula (5.1.42).

It must be remarked in the conclusion of this section that the periodic Dirac-type potential $U(x)$ is not a very realistic model for periodic potentials in crystal solids. Nevertheless, this model captures the main features of band structures in solids. This is because these features are consequences of **periodicity** of crystal structures. This model reveals the energy gaps existing between energy bands, and these are actually direct band gaps when energy minima and maxima in adjacent bands occur for the same zero value of k.

It is also important to point out that, despite infinite potential barriers represented by δ-functions, electrons with energies corresponding to energy bands move through these barriers without any scattering. This is because these barriers are identical, and electron tunneling through them can be viewed as resonant tunneling (see Chapter 3). Deviations from ideal periodicity of $U(x)$ will cause electron scattering.

5.3 Effective Mass Schrödinger Equation and Semiclassical Transport

In the last two sections, the motion of electrons in periodic potentials of crystal lattices has been studied and energy bands have been introduced. The actual motion of electrons is more complicated than that. This is due to the presence of macroscopic electric fields as well as due to deviations from ideal periodicity of lattice potentials. These deviations may occur due to the presence of impurities when not all lattice sites are occupied by identical ionized atoms. These deviations may also happen due to random thermal vibrations of crystal lattices when atoms are displaced from periodically spaced lattice sites. This suggests that the actual potential energy $U(\mathbf{r}, t)$ of the Hamiltonian can be represented as the sum of three distinct terms

$$U(\mathbf{r}, t) = U_L(\mathbf{r}) + U_\mathbf{E}(\mathbf{r}, t) + U_S(\mathbf{r}, t). \tag{5.3.1}$$

Here: $U_L(\mathbf{r})$ is the perfectly periodic lattice potential, $U_\mathbf{E}(\mathbf{r}, t)$ is the potential due to applied macroscopic electric fields, while $U_S(\mathbf{r}, t)$ is the scattering potential due to the presence of ionized impurities and thermal vibrations of the crystal lattice. The latter potential is random in nature because of the randomness of impurity locations as well as the randomness of thermal lattice vibrations. These random deviations from lattice periodicity result in electron scattering.

The time-dependent Schrödinger equation with this potential energy $U(\mathbf{r}, t)$ can be written as follows

$$i\hbar\frac{\partial\psi(\mathbf{r}, t)}{\partial t} = -\frac{\hbar^2}{2m}\nabla^2\psi(\mathbf{r}, t) + [U_L(\mathbf{r}) + U_E(\mathbf{r}, t) + U_S(\mathbf{r}, t)]\psi(\mathbf{r}, t). \quad (5.3.2)$$

The solution of this equation is a very difficult task. This task can be appreciably simplified by using the observation that $U_L(\mathbf{r})$ varies on a very different spatial scale than $U_E(\mathbf{r}, t)$ and $U_S(\mathbf{r}, t)$. Indeed, $U_L(\mathbf{r})$ oscillates very fast by varying on the atomic scale, i.e., on the scale defined by the spatial separation of adjacent crystal lattice sites, while $U_E(\mathbf{r}, t)$ and $U_S(\mathbf{r}, t)$ are almost constant on this spatial scale. This allows the reduction of the Schrödinger equation (5.3.2) to the so-called "effective mass" Schrödinger equation. The latter equation is constructed as follows. First, we consider the Hamiltonian $\hat{H}_0$ with the fast varying lattice potential

$$\hat{H}_0 = -\frac{\hbar^2}{2m}\nabla^2 + U_L(\mathbf{r}) \quad (5.3.3)$$

and find the band structure $\mathcal{E}_n(\mathbf{k})$ for this Hamiltonian by solving the following time-independent Schrödinger equation

$$\hat{H}_0\Phi_{n,\mathbf{k}}(\mathbf{r}) = \mathcal{E}_n(\mathbf{k})\Phi_{n,\mathbf{k}}(\mathbf{r}). \quad (5.3.4)$$

We shall limit our discussion to the "single-band effective mass" Schrödinger equation. In this case, we consider the highest energy conduction band $\mathcal{E}_c(\mathbf{k})$ and transform $\mathcal{E}_c(\mathbf{k})$ into the operator. This is done by replacing wave vector $\mathbf{k}$ by the operator according to the following rule

$$\mathbf{k} \to \hat{\mathbf{k}} = -i\nabla. \quad (5.3.5)$$

Consequently,

$$\mathcal{E}_c(\mathbf{k}) \to \hat{\mathcal{E}}_c(\hat{\mathbf{k}}) = \hat{\mathcal{E}}_c(-i\nabla). \quad (5.3.6)$$

Now, the single-band effective mass Schrödinger equation can be written as follows

$$\boxed{i\hbar\frac{\partial\tilde{\psi}(\mathbf{r}, t)}{\partial t} = \hat{\mathcal{E}}_c(-i\nabla)\tilde{\psi}(\mathbf{r}, t) + [U_E(\mathbf{r}, t) + U_S(\mathbf{r}, t)]\tilde{\psi}(\mathbf{r}, t).} \quad (5.3.7)$$

The wave function $\tilde{\psi}(\mathbf{r}, t)$ in the effective mass equation is frequently called an "envelope function." It can be viewed as some average over the lattice cell of the wave function $\psi(\mathbf{r}, t)$ in the original Schrödinger equation (5.3.2).

The transition from the original Schrödinger equation (5.3.2) to the single-band effective mass Schrödinger equation can be mathematically justified under some conditions by using the averaging (over the lattice cell) procedure. This averaging procedure is often called the homogenization technique. It removes the fast oscillating lattice potential $U_L(\mathbf{r})$, while the remaining potentials $U_\mathbf{E}(\mathbf{r}, t)$ and $U_S(\mathbf{r}, t)$ in the effective mass equation (5.3.7) vary on a much larger spatial scale than potential $U(\mathbf{r}, t)$ (see formula (5.3.1)). In this sense, the potentials $U_\mathbf{E}(\mathbf{r}, t)$ and $U_S(\mathbf{r}, t)$ are more "homogeneous" than the potential $U(\mathbf{r}, t)$ in equation (5.3.2). Physically, the single-band equation (5.3.7) is justified if $U_\mathbf{E}(\mathbf{r}, t)$ and $U_S(\mathbf{r}, t)$ are almost constant over a unit cell of the crystal lattice and there is no appreciable coupling to other bands. The latter means that this equation is not applicable, for instance, to valence band electrons because valence bands (representing "heavy" and "light" holes) may overlap in energy leading to easily occurring interband transitions. In this case, the multiband effective mass Schrödinger equations can be derived, but we shall not discuss this matter further. We shall also omit the mathematics leading to the derivation of the effective mass equation. Instead, we shall explain briefly why it is called the "effective mass" equation.

If the conduction band $\mathcal{E}_c(\mathbf{k})$ is isotropic in $\mathbf{k}$-space near its minimum $\mathcal{E}_c^{(0)}$, then it can be approximated as follows

$$\mathcal{E}_c(\mathbf{k}) = \mathcal{E}_c^{(0)} + \frac{\mathbf{k}^2}{2m^*}, \tag{5.3.8}$$

where m^* is called the effective mass. According to formulas (5.3.5) and (5.3.6), $\mathcal{E}(\mathbf{k})$ can be replaced by the following operator

$$\hat{\mathcal{E}}(\hat{\mathbf{k}}) = \hat{\mathcal{E}}(-i\nabla) = \mathcal{E}_c^{(0)} - \frac{\nabla^2}{2m^*}. \tag{5.3.9}$$

This means that the single-band effective mass Schrödinger equation can be written in the form

$$i\hbar \frac{\partial \tilde{\psi}(\mathbf{r}, t)}{\partial t} = -\frac{\hbar^2}{2m^*} \nabla^2 \tilde{\psi}(\mathbf{r}, t) + [U_\mathbf{E}^{(0)}(\mathbf{r}, t) + U_S(\mathbf{r}, t)]\tilde{\psi}(\mathbf{r}, t), \tag{5.3.10}$$

where

$$U_\mathbf{E}^{(0)}(\mathbf{r}, t) = \mathcal{E}_c^{(0)} + U_\mathbf{E}(\mathbf{r}, t) \tag{5.3.11}$$

and it can be viewed as the conduction band edge energy in the presence of an electric field.

Equation (5.3.10) is mathematically similar to the Schrödinger equation for an electron with mass m^* in the absence of periodic crystal potential. This means that the effect of the crystal lattice is taken into account by renormalizing the electron mass. It must be noted that in the case when the energy band function $\mathcal{E}_c(\mathbf{k})$ is not isotropic in $\mathbf{k}$-space near its minimum, an effective mass is a tensor rather than a scalar.

Next, we shall use the single-band effective mass Schrödinger equation (5.3.7) to derive the equations of semi-classical transport of conduction band electrons. These equations are extensively used in numerical simulations of electron transport in semiconductors. The central idea of this derivation is to replace the effective mass equation (5.3.7) by the differential equations for operators $\hat{\mathbf{r}}$ and $\hat{\mathbf{k}}$. These differential equations are somewhat similar to those used in the derivation of Ehrenfest's theorem. This theorem states that the dynamic equations for average values of quantum mechanical quantities are mathematically identical to dynamic equations for the corresponding classical quantities.

The operator-differential equations that we are going to derive have the following form

$$\frac{d\hat{\mathbf{r}}}{dt} = \hat{\boldsymbol{v}}(\hat{\mathbf{k}}), \qquad (5.3.12)$$

$$\hbar\frac{d\hat{\mathbf{k}}}{dt} = -\nabla U_{\mathbf{E}}(\mathbf{r}, t) - \nabla U_S(\mathbf{r}, t), \qquad (5.3.13)$$

where $\hat{\boldsymbol{v}}(\hat{\mathbf{k}})$ is obtained from the following formula

$$\boldsymbol{v}(\mathbf{k}) = \frac{1}{\hbar}\nabla_{\mathbf{k}}\mathcal{E}_c(\mathbf{k}) \qquad (5.3.14)$$

by replacing $\mathbf{k}$ by the operator $\hat{\mathbf{k}} = -i\nabla$.

To derive the above equations, we first recall that the derivative of operator $\hat{f}$ is understood in the sense of the following equality

$$\langle\psi|\frac{d\hat{f}}{dt}|\psi\rangle = \frac{d}{dt}\langle\psi|\hat{f}|\psi\rangle. \qquad (5.3.15)$$

If an operator $\hat{f}$ does not depend on time explicitly $\left(\frac{\partial \hat{f}}{\partial t} = 0\right)$, then the following formula is valid

$$\frac{d\hat{f}}{dt} = \frac{i}{\hbar}\left[\hat{H}, \hat{f}\right]. \qquad (5.3.16)$$

This formula can be established by using the same reasoning as in the Section 2.2 of Chapter 2 (see formulas (2.2.36)–(2.2.41)).

Now, by using formula (5.3.16), we shall derive equation (5.3.12). According to formula (5.3.16), we find

$$\frac{d\hat{\mathbf{r}}}{dt} = \frac{i}{\hbar}\left[\hat{H}, \hat{\mathbf{r}}\right]. \qquad (5.3.17)$$

The Hamiltonian in the effective mass equation (5.3.7) has the form

$$\hat{H} = \hat{\mathcal{E}}_c(-i\nabla) + U_{\mathbf{E}}(\mathbf{r}, t) + U_S(\mathbf{r}, t). \qquad (5.3.18)$$

It is clear from the last formula that

$$\left[\hat{H}, \hat{\mathbf{r}}\right] = \hat{\mathcal{E}}_c(-i\nabla)\mathbf{r} - \mathbf{r}\hat{\mathcal{E}}_c(-i\nabla) + [U_{\mathbf{E}}(\mathbf{r}, t) + U_S(\mathbf{r}, t)]\mathbf{r} - \mathbf{r}[U_{\mathbf{E}}(\mathbf{r}, t) + U_S(\mathbf{r}, t)],$$
(5.3.19)

which is obviously reduced to

$$\left[\hat{H}, \hat{\mathbf{r}}\right] = \hat{\mathcal{E}}_c(-i\nabla)\mathbf{r} - \mathbf{r}\hat{\mathcal{E}}_c(-i\nabla).$$
(5.3.20)

We consider the power series expansion of $\mathcal{E}_c(\mathbf{k})$:

$$\mathcal{E}_c(\mathbf{k}) = \sum_n a_n \mathbf{k}^n,$$
(5.3.21)

which in the operator form can be rewritten as follows

$$\hat{\mathcal{E}}_c(-i\nabla) = \sum_n a_n(-i\nabla)^n.$$
(5.3.22)

Next, we want to establish the commutation relation

$$(-i\nabla)^n \mathbf{r} - \mathbf{r}(-i\nabla)^n = -in(-i\nabla)^{n-1}.$$
(5.3.23)

To do this, we shall use the induction argument. First, it is easy to check that the commutation relation (5.3.23) is valid for $n = 1$, that is, when it has the form

$$(-i\nabla)\mathbf{r} - \mathbf{r}(-i\nabla) = -i\hat{I},$$
(5.3.24)

where $\hat{I}$ is the identity operator.

Next, we assume that the commutation relation (5.3.23) is valid for $n - 1$, that is

$$(-i\nabla)^{n-1}\mathbf{r} - \mathbf{r}(-i\nabla)^{n-1} = -i(n-1)(-i\nabla)^{n-2}.$$
(5.3.25)

Now, we shall demonstrate that from the validity of formulas (5.3.24) and (5.3.25) follows the validity of (5.3.23).

Indeed, we have

$$(-i\nabla)^n \mathbf{r} = (-i\nabla)^{n-1}(-i\nabla)\mathbf{r} = (-i\nabla)^{n-1}\left[-i\hat{I} + \mathbf{r}(-i\nabla)\right].$$
(5.3.26)

Consequently,

$$(-i\nabla)^n \mathbf{r} = -i(-i\nabla)^{n-1} + (-i\nabla)^{n-1}\mathbf{r}(-i\nabla).$$
(5.3.27)

According to formula (5.3.25), we have

$$(-i\nabla)^{n-1}\mathbf{r} = \mathbf{r}(-i\nabla)^{n-1} - i(n-1)(-i\nabla)^{n-2}.$$
(5.3.28)

By substituting the last formula into equation (5.3.27), we find

$$(-i\nabla)^n \mathbf{r} = -in(-i\nabla)^{n-1} + \mathbf{r}(-i\nabla)^n,$$
(5.3.29)

which is equivalent to (5.3.23).

Having established the commutation relation (5.3.23), the formula (5.3.20) can be transformed as follows

$$\left[\hat{H}, \hat{\mathbf{r}}\right] = \mathcal{E}_c(-i\nabla)\mathbf{r} - \mathbf{r}\mathcal{E}_c(-i\nabla)$$

$$= \sum_n a_n[(-i\nabla)^n\mathbf{r} - \mathbf{r}(-i\nabla)^n] = -i\sum_n na_n(-i\nabla)^{n-1}. \quad (5.3.30)$$

This means that according to formulas (5.3.14) and (5.3.22)

$$\left[\hat{H}, \hat{\mathbf{r}}\right] = -i\hat{\boldsymbol{v}}(\hat{\mathbf{k}}). \quad (5.3.31)$$

By substituting the last formula into equation (5.3.17), we arrive at equation (5.3.12).

Next, we shall derive equation (5.3.13). It is clear from formula (5.3.16) that

$$\frac{d\hat{\mathbf{k}}}{dt} = \frac{i}{\hbar}\left[\hat{H}, \hat{\mathbf{k}}\right]. \quad (5.3.32)$$

By using formula (5.3.18), we find

$$\left[\hat{H}, \hat{\mathbf{k}}\right] = \hat{\mathcal{E}}_c(\hat{\mathbf{k}})\hat{\mathbf{k}} - \hat{\mathbf{k}}\hat{\mathcal{E}}_c\hat{\mathbf{k}} + \tilde{U}(\mathbf{r}, t)\hat{\mathbf{k}} - \hat{\mathbf{k}}\tilde{U}(\mathbf{r}, t), \quad (5.3.33)$$

where for the sake of brevity we use the notation

$$\tilde{U}(\mathbf{r}, t) = U_{\mathbf{E}}(\mathbf{r}, t) + U_S(\mathbf{r}, t). \quad (5.3.34)$$

It is clear that the first two terms in the right-hand side of formula (5.3.29) are canceled and, consequently,

$$\left[\hat{H}, \hat{\mathbf{k}}\right] = \tilde{U}(\mathbf{r}, t)\hat{\mathbf{k}} - \hat{\mathbf{k}}\tilde{U}(\mathbf{r}, t). \quad (5.3.35)$$

It is apparent that for any function $\varphi(\mathbf{r}, t)$ we have

$$\hat{\mathbf{k}}\tilde{U}(\mathbf{r}, t)\varphi(\mathbf{r}, t) = -i\nabla[\tilde{U}(\mathbf{r}, t)\varphi(\mathbf{r}, t)]$$

$$= -i(\nabla\tilde{U}(\mathbf{r}, t))\varphi(\mathbf{r}, t) - i\tilde{U}(\mathbf{r}, t)\nabla\varphi(\mathbf{r}, t)$$

$$= \tilde{U}(\mathbf{r}, t)\hat{\mathbf{k}}\varphi(\mathbf{r}, t) - i(\nabla\tilde{U}(\mathbf{r}, t))\varphi(\mathbf{r}, t). \quad (5.3.36)$$

Consequently,

$$\tilde{U}(\mathbf{r}, t)\hat{\mathbf{k}} - \hat{\mathbf{k}}\tilde{U}(\mathbf{r}, t) = i\nabla\tilde{U}(\mathbf{r}, t), \quad (5.3.37)$$

and formula (5.3.35) can be written as

$$\left[\hat{H}, \hat{\mathbf{k}}\right] = i[\nabla U_{\mathbf{E}}(\mathbf{r}, t) + \nabla U_S(\mathbf{r}, t)]. \quad (5.3.38)$$

By substituting the last formula into equation (5.3.32), we arrive at equation (5.3.13). This completes the derivation of equations (5.3.12) and (5.3.13).

Operator equations (5.3.12) and (5.3.13) are naturally transformed into the following equations for average quantities

$$\frac{\overline{d\mathbf{r}}}{dt} = \overline{\mathbf{v}(\mathbf{k})}, \tag{5.3.39}$$

$$\hbar\frac{\overline{d\mathbf{k}}}{dt} = -\overline{\nabla U_{\mathbf{E}}(\mathbf{r}, t)} - \overline{\nabla U_S(\mathbf{r}, t)}. \tag{5.3.40}$$

If an electron is represented by a wave packet localized in a small region of $(\mathbf{r}, \mathbf{k})$-space and $\tilde{U}(\mathbf{r}, t)$ is a slowly varying function on the scale of spatial extension of the electron wave packet, the above equations can be with some accuracy written as the following equations for average electron position $\mathbf{r}$ and average wave vector $\mathbf{k}$:

$$\frac{d\mathbf{r}}{dt} = \mathbf{v}(\mathbf{k}), \tag{5.3.41}$$

$$\hbar\frac{d\mathbf{k}}{dt} = -e\mathbf{E}(\mathbf{r}, t) + \mathbf{F}_r, \tag{5.3.42}$$

where

$$\nabla U_{\mathbf{E}}(\mathbf{r}, t) = e\mathbf{E}(\mathbf{r}, t) \tag{5.3.43}$$

and

$$\mathbf{F}_r = -\nabla U_S(\mathbf{r}, t) \tag{5.3.44}$$

is the random force that accounts for electron scattering. Usually, this force is modeled as

$$\mathbf{F}_r = \hbar\sum_i \boldsymbol{\zeta}_i \delta(t - t_i), \tag{5.3.45}$$

where $\boldsymbol{\zeta}_i$ are random "jumps" of the wave vector

$$\mathbf{k}(t_{i+}) = \mathbf{k}(t_{i-}) + \boldsymbol{\zeta}_i \tag{5.3.46}$$

occurring at random times t_i due to electron scattering. Statistics of random times t_i and random jumps $\boldsymbol{\zeta}_i$ can be found by specifying the function

$$S(\mathbf{k}, \mathbf{k}'), \tag{5.3.47}$$

which is called the transition probability rate. This name implies that the quantity $S(\mathbf{k}, \mathbf{k}')dt$ is the probability of scattering from $\mathbf{k}$ to $\mathbf{k}'$ during the time dt. There are analytical expressions for $S(\mathbf{k}, \mathbf{k}')$ valid for different physical mechanisms of scattering such as, for example, optical phonon scattering, acoustic phonon scattering, etc.

By using $S(\mathbf{k}, \mathbf{k}')$, the scattering rate $\lambda(\mathbf{k})$ can be computed as follows

$$\lambda(\mathbf{k}) = \int S(\mathbf{k}, \mathbf{k}')d\mathbf{k}'. \tag{5.3.48}$$

It is apparent that $\lambda(\mathbf{k})dt$ has the physical meaning of probability of scattering from $\mathbf{k}$ to any $\mathbf{k}'$ during the time dt.

Now, the statistics of t_i and ζ_i are specified by the formulas

$$\text{Probability}(t_i - t_{i-1} > \tau) = e^{-\int_0^\tau \lambda(\mathbf{k}(t))dt}, \tag{5.3.49}$$

and

$$\rho_{\mathbf{k}_i}(\zeta_i) = \frac{S(\mathbf{k}_i, \mathbf{k}_i + \zeta_i)}{\lambda(\mathbf{k}_i)}, \tag{5.3.50}$$

where $\rho_{\mathbf{k}_i}(\zeta_i)$ is the probability density of random jump ζ_i from $\mathbf{k}_i$.

The equations (5.3.41)–(5.3.50) constitute the basis of semi-classical transport theory, and they are extensively used in the Monte Carlo simulations of electron transport in semiconductors. In these simulations, a self-scattering process is usually introduced to simplify computations of random scattering time on the basis of formula (5.3.49). The transition probability rate $\tilde{S}(\mathbf{k}, \mathbf{k}')$ for self-scattering is defined by the formula

$$\tilde{S}(\mathbf{k}, \mathbf{k}') = \tilde{\lambda}(\mathbf{k})\delta(\mathbf{k} - \mathbf{k}') \tag{5.3.51}$$

and the self-scattering rate $\tilde{\lambda}(\mathbf{k})$ is chosen in such a way that

$$\lambda(\mathbf{k}) + \tilde{\lambda}(\mathbf{k}) = \Gamma = \text{const.} \tag{5.3.52}$$

This leads to the following simplification of formula (5.3.49):

$$\text{Probability}(t_i - t_{i-1} > \tau) = e^{-\Gamma\tau}. \tag{5.3.53}$$

In Monte Carlo simulations, a large number of trajectories of electrons are computed, and then the desired average quantities are calculated. It is clear from equations (5.3.41)–(5.3.50) that these random trajectories consists of pieces of deterministic trajectories of electrons caused by applied electric field which are randomly interrupted by random jumps of the wave vector $\mathbf{k}$ caused by random scattering.

Equations (5.3.41)–(5.3.50) can be interpreted as stochastic differential equations driven by a jump process $\mathbf{F}_r$. Solutions of these equations are realizations of a Markovian process which are discontinuous in $\mathbf{k}$-space. Such a process can be characterized by a transition probability density function as well as by the distribution function

$$f(\mathbf{r}, \mathbf{k}, t). \tag{5.3.54}$$

It is proven in the theory of stochastic processes that this distribution function is a solution of the Kolmogorov-Feller equation which in physics is called the Boltzmann equation. This equation has the form

$$\frac{\partial f(\mathbf{r}, \mathbf{k}, t)}{\partial t} = -v(\mathbf{k}) \cdot \nabla_{\mathbf{r}} f(\mathbf{r}, \mathbf{k}, t) + e\mathbf{E}(\mathbf{r}, t) \cdot \nabla_{\mathbf{k}} f(\mathbf{r}, \mathbf{k}, t)$$

$$+ \int [S(\mathbf{k}', \mathbf{k}) f(\mathbf{r}, \mathbf{k}', t) - S(\mathbf{k}, \mathbf{k}') f(\mathbf{r}, \mathbf{k}, t)] d\mathbf{k}'. \quad (5.3.55)$$

This equation reflects the fact that the change in time of the distribution function is due to the deterministic motion of electrons from (or into) the infinitesimal volume of $(\mathbf{r}, \mathbf{k})$-space as well as due to the balance of scattering from and into this volume. The latter is described by the integral in equation (5.3.55) which is often called the collision integral. The Boltzmann equation (5.3.55) is also extensively used in numerical simulations of semi-classical electron transport in semiconductors. It is a deterministic partial differential-integral equation whose direct solution avoids numerous generation of stochastic trajectories. For this reason, the numerical solution of this equation is considered as an attractive alternative to the Monte Carlo simulations.

Problems

(1) Prove formula (5.1.14).

(2) Prove formula (5.1.23).

(3) Demonstrate that the primitive vectors $\mathbf{K}_1, \mathbf{K}_2$ and $\mathbf{K}_3$ of the reciprocal lattice given by formulas (5.1.43), (5.1.44), and (5.1.45) are solutions of equations (5.1.42).

(4) Establish the identity (5.1.47).

(5) Derive equation (5.1.60).

(6) Derive equations (5.2.19) and (5.2.20).

(7) Analyze the Kronig-Penney model in the case when the periodic potential $U(x)$ consists of periodically spaced rectangular barriers (see Figure below).

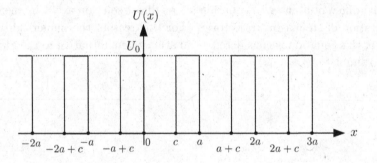

Fig. P.1

(8) Derive the formula for the effective mass tensor in the case when the conduction band is not isotropic in $\mathbf{k}$-space near its minimum.

(9) How can the single-band effective mass Schrödinger equation be written in terms of the effective mass tensor?

(10) Prove formula (5.3.16).

(11) Verify formula (5.3.24).

Chapter 6

Matrix Form of Quantum Mechanics. Perturbation Theory. Density Matrix

6.1 Matrix Form of Quantum Mechanics

In previous chapters, we have studied quantum mechanical problems by using the coordinate representation for wave functions and operators. In this representation, the wave function $\psi(\mathbf{r}, t)$ is a function of coordinates and time, while quantum mechanical operators are operators with respect to spatial coordinates. Namely, we have used the following operators

$$\hat{\mathbf{r}}\psi(\mathbf{r}, t) = \mathbf{r}\psi(\mathbf{r}, t), \tag{6.1.1}$$

$$\hat{\mathbf{p}}\psi(\mathbf{r}, t) = -i\hbar\nabla_{\mathbf{r}}\psi(\mathbf{r}, t), \tag{6.1.2}$$

$$\hat{\mathbf{L}}\psi(\mathbf{r}, t) = (\mathbf{r} \times \hat{\mathbf{p}})\psi(\mathbf{r}, t) = -i\hbar(\mathbf{r} \times \nabla_{\mathbf{r}})\psi(\mathbf{r}, t), \tag{6.1.3}$$

$$\hat{H}\psi(\mathbf{r}, t) = -\frac{\hbar^2}{2m}\nabla^2\psi(\mathbf{r}, t) + U(\mathbf{r}, t)\psi(\mathbf{r}, t) \tag{6.1.4}$$

and

$$\hat{H}\psi(\mathbf{r}, t) = \frac{[\hat{\mathbf{p}} - q\mathbf{A}(\mathbf{r}, t)]^2\psi(\mathbf{r}, t)}{2m} + q\varphi(\mathbf{r}, t)\psi(\mathbf{r}, t). \tag{6.1.5}$$

It turns out that in many applications of quantum mechanics other representations of quantum mechanical states are useful. In this chapter, we shall deal with representations of quantum mechanical states that lead to the matrix form of quantum mechanics. Historically, the wave form and the matrix form of quantum mechanics were almost simultaneously developed.

To start the discussion, consider some physical quantity f and its corresponding Hermitian operator $\hat{f}$ with discrete spectrum. The latter implies that

$$\hat{f}\psi_n = f_n\psi_n \tag{6.1.6}$$

and

$$\langle \psi_n | \psi_k \rangle = \delta_{nk}. \tag{6.1.7}$$

It is assumed that the set of eigenfunctions ψ_n is complete and can be used as an orthonormal basis for expansion of any square-integrable wave function ψ:

$$\psi = \sum_n a_n \psi_n, \tag{6.1.8}$$

where

$$a_n = \langle \psi_n | \psi \rangle. \tag{6.1.9}$$

Now, we consider the infinite-dimensional vector

$$\mathbf{a} = \begin{pmatrix} a_1 \\ a_2 \\ \vdots \\ a_n \\ \vdots \end{pmatrix} \tag{6.1.10}$$

which will be regarded as the $\hat{f}$-representation of the quantum mechanical state $\psi(\mathbf{r}, t)$. It is apparent that there is one-to-one correspondence between $\psi(\mathbf{r}, t)$ and $\mathbf{a}$. For another wave function $\tilde{\psi}(\mathbf{r}, t)$, its $\hat{f}$-representation is the infinite-dimensional vector

$$\mathbf{b} = \begin{pmatrix} b_1 \\ b_2 \\ \vdots \\ b_n \\ \vdots \end{pmatrix} \tag{6.1.11}$$

whose components are the expansion coefficients in the formula

$$\tilde{\psi} = \sum_n b_n \psi_n. \tag{6.1.12}$$

As discussed in Section 2.1 of Chapter 2, we can consider two Hilbert spaces: the Hilbert space L_2 of square-integrable wave functions in coordinate representation with inner product

$$\langle \tilde{\psi} | \psi \rangle_{L_2} = \int \tilde{\psi}^* \psi \, dv \tag{6.1.13}$$

and norm

$$\|\psi\|_{L_2} = \sqrt{\langle\psi|\psi\rangle_{L_2}}, \qquad (6.1.14)$$

and the Hilbert space ℓ_2 of infinite-dimensional vectors, serving as $\hat{f}$-representations of wave functions with inner product

$$\langle\mathbf{b}|\mathbf{a}\rangle_{\ell_2} = \sum_n b_n^* a_n \qquad (6.1.15)$$

and norm

$$\|\mathbf{a}\|_{\ell_2} = \sqrt{\langle\mathbf{a}|\mathbf{a}\rangle_{\ell_2}}. \qquad (6.1.16)$$

It is apparent from formulas (6.1.8) and (6.1.9) that

$$\|\psi\|_{L_2}^2 = \langle\psi|\psi\rangle_{L_2} = \sum_n a_n^* a_n = \langle\mathbf{a}|\mathbf{a}\rangle_{\ell_2} = \|\mathbf{a}\|_{\ell_2}. \qquad (6.1.17)$$

Similarly, it can be established that

$$\langle\tilde{\psi}|\psi\rangle_{L_2} = \langle\mathbf{b}|\mathbf{a}\rangle_{\ell_2}. \qquad (6.1.18)$$

Thus, there is one-to-one correspondence between coordinate representation and $\hat{f}$-representation of quantum mechanical states, and this correspondence preserves the norm and inner products. In mathematics, this correspondence is called isometric isomorphism of L_2 and ℓ_2.

Next, consider an operator $\hat{g}$:

$$\chi = \hat{g}\psi. \qquad (6.1.19)$$

We want to find the $\hat{f}$-representation of operator $\hat{g}$. To this end, we shall use the expansion

$$\chi = \sum_k c_k \psi_k \qquad (6.1.20)$$

and introduce the infinite-dimensional vector $\mathbf{c}$ as the $\hat{f}$-representation of χ:

$$\mathbf{c} = \begin{pmatrix} c_1 \\ c_2 \\ \vdots \\ c_n \\ \vdots \end{pmatrix}. \qquad (6.1.21)$$

By substituting formulas (6.1.20) and (6.1.8) into equation (6.1.19), we derive

$$\sum_k c_k \psi_k = \sum_n a_n \hat{g}\psi_n. \qquad (6.1.22)$$

From the last equation it follows that

$$c_k = \sum_n \langle \psi_k | \hat{g} | \psi_n \rangle a_n, \quad (k = 1, 2, \ldots).$$ (6.1.23)

By using the notation

$$g_{kn} = \langle \psi_k | \hat{g} | \psi_n \rangle$$ (6.1.24)

the last formula can be written as

$$c_k = \sum_n g_{kn} a_n, \quad (k = 1, 2, \ldots).$$ (6.1.25)

By introducing the infinite-dimensional matrix

$$\hat{g} = \begin{pmatrix} g_{11} & g_{12} & g_{13} & \cdots \\ g_{21} & g_{22} & g_{23} & \cdots \\ g_{31} & g_{32} & g_{33} & \cdots \\ \vdots & \vdots & \vdots & \vdots \end{pmatrix},$$ (6.1.26)

or, in abbreviated form,

$$\hat{g} = \{g_{kn}\},$$ (6.1.27)

formula (6.1.25) can be represented as follows

$$\mathbf{c} = \hat{g}\mathbf{a}.$$ (6.1.28)

The last equation is the $\hat{f}$-representation of equation (6.1.19) and the matrix (6.1.27) is the $\hat{f}$-representation of operator $\hat{g}$.

If operator $\hat{g}$ is Hermitian, then it is easy to see that

$$g_{kn} = g_{nk}^*,$$ (6.1.29)

where, as before, superscript "∗" indicates a complex conjugate quantity.

It is apparent that by choosing different representations of quantum mechanical states we end up with different matrix representations of operators.

Next, we shall discuss how these different matrix representations of operators are related. To this end, consider some physical quantity f' (different from f) and its corresponding Hermitian operator $\hat{f}'$ with discrete spectrum. The latter implies that

$$\hat{f}' \psi_k' = f_n' \psi_k',$$ (6.1.30)

and

$$\psi = \sum_k a_k' \psi_k'.$$ (6.1.31)

The infinite-dimensional vector

$$\mathbf{a}' = \begin{pmatrix} a'_1 \\ a'_2 \\ \vdots \\ a'_n \\ \vdots \end{pmatrix} \tag{6.1.32}$$

can be viewed as the f'-representation of state ψ. We shall now find the relation between vectors $\mathbf{a}$ and $\mathbf{a}'$, i.e., between $\hat{f}$- and $\hat{f}'$-representations of the same quantum mechanical state. To do this, we shall first expand the basis functions ψ_n in terms of basis functions ψ'_k:

$$\psi_n = \sum_k u_{kn} \psi'_k, \tag{6.1.33}$$

where

$$u_{kn} = \langle \psi'_k | \psi_n \rangle. \tag{6.1.34}$$

By substituting formula (6.1.33) into equation (6.1.8), we find

$$\psi = \sum_n a_n \left(\sum_k u_{kn} \psi'_k \right) = \sum_k \left(\sum_n u_{kn} a_n \right) \psi'_k. \tag{6.1.35}$$

By comparing formulas (6.1.31) and (6.1.35), we conclude that

$$a'_k = \sum_n u_{kn} a_n, \quad (k = 1, 2, \dots). \tag{6.1.36}$$

The last equation can be written in the matrix form as follows

$$\mathbf{a}' = \hat{U} \mathbf{a}, \tag{6.1.37}$$

where $\hat{U}$ is the infinite-dimensional matrix

$$\hat{U} = \{u_{kn}\}. \tag{6.1.38}$$

Next, we shall prove that this matrix $\hat{U}$, which relates different representations of the same quantum mechanical state, has special properties. To do this, we expand basis functions ψ'_k in terms of basis function ψ_n:

$$\psi'_k = \sum_n v_{nk} \psi_n, \tag{6.1.39}$$

where

$$v_{nk} = \langle \psi_n | \psi'_k \rangle. \tag{6.1.40}$$

By substituting formula (6.1.39) into equation (6.1.31), we find

$$\psi = \sum_k a_k' \left(\sum_n v_{nk} \psi_n \right) = \sum_n \left(\sum_k v_{nk} a_k' \right) \psi_n. \qquad (6.1.41)$$

By comparing formulas (6.1.41) and (6.1.8), we conclude that

$$a_n = \sum_k v_{nk} a_k', \quad (n = 1, 2, \dots). \qquad (6.1.42)$$

The last equation can be written in the matrix form as follows

$$\mathbf{a} = \hat{V} \mathbf{a}', \qquad (6.1.43)$$

where $\hat{V}$ is the infinite-dimensional matrix

$$\hat{V} = \{v_{nk}\}. \qquad (6.1.44)$$

From formulas (6.1.37) and (6.1.44), we find that

$$\hat{V} = \hat{U}^{-1}. \qquad (6.1.45)$$

Furthermore, from formulas (6.1.34) and (6.1.40) it follows that

$$v_{nk} = u_{kn}^*. \qquad (6.1.46)$$

This implies that

$$\hat{V} = \hat{U}^*, \qquad (6.1.47)$$

where $\hat{U}^*$ is the Hermitian conjugate (also called Hermitian adjoint) of matrix $\hat{U}$.

From formulas (6.1.45) and (6.1.47), it follows that

$$\boxed{\hat{U}^{-1} = \hat{U}^*,} \qquad (6.1.48)$$

or, equivalently,

$$\boxed{\hat{U}^* \hat{U} = \hat{U} \hat{U}^* = \hat{I},} \qquad (6.1.49)$$

where $\hat{I}$ is the identity matrix.

A matrix whose inverse coincides with its Hermitian conjugate is called **unitary**. By using formulas (6.1.37) and (6.1.49), it is easy to establish that

$$\langle \mathbf{c}' | \mathbf{a}' \rangle = \langle \mathbf{c} | \mathbf{a} \rangle, \qquad (6.1.50)$$

and, consequently,

$$\| \mathbf{a}' \| = \| \mathbf{a} \| = 1. \qquad (6.1.51)$$

Thus, it can be concluded that **different representations of the same quantum mechanical state are related by unitary matrices (unitary operators) which preserve the inner products and norms.**

Next, by using the derived formulas, we find the relation between different representations of operators. For this purpose, we shall write formula (6.1.19) in f'-representation as follows

$$\mathbf{c}' = \hat{g}'\mathbf{a}', \tag{6.1.52}$$

where $\mathbf{c}'$, $\mathbf{a}'$ and $\hat{g}'$ are f'-representations of χ, ψ and operator $\hat{g}$, respectively. It is clear that

$$\hat{g}' = \{g'_{kn}\} \tag{6.1.53}$$

and

$$g'_{kn} = \langle \psi'_k | \hat{g} | \psi'_n \rangle. \tag{6.1.54}$$

It is also clear that

$$\mathbf{c}' = \hat{U}\mathbf{c}, \tag{6.1.55}$$

where, as before, $\hat{U}$ is the unitary matrix that relates f'- and f-representations of the same quantum mechanical states.

By substituting formulas (6.1.55) and (6.1.37) into equation (6.1.52), we obtain

$$\hat{U}\mathbf{c} = \hat{g}'\hat{U}\mathbf{a}. \tag{6.1.56}$$

Now, by applying $\hat{U}^{-1}$ to both sides of the last equation, we arrive at:

$$\mathbf{c} = \hat{U}^{-1}\hat{g}'\hat{U}\mathbf{a}. \tag{6.1.57}$$

By comparing formulas (6.1.28) and (6.1.57) we find that

$$\boxed{\hat{g} = \hat{U}^{-1}\hat{g}'\hat{U}.} \tag{6.1.58}$$

From the last formula, it also follows that

$$\boxed{\hat{g}' = \hat{U}\hat{g}\hat{U}^{-1}.} \tag{6.1.59}$$

Thus, it can be concluded that **different matrix representations of operators are related by similarity transformations** (6.1.58) **and** (6.1.59). It can also be said that similar matrices $\hat{g}$ and $\hat{g}'$ represent the same operator under two different orthonormal bases $\{\psi_n\}$ and $\{\psi'_k\}$, with unitary matrix $\hat{U}$ being the change of basis matrix.

Next, we consider two useful examples of matrix representations of operators. The first example deals with the matrix form of the coordinate operator in the energy representation of the one-dimensional harmonic oscillator. The basis functions in this case are the eigenfunctions Φ_n of the harmonic oscillator Hamiltonian (see Section 4.2 of Chapter 4):

$$\hat{H}\Phi_n = \hbar\omega\left(n + \frac{1}{2}\right)\Phi_n, \tag{6.1.60}$$

where

$$\hat{H} = -\frac{\hbar^2}{2m}\frac{d^2}{dx^2} + \frac{m\omega^2}{2}x^2. \tag{6.1.61}$$

By using formulas (4.2.64) and the recursion relation (4.2.69) it is easy to show the validity of the following recursion relation for eigenfunctions $\Phi_n(x)$:

$$\Phi_{n+1}(X) = \sqrt{\frac{2}{n+1}}X\Phi_n(X) - \sqrt{\frac{n}{n+1}}\Phi_{n-1}(X), \tag{6.1.62}$$

where (see formula (4.2.15))

$$X = x\sqrt{\frac{m\omega}{\hbar}}. \tag{6.1.63}$$

Since eigenfunctions $\Phi_n(X)$ are orthonormal, from the recursion relation (6.1.62) it follows that

$$\langle\Phi_{n+1}|X\Phi_n\rangle = \sqrt{\frac{n+1}{2}} \tag{6.1.64}$$

and

$$\langle\Phi_{n-1}|X\Phi_n\rangle = \sqrt{\frac{n}{2}}. \tag{6.1.65}$$

It is also clear from formula (6.1.62) that

$$\langle\Phi_k|X\Phi_n\rangle = 0 \tag{6.1.66}$$

if $k \neq n+1$ and $k \neq n-1$.

By using formulas (6.1.63)–(6.1.66), it can be concluded that the coordinate operator $\hat{x}$ in the energy representation is the infinite matrix

$$\hat{x} = \{x_{nk}\} \tag{6.1.67}$$

whose matrix elements are given by the formula

$$x_{nk} = \sqrt{\frac{\hbar}{2m\omega}}(\sqrt{k}\delta_{n,k-1} + \sqrt{k+1}\delta_{n,k+1}),$$
$$(n = 0, 1, 2, \ldots; k = 0, 1, 2, \ldots). \tag{6.1.68}$$

It is easy to see from the last formula that $\hat{x}$ is the infinite-dimensional "two diagonal" matrix of the form

$$\hat{x} = \sqrt{\frac{\hbar}{2m\omega}} \begin{pmatrix} 0 & 1 & 0 & 0 & \cdots \\ 1 & 0 & \sqrt{2} & 0 & \cdots \\ 0 & \sqrt{2} & 0 & \sqrt{3} & \cdots \\ 0 & 0 & \sqrt{3} & 0 & \cdots \\ \vdots & \vdots & \vdots & \vdots & \vdots \end{pmatrix}. \tag{6.1.69}$$

We have derived the formula (6.1.68) for the matrix elements x_{nk} by using the recursion relation (6.1.62). The same formula can be also derived by using raising $\hat{a}^+$ and lowering $\hat{a}$ operators. This is left as an exercise for the reader at this point. However, in the next section, we shall use this operator technique for the evaluation of matrix elements of $\hat{x}^\nu$-operators.

Next, we shall discuss the matrix forms of the angular momentum operator in a basis of eigenfunctions of $\widehat{\mathbf{L}}^2$ and $\hat{L}_z$ operators. Consequently, the following formulas are valid for the matrix elements of these operators

$$\widehat{\mathbf{L}}^2_{lm,l'm'} = \hbar^2 l(l+1)\delta_{ll'}\delta_{mm'}, \tag{6.1.70}$$

$$(\hat{L}_z)_{lm,l'm'} = \hbar m\delta_{ll'}\delta_{mm'}. \tag{6.1.71}$$

It is apparent from the above formulas that the matrix elements depend on four indices. For this reason, these matrix elements are usually displayed in the matrix $\widehat{\mathbf{L}}^2$ as follows. The rows and columns are ordered in such a way that for every value of l, m runs from $-l$ to $+l$ and for all these $2l+1$ consecutive values of m the diagonal elements are the same and equal to $\hbar^2 l(l+1)$. This structure is repeated for increasing values of l. Thus, it is clear that the diagonal matrix $\widehat{\mathbf{L}}^2$ consists of diagonal blocks with equal diagonal elements within each block. This is illustrated below.

$$\widehat{\mathbf{L}}^2 = \hbar^2 \left[\begin{array}{c|ccc|ccccc|c} 0 & 0 & & 0 & & & & & & \cdots \\ \hline & 2 & 0 & 0 & & & & & & \\ 0 & 0 & 2 & 0 & & & 0 & & & \cdots \\ & 0 & 0 & 2 & & & & & & \\ \hline & & & & 6 & 0 & 0 & 0 & 0 & \\ & & & & 0 & 6 & 0 & 0 & 0 & \\ 0 & & 0 & & 0 & 0 & 6 & 0 & 0 & \cdots \\ & & & & 0 & 0 & 0 & 6 & 0 & \\ & & & & 0 & 0 & 0 & 0 & 6 & \\ \hline \vdots & & \vdots & & & \vdots & & & & \vdots \end{array} \right]. \tag{6.1.72}$$

The matrix $\hat{L}_z$ is constructed in a similar way as shown below.

$$\hat{L}_z = \hbar \begin{bmatrix} 0 & 0 & 0 & \cdots \\ & \begin{matrix} 1 & 0 & 0 \\ 0 & 0 & 0 \\ 0 & 0 & -1 \end{matrix} & 0 & \cdots \\ 0 & 0 & \begin{matrix} 2 & 0 & 0 & 0 & 0 \\ 0 & 1 & 0 & 0 & 0 \\ 0 & 0 & 0 & 0 & 0 \\ 0 & 0 & 0 & -1 & 0 \\ 0 & 0 & 0 & 0 & -2 \end{matrix} & \cdots \\ \vdots & \vdots & \vdots & \end{bmatrix}. \qquad (6.1.73)$$

Now, we shall discuss the matrix form of $\hat{L}_+$, $\hat{L}_-$, $\hat{L}_x$ and $\hat{L}_y$ operators in the basis consisting of eigenfunctions $\psi_{lm}(\theta,\varphi)$ of $\widehat{\mathbf{L}}^2$ and $\hat{L}_z$ operators. According to formula (2.4.68), we have

$$\widehat{\mathbf{L}}^2\psi_{lm}(\theta,\varphi) = \hat{L}_-\hat{L}_+\psi_{lm}(\theta,\varphi) + \hat{L}_z^2\psi_{lm}(\theta,\varphi) + \hbar\hat{L}_z\psi_{lm}(\theta,\varphi). \quad (6.1.74)$$

Furthermore, since $\hat{L}_+$ and $\hat{L}_-$ are promotion (raising) and demotion (lowering) operators, respectively, the following formulas are valid

$$\hat{L}_+\psi_{lm}(\theta,\varphi) = \alpha_{lm}^+\psi_{l,m+1}(\theta,\varphi), \qquad (6.1.75)$$

$$\hat{L}_-\psi_{lm}(\theta,\varphi) = \alpha_{lm}^-\psi_{l,m-1}(\theta,\varphi). \qquad (6.1.76)$$

The above two formulas imply that the only nonzero matrix elements for operators $\hat{L}_+$ and $\hat{L}_-$ are $(\hat{L}_+)_{lm,l(m+1)}$ and $(\hat{L}_-)_{lm,l(m-1)}$, respectively. By using this fact, we derive from formula (6.1.74) that

$$\hbar^2 l(l+1) = (\hat{L}_+)_{lm,l(m+1)}(\hat{L}_-)_{l(m+1),lm} + \hbar^2 m^2 + \hbar^2 m. \qquad (6.1.77)$$

According to formulas (see Section 2.4 of Chapter 2)

$$\hat{L}_+ = \hat{L}_x + i\hat{L}_y, \quad \hat{L}_- = (\hat{L}_x - i\hat{L}_y), \qquad (6.1.78)$$

operators $\hat{L}_+$ and $\hat{L}_-$ are Hermitian adjoint. Consequently,

$$(\hat{L}_+)_{lm,l(m+1)} = (\hat{L}_-)_{l(m+1),lm}^*. \qquad (6.1.79)$$

By using this fact, we derive from formula (6.1.77) that

$$\left|(\hat{L}_+)_{lm,l(m+1)}\right|^2 = \hbar^2[l(l+1) - m(m+1)]. \qquad (6.1.80)$$

It can be shown that by properly scaling eigenfunctions ψ_{lm} matrix elements $(\hat{L}_+)_{lm,l(m+1)}$ can be made real and positive. This implies that

$$\boxed{(\hat{L}_+)_{lm,l(m+1)} = (\hat{L}_-)_{l(m+1),lm} = \hbar\sqrt{(l-m)(l+m+1)}.} \qquad (6.1.81)$$

Thus, the operators $\hat{L}_+$ and $\hat{L}_-$ can be represented in the matrix form as follows:

$$\hat{L}_+ = \hbar \begin{bmatrix} 0 & 0 & 0 & \cdots \\ & \begin{matrix} 0 & \sqrt{2} & 0 \\ 0 & 0 & \sqrt{2} \\ 0 & 0 & 0 \end{matrix} & 0 & \cdots \\ 0 & 0 & \begin{matrix} 0 & 2 & 0 & 0 & 0 \\ 0 & 0 & \sqrt{6} & 0 & 0 \\ 0 & 0 & 0 & \sqrt{6} & 0 \\ 0 & 0 & 0 & 0 & 2 \\ 0 & 0 & 0 & 0 & 0 \end{matrix} & \cdots \\ & \vdots & \vdots & \vdots & \end{bmatrix}, \tag{6.1.82}$$

$$\hat{L}_- = \hbar \begin{bmatrix} 0 & 0 & 0 & \cdots \\ & \begin{matrix} 0 & 0 & 0 \\ \sqrt{2} & 0 & 0 \\ 0 & \sqrt{2} & 0 \end{matrix} & 0 & \cdots \\ 0 & 0 & \begin{matrix} 0 & 0 & 0 & 0 & 0 \\ 2 & 0 & 0 & 0 & 0 \\ 0 & \sqrt{6} & 0 & 0 & 0 \\ 0 & 0 & \sqrt{6} & 0 & 0 \\ 0 & 0 & 0 & 2 & 0 \end{matrix} & \cdots \\ & \vdots & \vdots & \vdots & \end{bmatrix}. \tag{6.1.83}$$

From formula (6.1.78), it follows that

$$\hat{L}_x = \frac{1}{2}(\hat{L}_+ + \hat{L}_-), \quad \hat{L}_y = \frac{1}{2i}(\hat{L}_+ - \hat{L}_-). \tag{6.1.84}$$

This suggests that operators $\hat{L}_x$ and $\hat{L}_y$ can be represented in the matrix form as follows:

$$\hat{L}_x = \frac{\hbar}{2} \begin{bmatrix} 0 & 0 & 0 & \cdots \\ & \begin{matrix} 0 & \sqrt{2} & 0 \\ \sqrt{2} & 0 & \sqrt{2} \\ 0 & \sqrt{2} & 0 \end{matrix} & 0 & \cdots \\ 0 & 0 & \begin{matrix} 0 & 2 & 0 & 0 & 0 \\ 2 & 0 & \sqrt{6} & 0 & 0 \\ 0 & \sqrt{6} & 0 & \sqrt{6} & 0 \\ 0 & 0 & \sqrt{6} & 0 & 2 \\ 0 & 0 & 0 & 2 & 0 \end{matrix} & \cdots \\ & \vdots & \vdots & \vdots & \end{bmatrix}, \tag{6.1.85}$$

$$\hat{L}_y = \frac{\hbar}{2i} \begin{bmatrix} 0 & 0 & 0 & \cdots \\ & \begin{matrix} 0 & \sqrt{2} & 0 \\ -\sqrt{2} & 0 & \sqrt{2} \\ 0 & -\sqrt{2} & 0 \end{matrix} & & \\ 0 & & \begin{matrix} 0 & 2 & 0 & 0 & 0 \\ -2 & 0 & \sqrt{6} & 0 & 0 \\ 0 & -\sqrt{6} & 0 & \sqrt{6} & 0 \\ 0 & 0 & -\sqrt{6} & 0 & 2 \\ 0 & 0 & 0 & -2 & 0 \end{matrix} & \cdots \\ \vdots & \vdots & \vdots & \vdots \end{bmatrix} . \qquad (6.1.86)$$

Having discussed the matrix form of quantum mechanical operators, we shall conclude this section by deriving the matrix form of Schrödinger equations. We start with the time-dependent Schrödinger equation

$$i\hbar \frac{\partial \psi}{\partial t} = \hat{H}\psi. \qquad (6.1.87)$$

By expanding $\psi(\mathbf{r}, t)$ in some orthonormal basis $\{\psi_n(\mathbf{r})\}$, we obtain

$$\psi(\mathbf{r}, t) = \sum_n a_n(t)\psi_n(\mathbf{r}). \qquad (6.1.88)$$

By substituting the last formula into equation (6.1.87), we derive

$$i\hbar \sum_k \frac{da_k(t)}{dt} \psi_k(\mathbf{r}) = \sum_n a_n(t)\hat{H}\psi_n(\mathbf{r}), \qquad (6.1.89)$$

which leads to

$$i\hbar \frac{da_k(t)}{dt} = \sum_n \langle \psi_k | \hat{H}\psi_n \rangle a_n(t), \quad (k = 1, 2, \ldots). \qquad (6.1.90)$$

By introducing the notation

$$H_{kn} = \langle \psi_k | \hat{H} | \psi_n \rangle, \qquad (6.1.91)$$

formula (6.1.90) can be rewritten as follows

$$\boxed{i\hbar \frac{da_k(t)}{dt} = \sum_n H_{kn} a_n(t), \quad (k = 1, 2, \ldots).} \qquad (6.1.92)$$

The last equation can be written in compact form by introducing the matrix representation of the Hamiltonian in the basis $\{\psi_n(\mathbf{r})\}$

$$\hat{H} = \{H_{kn}\} \qquad (6.1.93)$$

and the infinite-dimensional vector representation $\mathbf{a}(t)$ of the wave function $\psi(\mathbf{r}, t)$:

$$\mathbf{a}(t) = \begin{pmatrix} a_1(t) \\ a_2(t) \\ \vdots \end{pmatrix}. \tag{6.1.94}$$

Now, the equation (6.1.92) can be written as follows

$$\boxed{i\hbar \frac{d\mathbf{a}(t)}{dt} = \hat{H}\mathbf{a}(t).} \tag{6.1.95}$$

Equations (6.1.92) and (6.1.95) can be viewed as the matrix form of the time-dependent Schrödinger equation (6.1.87).

Finally, consider the time-independent Schrödinger equation

$$\hat{H}\Phi_{\mathcal{E}} = \mathcal{E}\Phi_{\mathcal{E}}. \tag{6.1.96}$$

By using the expansion

$$\Phi_{\mathcal{E}} = \sum_n a_n \psi_n, \tag{6.1.97}$$

equation (6.1.96) can be written as follows

$$\sum_n a_n \hat{H}\psi_n = \mathcal{E}\sum_k a_k \psi_k, \tag{6.1.98}$$

which leads to

$$\sum_n \langle \psi_k | \hat{H} | \psi_n \rangle a_n = \mathcal{E} a_k, \quad (k = 1, 2, \dots). \tag{6.1.99}$$

By using the notation (6.1.91), equation (6.1.99) can be represented in the form

$$\boxed{\sum_n H_{kn} a_n = \mathcal{E} a_k, \quad (k = 1, 2, \dots),} \tag{6.1.100}$$

or in compact form

$$\boxed{\hat{H}\mathbf{a} = \mathcal{E}\mathbf{a}.} \tag{6.1.101}$$

Thus, the time-independent Schrödinger equation (6.1.96) is reduced to the eigenvalue problem for the matrix Hamiltonian.

6.2 Time-Independent Perturbation Theory

The perturbation technique is ubiquitous in the structure of quantum mechanics, and it represents an invaluable computational tool. There are two versions of the perturbation technique: time-independent perturbation theory and time-dependent perturbation theory. The time-independent perturbation theory was first **initiated by E. Schrödinger**, and it is used for the calculation of energy spectrum. The time-independent perturbation problem can be stated as follows.

Suppose we know the energy spectrum $\mathcal{E}_n^{(0)}$ and the corresponding eigenstates $\Phi_n^{(0)}$ for the Hamiltonian $\hat{H}_0$:

$$\hat{H}_0 \Phi_n^{(0)} = \mathcal{E}_n^{(0)} \Phi_n^{(0)}, \tag{6.2.1}$$

$$\langle \Phi_k^{(0)} | \Phi_n^{(0)} \rangle = \delta_{kn}, \tag{6.2.2}$$

and we want to find the energy spectrum $\mathcal{E}_n$ and the corresponding eigenstates Φ_n for the perturbed Hamiltonian

$$\hat{H} = \hat{H}_0 + \lambda \hat{H}', \tag{6.2.3}$$

where λ is a small parameter.

In other words, we want to find the solutions of the equation

$$\hat{H} \Phi_n = \mathcal{E}_n \Phi_n. \tag{6.2.4}$$

We first consider the nondegenerate case when all eigenvalues $\mathcal{E}_n^{(0)}$ are simple. The latter implies that there is only one linearly independent eigenstate $\Phi_n^{(0)}$ for each eigenvalue $\mathcal{E}_n^{(0)}$.

We shall look for solutions of equation (6.2.4) in the form of the following power series

$$\mathcal{E}_n = \mathcal{E}_n^{(0)} + \lambda \mathcal{E}_n^{(1)} + \lambda^2 \mathcal{E}_n^{(2)} + \cdots = \sum_{k=0} \lambda^k \mathcal{E}_n^{(k)}, \tag{6.2.5}$$

$$\Phi_n = \Phi_n^{(0)} + \lambda \Phi_n^{(1)} + \lambda^2 \Phi_n^{(2)} + \cdots = \sum_{k=0} \lambda^k \Phi_n^{(k)}. \tag{6.2.6}$$

By substituting formulas (6.2.3), (6.2.5) and (6.2.6) into equation (6.2.4), we arrive at:

$$(\hat{H}_0 + \lambda \hat{H}') \left(\sum_{k=0} \lambda^k \Phi_n^{(k)} \right) = \left(\sum_{k=0} \lambda^k \mathcal{E}_n^{(k)} \right) \left(\sum_{k=0} \lambda^k \Phi_n^{(k)} \right). \tag{6.2.7}$$

By equating the terms of the same powers in λ on both sides of equation (6.2.7), we obtain

$$\hat{H}_0 \Phi_n^{(0)} = \mathcal{E}_n^{(0)} \Phi_n^{(0)}, \tag{6.2.8}$$

$$\hat{H}_0 \Phi_n^{(1)} + \hat{H}' \Phi_n^{(0)} = \mathcal{E}_n^{(0)} \Phi_n^{(1)} + \mathcal{E}_n^{(1)} \Phi_n^{(0)}, \tag{6.2.9}$$

$$\hat{H}_0 \Phi_n^{(2)} + \hat{H}' \Phi_n^{(1)} = \mathcal{E}_n^{(0)} \Phi_n^{(2)} + \mathcal{E}_n^{(1)} \Phi_n^{(1)} + \mathcal{E}_n^{(2)} \Phi_n^{(0)}, \tag{6.2.10}$$

$$\hat{H}_0 \Phi_n^{(3)} + \hat{H}' \Phi_n^{(2)} = \mathcal{E}_n^{(0)} \Phi_n^{(3)} + \mathcal{E}_n^{(1)} \Phi_n^{(2)} + \mathcal{E}_n^{(2)} \Phi_n^{(1)} + \mathcal{E}_n^{(3)} \Phi_n^{(0)}. \tag{6.2.11}$$

The pattern of these equations is easily recognizable, and it can be expressed in the general form as follows:

$$\hat{H}_0 \Phi_n^{(k)} + \hat{H}' \Phi_n^{(k-1)} = \mathcal{E}_n^{(0)} \Phi_n^{(k)} + \sum_{j=1}^{k} \mathcal{E}_n^{(j)} \Phi_n^{(k-j)}. \tag{6.2.12}$$

Equation (6.2.8) coincides with equation (6.2.1) and its solutions are presumed to be known. By using these solutions, we can solve equation (6.2.9) to find $\mathcal{E}_n^{(1)}$ and $\Phi_n^{(1)}$. Then, we can use equation (6.2.10), to find $\mathcal{E}_n^{(2)}$ and $\Phi_n^{(2)}$, and so on. In this way, higher-order corrections to the eigenvalues $\mathcal{E}_n^{(0)}$ and eigenstates $\Phi_n^{(0)}$ can be sequentially computed. For the purposes of computations, it is convenient to rewrite the above equations as follows

$$\hat{H}_0 \Phi_n^{(1)} - \mathcal{E}_n^{(0)} \Phi_n^{(1)} = -\hat{H}' \Phi_n^{(0)} + \mathcal{E}_n^{(1)} \Phi_n^{(0)}, \tag{6.2.13}$$

$$\hat{H}_0 \Phi_n^{(2)} - \mathcal{E}_n^{(0)} \Phi_n^{(2)} = -\hat{H}' \Phi_n^{(1)} + \mathcal{E}_n^{(1)} \Phi_n^{(1)} + \mathcal{E}_n^{(2)} \Phi_n^{(0)}, \tag{6.2.14}$$

$$\hat{H}_0 \Phi_n^{(3)} - \mathcal{E}_n^{(0)} \Phi_n^{(3)} = -\hat{H}' \Phi_n^{(2)} + \mathcal{E}_n^{(1)} \Phi_n^{(2)} + \mathcal{E}_n^{(2)} \Phi_n^{(1)} + \mathcal{E}_n^{(3)} \Phi_n^{(0)}, \tag{6.2.15}$$

or, in general form, as

$$\boxed{\hat{H}_0 \Phi_n^{(k)} - \mathcal{E}_n^{(0)} \Phi_n^{(k)} = -\hat{H}' \Phi_n^{(k-1)} + \sum_{j=1}^{k} \mathcal{E}_n^{(j)} \Phi_n^{(k-j)}.} \tag{6.2.16}$$

All the above equations have the form

$$\hat{H}_0 \varphi - \mathcal{E}_n^{(0)} \varphi = f, \tag{6.2.17}$$

where the specific expressions for f depend on the order of corrections. The above equation is inhomogeneous and its solution does not exist for any f. This is because $\mathcal{E}_n^{(0)}$ is the eigenvalue of $\hat{H}_0$. It turns out, and this is demonstrated below, that equation (6.2.17) has a solution only for such right-hand sides f that are orthogonal to the eigenfunction $\Phi_n^{(0)}$, i.e., orthogonal to the solution of the corresponding homogeneous equation. This is the so-called **normal solvability** condition which is encountered in many areas of differential and integral equation theory (Fredholm theory is one example).

To establish the normal solvability of equation (6.2.17), we shall use the following expansions

$$\varphi = \sum_r a_r \Phi_r^{(0)}, \tag{6.2.18}$$

$$f = \sum_r b_r \Phi_r^{(0)}. \tag{6.2.19}$$

By substituting the last two formulas into equation (6.2.17) and taking into account that $\hat{H}\Phi_r^{(0)} = \mathcal{E}_r^{(0)}\Phi_r^{(0)}$, we obtain

$$\sum_{r \neq n} a_r(\mathcal{E}_r^{(0)} - \mathcal{E}_n^{(0)})\Phi_r^{(0)} = \sum_r b_r \Phi_r^{(0)}. \tag{6.2.20}$$

It is apparent from the last equation that

$$b_n = 0, \tag{6.2.21}$$

and, consequently,

$$f = \sum_{r \neq n} b_r \Phi_r^{(0)}. \tag{6.2.22}$$

From the last formula, we find that

$$\boxed{\langle \Phi_n^{(0)} | f \rangle = 0,} \tag{6.2.23}$$

which is the normal solvability condition.

Furthermore, from formula (6.2.20), it follows that

$$a_r = \frac{b_r}{\mathcal{E}_r^{(0)} - \mathcal{E}_n^{(0)}}. \tag{6.2.24}$$

On the other hand, the expansion (6.2.19) implies that

$$b_r = \langle \Phi_r^{(0)} | f \rangle. \tag{6.2.25}$$

Consequently,

$$a_r = \frac{\langle \Phi_r^{(0)} | f \rangle}{\mathcal{E}_r^{(0)} - \mathcal{E}_n^{(0)}}, \tag{6.2.26}$$

and, according to formula (6.2.18), a solution of inhomogeneous equation (6.2.17) can be written in the form

$$\boxed{\varphi = \sum_{r \neq n} \frac{\langle \Phi_r^{(0)} | f \rangle}{\mathcal{E}_r^{(0)} - \mathcal{E}_n^{(0)}} \Phi_r^{(0)}.} \tag{6.2.27}$$

It is apparent that this solution is not unique and that a general solution of equation (6.2.17) can be written as

$$\tilde{\varphi} = \alpha \Phi_n^{(0)} + \sum_{r \neq n} \frac{\langle \Phi_r^{(0)} | f \rangle}{\mathcal{E}_r^{(0)} - \mathcal{E}_n^{(0)}} \Phi_r^{(0)}, \qquad (6.2.28)$$

where α is an arbitrary constant.

Within the framework of the perturbation theory, the term $\alpha \Phi_n^{(0)}$ is usually neglected because when it is added to the first term in the perturbation expansion (6.2.6), it only increases the norm of Φ_n beyond the perturbation order in λ. Thus, we shall use formulas (6.2.23) and (6.2.27) for the calculation of perturbation corrections.

According to equation (6.2.13) for the first-order perturbations, we have

$$f = -\hat{H}' \Phi_n^{(0)} + \mathcal{E}_n^{(1)} \Phi_n^{(0)}. \qquad (6.2.29)$$

By using the normal solvability condition (6.2.23), from the last formula we derive

$$-\langle \Phi_n^{(0)} | \hat{H}' \Phi_n^{(0)} \rangle + \mathcal{E}_n^{(1)} = 0, \qquad (6.2.30)$$

which leads to

$$\boxed{\mathcal{E}_n^{(1)} = H'_{nn}.} \qquad (6.2.31)$$

This describes the first-order corrections for energy levels.

To find the first-order corrections for the corresponding eigenstates, we shall use formula (6.2.27) in which φ is identified with $\Phi_n^{(1)}$:

$$\Phi_n^{(1)} = \sum_{r \neq n} \frac{\langle \Phi_r^{(0)} | f \rangle}{\mathcal{E}_r^{(0)} - \mathcal{E}_n^{(0)}} \Phi_r^{(0)}. \qquad (6.2.32)$$

Next, by using formula (6.2.29), we find

$$\langle \Phi_r^{(0)} | f \rangle = -\langle \Phi_r^{(0)} | \hat{H}' | \Phi_n^{(0)} \rangle + \mathcal{E}_n^{(1)} \langle \Phi_r^{(0)} | \Phi_n^{(0)} \rangle, \qquad (6.2.33)$$

which leads to

$$\langle \Phi_r^{(0)} | f \rangle = -H'_{nr}. \qquad (6.2.34)$$

By substituting the last formula into equation (6.2.32), we arrive at the following expression for the first-order corrections for the eigenstates

$$\boxed{\Phi_n^{(1)} = \sum_{r \neq n} \frac{H'_{nr}}{\mathcal{E}_n^{(0)} - \mathcal{E}_r^{(0)}} \Phi_r^{(0)}.} \qquad (6.2.35)$$

Thus, within the framework of the first-order approximation of the perturbation theory, we have obtained the following expressions for energy levels and the corresponding eigenstates:

$$\boxed{\mathcal{E}_n = \mathcal{E}_n^{(0)} + \lambda H'_{nn},} \tag{6.2.36}$$

$$\boxed{\Phi_n = \Phi_n^{(0)} + \lambda \sum_{r \neq n} \frac{H'_{nr}}{\mathcal{E}_n^{(0)} - \mathcal{E}_r^{(0)}} \Phi_r^{(0)}.} \tag{6.2.37}$$

Next, we turn to the calculation of second-order corrections. According to equation (6.2.14), for the second-order corrections we have the following expressions for f:

$$f = -\hat{H}' \Phi_n^{(1)} + \mathcal{E}_n^{(1)} \Phi_n^{(1)} + \mathcal{E}_n^{(2)} \Phi_n^{(0)}. \tag{6.2.38}$$

By using the normal solvability condition (6.2.23), from the last formula we derive

$$\langle \Phi_n^{(0)} | f \rangle = -\langle \Phi_n^{(0)} | \hat{H}' \Phi_n^{(1)} \rangle + \mathcal{E}_n^{(1)} \langle \Phi_n^{(0)} | \Phi_n^{(1)} \rangle + \mathcal{E}_n^{(2)} \langle \Phi_n^{(0)} | \Phi_n^{(0)} \rangle = 0. \tag{6.2.39}$$

It is apparent according to formula (6.2.35) that

$$\langle \Phi_n^{(0)} | \Phi_n^{(1)} \rangle = 0. \tag{6.2.40}$$

Consequently, from equation (6.2.39), we obtain

$$\mathcal{E}_n^{(2)} = \langle \Phi_n^{(0)} | \hat{H}' \Phi_n^{(1)} \rangle. \tag{6.2.41}$$

It is clear from formula (6.2.35) that

$$\hat{H}' \Phi_n^{(1)} = \sum_{r \neq n} \frac{H'_{nr}}{\mathcal{E}_n^{(0)} - \mathcal{E}_r^{(0)}} \hat{H}' \Phi_r^{(0)}. \tag{6.2.42}$$

By substituting the last formula into equation (6.2.41), we arrive at the following second-order corrections for energy levels

$$\boxed{\mathcal{E}_n^{(2)} = \sum_{r \neq n} \frac{|H'_{nr}|^2}{\mathcal{E}_n^{(0)} - \mathcal{E}_r^{(0)}}.} \tag{6.2.43}$$

Thus, within the framework of the second-order approximation of the perturbation theory, we have obtained the following expressions for energy levels

$$\boxed{\mathcal{E}_n = \mathcal{E}_n^{(0)} + \lambda H'_{nn} + \lambda^2 \sum_{r \neq n} \frac{|H'_{nr}|^2}{\mathcal{E}_n^{(0)} - \mathcal{E}_r^{(0)}}.} \tag{6.2.44}$$

It is interesting to point out that the second-order correction for the ground state energy is always negative. Indeed, from formula (6.2.43) we find

$$\mathcal{E}_0^{(2)} = \sum_{n \neq r} \frac{|H'_{0r}|^2}{\mathcal{E}_0^{(0)} - \mathcal{E}_r^{(0)}}. \tag{6.2.45}$$

Since for the ground state energy level we have $\mathcal{E}_0^{(0)} < \mathcal{E}_r^{(0)}$, this means according to the last formula that

$$\boxed{\mathcal{E}_0^{(2)} < 0.} \tag{6.2.46}$$

By using formulas (6.2.38) and (6.2.27) and the same line of reasoning as in the derivation of formula (6.2.35), the following second-order correction for the eigenstates can be derived:

$$\boxed{\begin{aligned} \Phi_n^{(2)} = \sum_{r \neq n} \left(\sum_{q \neq n} \frac{H'_{rq} H'_{qn}}{(\mathcal{E}_r^{(0)} - \mathcal{E}_n^{(0)})(\mathcal{E}_q^{(0)} - \mathcal{E}_n^{(0)})} \right) \Phi_r^{(0)} - \\ \sum_{r \neq n} \frac{H'_{nn} H'_{rn}}{(\mathcal{E}_r^{(0)} - \mathcal{E}_n^{(0)})^2} \Phi_r^{(0)} - \frac{\Phi_n^{(0)}}{2} \sum_{r \neq n} \frac{|H'_{rn}|^2}{(\mathcal{E}_r^{(0)} - \mathcal{E}_n^{(0)})^2}. \end{aligned}} \tag{6.2.47}$$

The details of the derivation of the last formula are left as an exercise for the reader.

Finally, consider the calculation of the third-order corrections for energy levels. According to formula (6.2.15), we have the following expression for f in the case of third-order corrections

$$f = -\hat{H}' \Phi_n^{(2)} + \mathcal{E}_n^{(1)} \Phi_n^{(2)} + \mathcal{E}_n^{(2)} \Phi_n^{(1)} + \mathcal{E}_n^{(3)} \Phi_n^{(0)}. \tag{6.2.48}$$

By using the normal solvability condition (6.2.23) and the same line of reasoning as in the derivation of formula (6.2.43), the following formula can be obtained

$$\boxed{\mathcal{E}_n^{(3)} = \sum_{q \neq n} \sum_{r \neq n} \frac{H'_{nr} H'_{rq} H'_{qn}}{(\mathcal{E}_r^{(0)} - \mathcal{E}_n^{(0)})(\mathcal{E}_q^{(0)} - \mathcal{E}_n^{(0)})} - H'_{nn} \sum_{r \neq n} \frac{|H'_{nr}|^2}{(\mathcal{E}_r^{(0)} - \mathcal{E}_n^{(0)})^2}.} \tag{6.2.49}$$

Again, the details of the derivation of this formula are left as an exercise for the reader.

Now, we shall illustrate the presented time-independent perturbation technique by solving two problems related to the quantum mechanical harmonic oscillator. The first problem is the harmonic oscillator in the presence of an electric field. The Hamiltonian for this problem is given by the following formula

$$\hat{H} = -\frac{\hbar^2}{2m} \frac{d^2}{dx^2} + \frac{m\omega^2}{2} x^2 - eEx, \tag{6.2.50}$$

where E is the applied electric field.

This Hamiltonian can be represented as

$$\hat{H} = \hat{H}_0 + \hat{H}', \tag{6.2.51}$$

where

$$\hat{H}_0 = -\frac{\hbar^2}{2m}\frac{d^2}{dx^2} + \frac{m\omega^2}{2}x^2, \tag{6.2.52}$$

while

$$\hat{H}' = -eEx \tag{6.2.53}$$

and is regarded as a small perturbation of $\hat{H}_0$, which is certainly true if electric field E is sufficiently small. As discussed in Section 4.2 of Chapter 4, the energy levels $\mathcal{E}_n^{(0)}$ of unperturbed Hamiltonian H_0 are

$$\mathcal{E}_n^{(0)} = \hbar\omega(n + \frac{1}{2}), \quad (n = 0, 1, 2, \dots), \tag{6.2.54}$$

and the corresponding eigenstates $\Phi_n^{(0)}$ are expressed in terms of Hermite polynomials. We want to find the corrections to energy levels $\mathcal{E}_n^{(0)}$ due to the presence of an electric field. In other words, we want to find the corrections to $\mathcal{E}_n^{(0)}$ due to the perturbation $\hat{H}'$ of the Hamiltonian $\hat{H}_0$.

According to formulas (6.2.31) and (6.2.53), the first-order corrections can be computed as follows

$$\mathcal{E}_n^{(1)} = H'_{nn} = -eEx_{nn}. \tag{6.2.55}$$

By recalling the formula (6.1.68), we conclude that

$$x_{nn} = 0, \tag{6.2.56}$$

and, consequently,

$$\mathcal{E}_n^{(1)} = 0. \tag{6.2.57}$$

Next, we calculate the second-order corrections to $\mathcal{E}_n^{(0)}$ by using formula (6.2.43). In our case, this formula can be written as follows

$$\mathcal{E}_n^{(2)} = (eE)^2 \sum_{r \neq n} \frac{|x_{nr}|^2}{\mathcal{E}_n^{(0)} - \mathcal{E}_r^{(0)}}. \tag{6.2.58}$$

According to formula (6.1.68), we find

$$|x_{nr}|^2 = \frac{\hbar}{2m\omega}[r\delta_{n,r-1} + (r+1)\delta_{n,r+1}]. \tag{6.2.59}$$

Consequently, by using equations (6.2.54), (6.2.58) and (6.2.59), we derive

$$\mathcal{E}_n^{(2)} = \frac{(eE)^2\hbar}{2m\omega}\left[-\frac{n+1}{\hbar\omega} + \frac{n}{\hbar\omega}\right], \tag{6.2.60}$$

which leads to

$$\mathcal{E}_n^{(2)} = -\frac{(eE)^2}{2m\omega^2}.$$ (6.2.61)

As expected, the second-order correction is negative. It is remarkable that it does not depend on n; it is the same for any n.

By using formulas (6.2.54) and (6.2.61), we arrive at the following expression for the perturbed energy levels $\mathcal{E}_n$:

$$\mathcal{E}_n = \mathcal{E}_n^{(0)} + \mathcal{E}_n^{(2)} = \hbar\omega\left(n + \frac{1}{2}\right) - \frac{(eE)^2}{2m\omega^2}.$$ (6.2.62)

It turns out that the last formula is absolutely accurate. This can be demonstrated as follows. By using a new variable

$$\eta = x - \frac{eE}{m\omega^2},$$ (6.2.63)

the Hamiltonian (6.2.50) can be transformed as

$$\hat{H} = -\frac{\hbar^2}{2m}\frac{d^2}{2\eta^2} + \frac{m\omega^2}{2}\eta^2 - \frac{(eE)^2}{2m\omega^2},$$ (6.2.64)

which is equivalent to

$$\hat{H} = \hat{H}_0 - \frac{(eE)^2}{2m\omega^2}.$$ (6.2.65)

Thus, it is apparent that the Hamiltonian $\hat{H}$ is identical up to the constant $-\frac{(eE)^2}{2m\omega^2}$ with the Hamiltonian $\hat{H}_0$. Consequently, all its eigenvalues $\mathcal{E}_n$ are shifted from $\mathcal{E}_n^{(0)}$ by this constant, and they are given by formula (6.2.62), which proves that the perturbation technique leads in this particular case to the absolutely accurate result. The shifting of energy levels due to the presence of electric field is called the Stark effect. When the energy levels are degenerate (which is not the case for one-dimensional problems), the Stark effect also manifests itself in splitting degenerate energy levels. This is because the presence of electric field breaks a symmetry which is usually responsible for the existence of degeneracy of energy levels.

In the above case of the harmonic oscillator, one deals with the second-order Stark effect because the shifting of energy levels is proportional to the square E^2 of the applied electric field. The Stark effect can be viewed as the electric analogue of the Zeeman effect when degenerate energy levels are split into several energy levels due to the presence of a magnetic field. Both the Stark and Zeeman effects are essential in spectroscopy studies of atoms and molecules.

Next, we shall use the time-independent perturbation technique for the calculation of the energy spectrum of the anharmonic oscillator. The Hamiltonian for this oscillator is given by the formula

$$\hat{H} = -\frac{\hbar^2}{2m}\frac{d^2}{dx^2} + \frac{m\omega^2}{2}x^2 + \lambda x^4, \tag{6.2.66}$$

where λ is a small parameter.

It is clear that this Hamiltonian can be written as

$$\hat{H} = \hat{H}_0 + \lambda \hat{H}', \tag{6.2.67}$$

where $\hat{H}_0$ is defined as before by formula (6.2.52), while

$$\hat{H}' = \hat{x}^4. \tag{6.2.68}$$

According to formula (6.2.31), the first-order corrections to the energy levels $\mathcal{E}_n^{(0)}$ specified by equation (6.2.54) can be computed as

$$\mathcal{E}_n^{(1)} = (\hat{x}^4)_{nn}. \tag{6.2.69}$$

Thus, we have to evaluate the diagonal elements of the infinite-dimensional matrix of operator $\hat{x}^4$ in the energy representation, i.e., in the representation of the basis function $\{\Phi_n^0\}$ which are the eigenfunctions of the Hamiltonian $\hat{H}_0$. We shall do this by using the raising $\hat{a}^+$ and lowering $\hat{a}$ operators. According to formulas (4.2.15), (4.2.17) and (4.2.18) from Chapter 4, we have

$$\hat{x} = \sqrt{\frac{\hbar}{2m\omega}}(\hat{a} + \hat{a}^+). \tag{6.2.70}$$

Furthermore, from formulas (4.2.48) and (4.2.55) follows that

$$\boxed{\hat{a}^+\Phi_n^{(0)} = \sqrt{n+1}\,\Phi_{n+1}^{(0)}.} \tag{6.2.71}$$

From the last equation and formula

$$\hat{H}_0\Phi_n^{(0)} = \hbar\omega\left[\hat{a}\hat{a}^+ - \frac{1}{2}\right]\Phi_n^{(0)} = \hbar\omega\left(n + \frac{1}{2}\right)\Phi_n^{(0)} \tag{6.2.72}$$

follows that

$$\boxed{\hat{a}\Phi_{n+1}^{(0)} = \sqrt{n+1}\,\Phi_n^{(0)}.} \tag{6.2.73}$$

Formulas (6.2.71) and (6.2.73) are central to our calculations. Indeed, according to formula (6.2.70), we have

$$(\hat{x}^4)_{nn} = \left(\frac{\hbar}{2m\omega}\right)^2 \langle\Phi_n^{(0)}|(\hat{a} + \hat{a}^+)^4\Phi_n^{(0)}\rangle. \tag{6.2.74}$$

Next, we present operator $(\hat{a} + \hat{a}^+)^4$ as a sum of products of raising and lowering operators. It is clear according to formulas (6.2.71) and (6.2.73) that **only terms in this sum which contain an equal number of raising $\hat{a}^+$ and lowering $\hat{a}$ operators provide nonzero contribution to the inner product in formula** (6.2.74). Consequently,

$$\langle \Phi_n^{(0)} | (\hat{a} + \hat{a}^+)^4 \Phi_n^{(0)} \rangle$$
$$= \langle \Phi_n^{(0)} | [\hat{a}^2(\hat{a}^+)^2 + (\hat{a}\hat{a}^+)^2 + \hat{a}(\hat{a}^+)^2\hat{a} + \hat{a}^+\hat{a}^2\hat{a}^+ + (\hat{a}^+\hat{a})^2$$
$$+ (\hat{a}^+)^2\hat{a}^2] \Phi_n^{(0)} \rangle. \tag{6.2.75}$$

By using formulas (6.2.71) and (6.2.73), we derive:

$$\hat{a}^2(\hat{a}^+)^2 \Phi_n^{(0)} = \sqrt{n+1} \hat{a}^2 \hat{a}^+ \Phi_{n+1}^{(0)} = \sqrt{(n+1)(n+2)} \hat{a}^2 \Phi_{n+2}^{(0)}$$
$$= \sqrt{n+1}(n+2)\hat{a}\Phi_{n+1}^{(0)} = (n+1)(n+2)\Phi_n^{(0)}. \tag{6.2.76}$$

Similarly, the following formulas can be derived

$$(\hat{a}\hat{a}^+)^2 \Phi_n^{(0)} = (n+1)^2 \Phi_n^{(0)}, \tag{6.2.77}$$

$$\hat{a}(\hat{a}^+)^2\hat{a} \Phi_n^{(0)} = n(n+1) \Phi_n^{(0)}, \tag{6.2.78}$$

$$\hat{a}^+\hat{a}^2\hat{a}^+ \Phi_n^{(0)} = n(n+1) \Phi_n^{(0)}, \tag{6.2.79}$$

$$(\hat{a}^+\hat{a})^2 \Phi_n^{(0)} = n^2 \Phi_n^{(0)}, \tag{6.2.80}$$

$$(\hat{a}^+)^2\hat{a}^2 \Phi_n^{(0)} = n(n-1) \Phi_n^{(0)}. \tag{6.2.81}$$

From formulas (6.2.75)–(6.2.81), it follows that

$$\langle \Phi_n^{(0)} | (\hat{a} + \hat{a}^+)^4 \Phi_n^{(0)} \rangle = (n+1)(n+2) + (n+1)^2$$
$$+ 2n(n+1) + n^2 + n(n-1) = 3(2n^2 + 2n + 1). \tag{6.2.82}$$

By combining (6.2.69), (6.2.74) and (6.2.82), we find

$$\mathcal{E}_n^{(1)} = 3 \left(\frac{\hbar}{2m\omega} \right)^2 (2n^2 + 2n + 1), \tag{6.2.83}$$

which leads to the following expression for energy levels of anharmonic oscillator (6.2.66):

$$\boxed{\mathcal{E}_n = \mathcal{E}_n^{(0)} + \lambda \mathcal{E}_n^{(1)} = \hbar\omega \left(n + \frac{1}{2} \right) + 3\lambda \left(\frac{\hbar}{2m\omega} \right)^2 (2n^2 + 2n + 1).}$$
$$\tag{6.2.84}$$

We have discussed the calculations of the energy spectrum of the anharmonic oscillator when the potential energy $U(x) = \frac{m\omega^2}{2}x^2 + \lambda x^4$ has even symmetry. The same technique can be used for the energy spectrum calculation of the anharmonic oscillator with the Hamiltonian

$$\hat{H} = -\frac{\hbar^2}{2m}\frac{d^2}{dx^2} + \frac{m\omega^2}{2}x^2 + \lambda x^3, \tag{6.2.85}$$

that is, when the potential energy is not symmetric. In this case

$$\hat{H}' = \hat{x}^3, \tag{6.2.86}$$

and

$$\mathcal{E}_n^{(1)} = (\hat{x}^3)_{nn}. \tag{6.2.87}$$

By using the formula (6.2.70) and the same line of reasoning as before, it can be shown that

$$(\hat{x}^3)_{nn} = 0, \tag{6.2.88}$$

and, consequently,

$$\mathcal{E}_n^{(1)} = 0. \tag{6.2.89}$$

To calculate the second-order corrections, the formula

$$\mathcal{E}_n^{(2)} = \lambda^2 \sum_{r \neq n} \frac{|(\hat{x}^3)_{rn}|^2}{\mathcal{E}_n^{(0)} - \mathcal{E}_r^{(0)}} \tag{6.2.90}$$

must be used. Again, by using relation (6.2.70) and the same line of reasoning as before it can be demonstrated that all matrix elements $(\hat{x}^3)_{nr}$ are equal to zero, except for the following four which are given by the formulas

$$(\hat{x}^3)_{n-1,n} = (\hat{x}^3)_{n,n-1} = 3\left(\frac{\hbar}{2m\omega}\right)^{\frac{3}{2}} n^{\frac{3}{2}}, \tag{6.2.91}$$

$$(\hat{x}^3)_{n-3,n} = (\hat{x}^3)_{n,n-3} = \left(\frac{\hbar}{2m\omega}\right)^{\frac{3}{2}} \sqrt{n(n-1)(n-2)}. \tag{6.2.92}$$

This leads according to formula (6.2.90) to the following results

$$\mathcal{E}_n^{(2)} = -\frac{15}{4}\frac{\lambda^2}{\hbar\omega}\left(\frac{\hbar}{m\omega}\right)^3\left(n^2 + n + \frac{11}{30}\right) \tag{6.2.93}$$

and

$$\boxed{\mathcal{E}_n = \hbar\omega\left(n + \frac{1}{2}\right) - \frac{15}{4}\frac{\lambda^2}{\hbar\omega}\left(\frac{\hbar}{m\omega}\right)^3\left(n^2 + n + \frac{11}{30}\right).} \tag{6.2.94}$$

The details of derivations of formulas (6.2.88), (6.2.91), (6.2.92) and (6.2.93) are left as an exercise for the reader.

Up to this point, we have discussed the time-independent perturbation theory in the case when all eigenvalues $\mathcal{E}_n^{(0)}$ of the unperturbed Hamiltonian were assumed to be simple, i.e., non-degenerate. Now, we extend this theory to cover the situation when these eigenvalues can be degenerate. Namely, consider degenerate eigenvalue $\mathcal{E}_n^{(0)}$ and assume that there are N linearly independent eigenfunctions $\Phi_{n,r}^{(0)}$ corresponding to this eigenvalue:

$$\hat{H}_0\Phi_{n,r}^{(0)} = \mathcal{E}_n^{(0)}\Phi_{n,r}^{(0)}, \quad (r = 1, 2, \ldots N). \tag{6.2.95}$$

We are interested in the perturbed problem

$$(\hat{H}_0 + \lambda\hat{H}')\Phi_n = \mathcal{E}_n\Phi_n. \tag{6.2.96}$$

We look for solutions of this problem in the form

$$\Phi_n = \sum_{r=1}^{N} a_r\Phi_{n,r}^{(0)} + \lambda\Phi_n^{(1)} \tag{6.2.97}$$

and

$$\mathcal{E}_n = \mathcal{E}_n^{(0)} + \lambda\mathcal{E}_n^{(1)}. \tag{6.2.98}$$

It is apparent from formula (6.2.97) that in the limit of λ going to zero, Φ_n is reduced to

$$\Phi_n^{(0)} = \sum_{r=1}^{N} a_r\Phi_{n,r}^{(0)}. \tag{6.2.99}$$

This implies that $\Phi_n^{(0)}$ rather than $\Phi_{n,r}^{(0)}$ represents the zero order approximation for Φ_n. Our immediate goal is to discuss how to find coefficients a_r as well as $\mathcal{E}_n^{(1)}$. To this end, we substitute formulas (6.2.97) and (6.2.98) into equation (6.2.96) and equate the terms of first-order of smallness with respect to λ. This leads to the following equation

$$\hat{H}_0\Phi_n^{(1)} - \mathcal{E}_n^{(0)}\Phi_n^{(1)} = f, \tag{6.2.100}$$

where

$$f = -\hat{H}'\left(\sum_{r=1}^{N} a_r\Phi_{n,r}^{(0)}\right) + \mathcal{E}_n^{(1)}\sum_{r=1}^{N} a_r\Phi_{n,r}^{(0)}. \tag{6.2.101}$$

Since $\mathcal{E}_n^{(0)}$ is the eigenvalue of operator $\hat{H}_0$, equation (6.2.100) has solutions (i.e., is solvable) only for such f that are orthogonal to all linearly independent solutions of homogeneous equation (6.2.95):

$$\langle\Phi_{n,k}^{(0)}|f\rangle = 0, \quad (k = 1, 2, \ldots N). \tag{6.2.102}$$

This is the extension of the normal solvability principle to the case of degenerate eigenvalue. The proof of this generalization is very similar to the proof of normal solvabililty condition (6.2.23).

From formulas (6.2.101) and (6.2.102), we find

$$\langle \Phi_{n,k}^{(0)} | \sum_{r=1}^{N} a_r \hat{H}' \Phi_{n,r}^{(0)} \rangle = \mathcal{E}_n^{(1)} a_k, \quad (k = 1, 2, \ldots N). \tag{6.2.103}$$

By introducing, as before, the notation

$$H_{kr}'^{(n)} = \langle \Phi_{n,k}^{(0)} | \hat{H}' | \Phi_{n,r}^{(0)} \rangle \tag{6.2.104}$$

equations (6.2.103) can be written as follows

$$\boxed{\sum_{r=1}^{N} H_{kr}'^{(n)} a_r = \mathcal{E}_n^{(1)} a_k, \quad (k = 1, \ldots N).} \tag{6.2.105}$$

This is the eigenvalue problem for N-dimensional matrix

$$\hat{H}'^{(n)} = \{ H_{kr}'^{(n)} \} \tag{6.2.106}$$

and it can be written in the following compact form

$$\boxed{\hat{H}'^{(n)} \mathbf{a} = \mathcal{E}_n^{(1)} \mathbf{a}.} \tag{6.2.107}$$

This eigenvalue problem may have N distinct eigenvalues

$$\mathcal{E}_{n,m}^{(1)}, \quad (m = 1, 2, \ldots N), \tag{6.2.108}$$

which according to formula (6.2.98) lead to N distinct eigenvalues of the perturbed Hamiltonian

$$\mathcal{E}_{n,m} = \mathcal{E}_n^{(0)} + \lambda \mathcal{E}_{n,m}^{(1)}, \quad (m = 1, 2, \ldots N). \tag{6.2.109}$$

In physical language, this means that the perturbation results in energy level splitting.

The eigenvalue problem (6.2.107) may also have N distinct eigenvectors

$$\mathbf{a}^{(m)} = \begin{pmatrix} a_1^{(m)} \\ a_2^{(m)} \\ \vdots \\ a_N^{(m)} \end{pmatrix} \tag{6.2.110}$$

and this leads to N distinct zero-order eigenstates

$$\Phi_{n,m}^{(0)} = \sum_{r=1}^{N} a_r^{(m)} \Phi_{n,r}^{(0)}, \quad (m = 1, 2, \ldots N) \tag{6.2.111}$$

corresponding to eigenvalues $\mathcal{E}_{n,m}$. The eigenstates given by formula (6.2.97) are reduced to $\Phi_{n,m}^{(0)}$ in the limit of zero λ.

The perturbation technique for degenerate energy levels is extensively used in quantum mechanics to study the energy level splitting in atoms and molecules when perturbations break symmetries responsible for the degeneracy of energy levels. The Zeeman and Stark effects in atoms and molecules are, for instance, studied by using the degenerate perturbation theory. In this text, we have already used this theory in Section 4.5 of Chapter 4 when we discussed a simple model for energy gap formation in superconductors. This energy gap formation and the creation of the ground state were viewed as energy level splitting for a multiple degenerate energy level; the splitting is caused by a small attractive interaction between electrons in superconductors responsible for Cooper pair formation.

6.3 Time-Dependent Perturbations. Adiabatic Perturbations and Berry's Phase

In the previous section, we discussed time-independent perturbations which result in changes of energy spectrum of quantum mechanical systems. In this section, we shall discuss time-dependent perturbations which may cause the transitions in quantum mechanical systems. Time-dependent perturbation theory **was first initiated by P. Dirac** and applied to the study of interactions between atoms and electromagnetic radiation.

In time-dependent perturbation theory, the Hamiltonians of quantum mechanical systems are divided into two distinct parts

$$\hat{H} = \hat{H}_0 + \hat{V}(t). \tag{6.3.1}$$

The first part $\hat{H}_0$ is time-independent, and it is usually assumed that its eigenvalues (energy spectrum) and corresponding eigenstates (stationary states) are known. Those energy levels $\mathcal{E}_k$ and stationary states $\psi_k^{(0)}(\mathbf{r}, t)$ are defined by the equations

$$\hat{H}_0 \Phi_k^{(0)} = \mathcal{E}_k \Phi_k^{(0)}, \tag{6.3.2}$$

$$\psi_k^{(0)}(\mathbf{r}, t) = \Phi_k^{(0)}(\mathbf{r}) e^{-\frac{i}{\hbar}\mathcal{E}_k t}, \tag{6.3.3}$$

and

$$i\hbar \frac{\partial \psi_k^{(0)}(\mathbf{r}, t)}{\partial t} = \hat{H}_0 \psi_k^{(0)}(\mathbf{r}, t). \tag{6.3.4}$$

The second part $\hat{V}(t)$ of the Hamiltonian $\hat{H}$ (see formula (6.3.1)) is time-dependent, and it is assumed that it is equal to zero at $t < 0$. In other words, the physical interaction described by $\hat{V}(t)$ is turned on at $t = 0$. This time-dependent part of the Hamiltonian $\hat{H}$ is viewed as perturbation, and it is usually due to the presence of external fields. It is typically assumed that at $t < 0$, that is, before the interaction $\hat{V}(t)$ has been turned on, the quantum mechanical system was in one of the stationary states $\psi_l^{(0)}(\mathbf{r}, t)$, i.e.,

$$\psi(\mathbf{r}, t) = \psi_l^{(0)}(\mathbf{r}, t) \text{ for } t \leq 0, \tag{6.3.5}$$

and we want to study the time-evolution of the state $\psi(\mathbf{r}, t)$ of the quantum mechanical system under the perturbation $\hat{V}(t)$. It is easy to conclude that the time-evolving state $\psi(\mathbf{r}, t)$ is the solution of the following initial-value problem (Cauchy problem) for the time-dependent Schrödinger equation

$$i\hbar \frac{\partial \psi(\mathbf{r}, t)}{\partial t} = [\hat{H}_0 + \hat{V}(t)] \psi(\mathbf{r}, t), \quad (t > 0), \tag{6.3.6}$$

$$\psi(\mathbf{r}, t) \Big|_{t=0} = \Phi_l^{(0)}(\mathbf{r}). \tag{6.3.7}$$

To solve this initial-value problem, the set of stationary states of unperturbed Hamiltonian $\hat{H}_0$

$$\{\psi_k^{(0)}(\mathbf{r}, t)\} \tag{6.3.8}$$

will be used as the basis for the expansion of $\psi(\mathbf{r}, t)$. Namely, we represent $\psi(\mathbf{r}, t)$ as follows

$$\psi(\mathbf{r}, t) = \sum_k c_k(t) \psi_k^{(0)}(\mathbf{r}, t). \tag{6.3.9}$$

We are going to derive the coupled ordinary differential equations for the expansion coefficients $c_k(t)$. Before proceeding with this derivation, we first point out the physical meaning of the expansion coefficients $c_k(t)$. It is clear from the last formula that $|c_k(t)|^2$ can be viewed as the probability that the result of energy measurement at time t is $\mathcal{E}_k$. Namely,

$$|c_k(t)|^2 = \text{Probability } \{\mathcal{E}_{measured} = \mathcal{E}_k \text{ at } t\}. \tag{6.3.10}$$

This probability is denoted as $P_{lk}(t)$. Thus,

$$P_{lk}(t) = |c_k(t)|^2. \tag{6.3.11}$$

In literature, $P_{lk}(t)$ is usually called the transition probability from state l to state k during the time t. This terminology is somewhat ambiguous. Indeed, the actual state of the quantum mechanical system at time t is $\psi(\mathbf{r}, t)$ rather than the stationary state $\psi_k^{(0)}(\mathbf{r}, t)$. The state $\psi(\mathbf{r}, t)$ may randomly collapse to the stationary state $\psi_k^{(0)}(\mathbf{r}, t)$ with probability $P_{lk}(t)$ as a result of the measurement process, but not as a result of perturbation $\hat{V}(t)$. In this sense, the precise meaning of $P_{lk}(t)$ is specified by formula (6.3.10).

Now, we proceed to the derivation of differential equations for $c_k(t)$. To this end, we substitute formula (6.3.9) into equation (6.3.6) and, after simple transformations, we obtain

$$i\hbar \sum_k \frac{dc_k(t)}{dt} \psi_k^{(0)}(\mathbf{r}, t) + i\hbar \sum_k c_k(t) \frac{\partial \psi_k^{(0)}(\mathbf{r}, t)}{\partial t}$$
$$= \sum_k c_k(t) \hat{H}_0 \psi_k^{(0)}(\mathbf{r}, t) + \sum_k c_k(t) \hat{V}(t) \psi_k^{(0)}(\mathbf{r}, t). \qquad (6.3.12)$$

Next, we observe that according to equation (6.3.4) the second and third terms in the last formula cancel out. Consequently,

$$i\hbar \sum_k \frac{dc_k(t)}{dt} \psi_k^{(0)}(\mathbf{r}, t) = \sum_k c_k(t) \hat{V}(t) \psi_k^{(0)}(\mathbf{r}, t). \qquad (6.3.13)$$

Since

$$\langle \psi_n^{(0)}(\mathbf{r}, t) | \psi_k^{(0)}(\mathbf{r}, t) \rangle = \delta_{nk}, \qquad (6.3.14)$$

from equation (6.3.13) it follows that

$$i\hbar \frac{dc_n(t)}{dt} = \sum_k c_k(t) \langle \psi_n^{(0)}(\mathbf{r}, t) | \hat{V}(t) \psi_k^{(0)}(\mathbf{r}, t) \rangle, \quad (n = 1, 2, \dots). \qquad (6.3.15)$$

From formula (6.3.3), we find that

$$\langle \psi_n^{(0)}(\mathbf{r}, t) | \hat{V}(t) \psi_k^{(0)}(\mathbf{r}, t) \rangle = \langle \Phi_n^{(0)}(\mathbf{r}) | \hat{V}(t) \Phi_k^{(0)}(\mathbf{r}) \rangle e^{i \frac{\mathcal{E}_n - \mathcal{E}_k}{\hbar} t}. \qquad (6.3.16)$$

By introducing notations

$$\omega_{nk} = \frac{\mathcal{E}_n - \mathcal{E}_k}{\hbar}, \qquad (6.3.17)$$

and

$$V_{nk}(t) = \langle \Phi_n^{(0)}(\mathbf{r}) | \hat{V}(t) \Phi_k^{(0)}(\mathbf{r}) \rangle, \qquad (6.3.18)$$

from formulas (6.3.15), (6.3.16), (6.3.17) and (6.3.18), we derive the following Cauchy problem

$$\boxed{i\hbar \frac{dc_n(t)}{dt} = \sum_k e^{i\omega_{nk}t} V_{nk}(t) c_k(t), \quad (n = 1, 2, \dots),} \qquad (6.3.19)$$

$$\boxed{c_k(0) = \delta_{kl}.}$$ (6.3.20)

The initial condition (6.3.20) follows from the initial condition (6.3.7) and formula (6.3.9), which imply that

$$\sum_k c_k(0)\Phi_k^{(0)}(\mathbf{r}) = \Phi_l^{(0)}(\mathbf{r}).$$ (6.3.21)

The latter is equivalent to the initial condition (6.3.20).

It is worthwhile to point out that the transformation of the Cauchy problem (6.3.6)–(6.3.7) for the wave function $\psi(\mathbf{r}, t)$ into the Cauchy problem (6.3.19)–(6.3.20) for the expansion coefficients $c_k(t)$ is exact, i.e., it is not based on neglecting any small terms or quantities. Equations (6.3.19) are a set of coupled linear ordinary differential equations with variable in time coefficients. The latter makes the analytical solution of these equations by and large impossible. There is one special case when the exact analytical solution of these equations can be carried out. This is the case of **two-level** (also known as **two-state**) quantum mechanical systems. A multi-level quantum mechanical system can be approximately treated as a two-level system when the two energy levels of interest are appreciably separated (in energy) from other energy levels. The model of two-level systems is extensively used in quantum optics and lasers. Furthermore, two-level quantum mechanical systems are at the very foundation of quantum computing. Indeed, qubits, which are the main building blocks of a quantum computer are two-state quantum mechanical systems.

For two-level systems, the differential equations (6.3.19) can be written as follows

$$i\hbar\frac{dc_1(t)}{dt} = V_{11}(t)c_1(t) + e^{-i\omega_0 t}V_{12}(t)c_2(t),$$ (6.3.22)

$$i\hbar\frac{dc_2(t)}{dt} = e^{i\omega_0 t}V_{21}(t)c_1(t) + V_{22}(t)c_2(t),$$ (6.3.23)

where

$$\omega_0 = \frac{\mathcal{E}_2 - \mathcal{E}_1}{\hbar},$$ (6.3.24)

and

$$V_{12}(t) = V_{21}^*(t),$$ (6.3.25)

because $\hat{V}(t)$ is a Hermitian operator.

It is often the case that due to symmetry considerations, diagonal matrix elements naturally vanish

$$V_{11} = V_{22} = 0,$$ (6.3.26)

or they can be treated in some other ways. Furthermore, we shall assume that off-diagonal matrix elements have the form

$$V_{12}(t) = \frac{V_0}{2} e^{i\omega t}, \tag{6.3.27}$$

$$V_{21}(t) = \frac{V_0}{2} e^{-i\omega t}. \tag{6.3.28}$$

This assumption is the so-called **rotating wave approximation**. Under this assumption, the coupled differential equations (6.3.22)–(6.3.23) can be written as follows

$$\frac{dc_1(t)}{dt} = -\frac{i}{2\hbar} V_0 e^{i(\omega-\omega_0)t} c_2(t), \tag{6.3.29}$$

$$\frac{dc_2(t)}{dt} = -\frac{i}{2\hbar} V_0 e^{-i(\omega-\omega_0)t} c_1(t), \tag{6.3.30}$$

and the exact analytical solution of these equations can be carried out. Before proceeding with this solution, it is worthwhile to mention that $\Delta = \omega - \omega_0$ is often called detuning frequency. This terminology becomes apparent from Figure 6.1, representing the two-level system under consideration. The smallness of detuning frequency is required for accurate treatment of multi-level systems as two-level systems. To find the analyt-

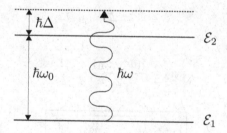

Fig. 6.1

ical solution of equations (6.3.29)–(6.3.30), we shall first rewrite equation (6.3.29) as follows:

$$\frac{dc_1(t)}{dt} e^{-i(\omega-\omega_0)t} = -\frac{i}{2\hbar} V_0 c_2(t). \tag{6.3.31}$$

By differentiating the last equation with respect to time, we find

$$\frac{d^2c_1(t)}{dt^2} e^{-i(\omega-\omega_0)t} - i(\omega - \omega_0)\frac{dc_1(t)}{dt} e^{-i(\omega-\omega_0)t} = -\frac{i}{2\hbar} V_0 \frac{dc_2(t)}{dt}. \tag{6.3.32}$$

By substituting formula (6.3.30) for $\frac{dc_2(t)}{dt}$ into the last equation and by subsequently canceling the exponential factor $e^{i(\omega-\omega_0)t}$ we arrive at the following second-order linear homogeneous differential equation with constant coefficients for $c_1(t)$:

$$\frac{d^2c_1(t)}{dt^2} - i(\omega - \omega_0)\frac{dc_1(t)}{dt} + \frac{V_0^2}{4\hbar^2}c_1(t) = 0. \qquad (6.3.33)$$

We want to find the solution of the above equation subject to the initial conditions

$$c_1(0) = 1, \qquad (6.3.34)$$

$$\left.\frac{dc_1}{dt}\right|_{t=0} = 0. \qquad (6.3.35)$$

These initial conditions follow from formulas (6.3.20) and (6.3.29).

By looking for a solution to equation (6.3.33) in the form

$$c_1(t) = Ae^{\alpha t}, \qquad (6.3.36)$$

we arrive at the following quadratic equation for α:

$$\alpha^2 - i(\omega - \omega_0)\alpha + \frac{V_0^2}{4\hbar^2} = 0. \qquad (6.3.37)$$

This equation has two solutions

$$\alpha_1 = \frac{i(\omega - \omega_0)}{2} + i\frac{\Omega}{2}, \qquad (6.3.38)$$

and

$$\alpha_2 = \frac{i(\omega - \omega_0)}{2} - i\frac{\Omega}{2}, \qquad (6.3.39)$$

where

$$\boxed{\Omega = \sqrt{(\omega - \omega_0)^2 + \frac{V_0^2}{\hbar^2}},} \qquad (6.3.40)$$

and it is called the **Rabi flopping frequency**. Thus a general solution of differential equation (6.3.33) can be written as follows

$$c_1(t) = e^{\frac{i(\omega-\omega_0)}{2}t}[A_1 e^{i\frac{\Omega}{2}t} + A_2 e^{-i\frac{\Omega}{2}t}]. \qquad (6.3.41)$$

Coefficients A_1 and A_2 can be found by using initial conditions (6.3.34) and (6.3.35). This is left as an exercise for the reader. Here, we just present the final expressions for $|c_1(t)|^2$ and $|c_2(t)|^2$. These expressions are as follows

$$|c_1(t)|^2 = 1 - \frac{V_0^2}{\hbar^2\Omega^2}\sin^2\frac{\Omega}{2}t, \qquad (6.3.42)$$

$$P_{12} = |c_2(t)|^2 = \frac{V_0^2}{\hbar^2 \Omega^2} \sin^2 \frac{\Omega}{2} t. \tag{6.3.43}$$

It is clear from the last formula that P_{12} oscillates with frequency Ω. These oscillations are called **Rabi oscillations**. It is also apparent according to formula (6.3.40) that

$$\frac{V_0^2}{\hbar^2 \Omega^2} = 1, \tag{6.3.44}$$

when the detuning $\Delta = \omega - \omega_0$ is equal to zero. In this case, P_{12} oscillates between zero and one. As the detuning is increased, the amplitudes of oscillations are decreased, and the probability of transition gets smaller and smaller.

Now, we shall turn to equations (6.3.19). As mentioned before, in general it is not possible to solve these equations analytically. However, if the time-dependent perturbations $\hat{V}(t)$ are small and can be represented as

$$\hat{V}(t) = \lambda \hat{H}'(t), \tag{6.3.45}$$

where λ is a small parameter, then a general approximate technique can be developed.

By using formula (6.3.45), differential equations (6.3.19) can be written in the form

$$i\hbar \frac{dc_n(t)}{dt} = \lambda \sum_k e^{i\omega_{nk}t} H'_{nk}(t) c_k(t), \quad (n = 1, 2, \ldots), \tag{6.3.46}$$

where, according to formula (6.3.18),

$$H'_{nk} = \langle \Phi_n^{(0)}(\mathbf{r}) | \hat{H}' \Phi_k^{(0)}(\mathbf{r}) \rangle. \tag{6.3.47}$$

Now, we look for a solution of equations (6.3.46) in the form of the following power series

$$c_k(t) = c_k^{(0)}(t) + \lambda c_k^{(1)}(t) + \lambda^2 c_k^{(2)}(t) + \cdots = \sum_m \lambda^m c_k^{(m)}(t). \tag{6.3.48}$$

By substituting formula (6.3.48) into equations (6.3.46) and equating the terms of the same powers in λ, we end up with

$$i\hbar \frac{dc_n^{(0)}}{dt} = 0, \quad (n = 1, 2, \ldots), \tag{6.3.49}$$

$$i\hbar \frac{dc_n^{(1)}}{dt} = \sum_k e^{i\omega_{nk}t} H'_{nk}(t) c_k^{(0)}(t), \quad (n = 1, 2, \ldots), \tag{6.3.50}$$

$$i\hbar\frac{dc_n^{(m)}(t)}{dt} = \sum_k e^{i\omega_{nk}t}H'_{nk}(t)c_k^{(m-1)}(t), \quad (n = 1, 2, \dots). \tag{6.3.51}$$

It is apparent from formulas (6.3.20) and (6.3.48) that the following initial conditions are valid:

$$c_k^{(0)}(0) = \delta_{kl}, \quad (k = 1, 2, \dots), \tag{6.3.52}$$

$$c_k^{(m)}(0) = 0, \quad (m = 1, 2, \dots; k = 1, 2, \dots). \tag{6.3.53}$$

From equations (6.3.49) and initial conditions (6.3.52) it follows that

$$c_k^{(0)}(t) = \delta_{kl}, \quad (k = 1, 2, \dots). \tag{6.3.54}$$

By using the last formula in equations (6.3.50), we arrive at the following equations for $c_n^{(1)}(t)$

$$i\hbar\frac{dc_n^{(1)}(t)}{dt} = e^{i\omega_{nl}t}H'_{nl}(t), \tag{6.3.55}$$

which can be easily integrated with account of the initial conditions (6.3.53). The final result is

$$c_n^{(1)}(t) = -\frac{i}{\hbar}\int_0^t H'_{nl}(\tau)e^{i\omega_{nl}\tau}d\tau, \quad (n = 1, 2, \dots). \tag{6.3.56}$$

By limiting our calculations only to the first two terms in expansion (6.3.48), from formulas (6.3.45), (6.3.54) and (6.3.56) we obtain

$$c_n(t) = -\frac{i}{\hbar}\int_0^t V_{nl}(\tau)e^{i\omega_{nl}\tau}d\tau, \quad (n = 1, 2, \dots), \tag{6.3.57}$$

which leads to the following final expressions for transition probabilities

$$\boxed{P_{ln} = \frac{1}{\hbar^2}\left|\int_0^t V_{nl}(\tau)e^{i\omega_{nl}\tau}d\tau\right|^2, \quad (n = 1, 2, \dots).} \tag{6.3.58}$$

Next, we shall apply the last formula to two specific cases. First, we consider a short pulse of $\hat{V}(t)$, which implies short pulses of $V_{nl}(t)$ (see Figure 6.2). We want to find transition probabilities P_{ln} for $t > T$. According to the last formula, we have

$$P_{ln} = \frac{1}{\hbar^2}\left|\int_0^T V_{nl}(\tau)e^{i\omega_{nl}\tau}d\tau\right|^2 = \frac{2\pi}{\hbar^2}\left|\frac{1}{\sqrt{2\pi}}\int_{-\infty}^{\infty} V_{nl}(\tau)e^{i\omega_{nl}\tau}d\tau\right|^2, \tag{6.3.59}$$

which can be also written as

$$\boxed{P_{ln} = \frac{2\pi}{\hbar^2}\left|V_{nl}^F(\omega_{nl})\right|^2.} \tag{6.3.60}$$

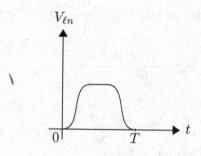

Fig. 6.2

Here, $V_{nl}^F(\omega_{nl})$ is the Fourier transform of $V_{nl}(t)$ evaluated at the transition frequency. One may say that formula (6.3.60) reflects the resonance nature of transitions. Indeed, if the Fourier spectrum of $V_{nl}(t)$ is changed, but the value of $V_{nl}^F(\omega)$ at the transition frequency ω_{nl} remains the same, the transition probability P_{ln} remains unchanged.

Next, we consider the case of time-harmonic perturbation

$$\hat{V}(\mathbf{r}, t) = 2\hat{W}(\mathbf{r}) \cos \omega t. \qquad (6.3.61)$$

According to formula (6.3.18), we find

$$V_{nl}(t) = 2W_{nl} \cos \omega t, \qquad (6.3.62)$$

where

$$W_{nl} = \langle \Phi_n^{(0)}(\mathbf{r})|\hat{W}(\mathbf{r})\Phi_l^{(0)}(\mathbf{r})\rangle. \qquad (6.3.63)$$

From formulas (6.3.57) and (6.3.62), it follows that

$$c_n(t) = -\frac{2i}{\hbar}W_{nl}\int_0^t e^{i\omega_{nl}\tau}\cos\omega\tau d\tau = -\frac{i}{\hbar}W_{nl}\int_0^t [e^{i(\omega_{nl}+\omega)\tau}+e^{i(\omega_{nl}-\omega)\tau}]d\tau.$$
$$(6.3.64)$$

By performing the integration, we find

$$c_n(t) = -\frac{W_{nl}}{\hbar}\left(\frac{e^{i(\omega_{nl}+\omega)\tau}-1}{\omega_{nl}+\omega}+\frac{e^{i(\omega_{nl}-\omega)t}-1}{\omega_{nl}-\omega}\right). \qquad (6.3.65)$$

Consider the practically interesting case of small detuning

$$\omega \approx \omega_{nl}. \qquad (6.3.66)$$

In this case, the first term in the brackets can be neglected, which results in the following expression for $c_n(t)$:

$$c_n(t) = -\frac{W_{nl}}{\hbar(\omega_{nl}-\omega)}[\cos(\omega_{nl}-\omega)t - 1 + i\sin(\omega_{nl}-\omega)t]. \qquad (6.3.67)$$

The last formula leads to the following expression for the transition probability

$$P_{ln} = \frac{|W_{nl}|^2}{\hbar^2} \frac{\sin^2 \frac{(\omega_{nl}-\omega)t}{2}}{\left(\frac{\omega_{nl}-\omega}{2}\right)^2}. \tag{6.3.68}$$

By introducing the notation

$$\beta = \frac{\omega_{nl} - \omega}{2}, \tag{6.3.69}$$

formula (6.3.68) can be written as follows

$$P_{ln}(t) = \frac{|W_{nl}|^2}{\hbar^2} t \frac{\sin^2 \beta t}{\beta^2 t}. \tag{6.3.70}$$

It can be noted that function $\frac{\sin^2 \beta t}{\beta^2 t}$ is narrowly peaked at zero as t is increased (see Figure 6.3). It can be shown that

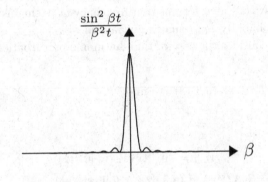

Fig. 6.3

$$\lim_{t\to\infty} \frac{\sin^2 \beta t}{\beta^2 t} = \pi \delta(\beta). \tag{6.3.71}$$

Consequently, for large times the transition probability P_{ln} is given by the formula

$$P_{ln} = \frac{\pi |W_{nl}|^2}{\hbar^2} t \delta\left(\frac{\omega_{nl} - \omega}{2}\right). \tag{6.3.72}$$

Taking into account that

$$\delta\left(\frac{\omega_{nl} - \omega}{2}\right) = \delta\left(\frac{\mathcal{E}_n - \mathcal{E}_l - \hbar\omega}{2\hbar}\right) = 2\hbar\delta(\mathcal{E}_n - \mathcal{E}_l - \hbar\omega), \tag{6.3.73}$$

formula (6.3.72) can be written as follows

$$P_{ln} = \frac{2\pi |W_{nl}|^2}{\hbar} t\delta(\mathcal{E}_n - \mathcal{E}_l - \hbar\omega). \tag{6.3.74}$$

The appearance of the delta function in the last formula may cause some concern. However, in most applications of this formula, the delta function is integrated. As an example, consider the case when energy levels are closely bunched around the energy level $\mathcal{E}_n$ (see Figure 6.4). In this situation, it

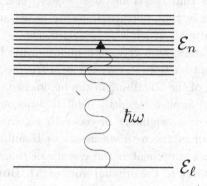

Fig. 6.4

is natural to talk about the density $\rho(\mathcal{E}_n)$ of energy states around $\mathcal{E}_n$. This density of states is defined by the formula

$$\Delta N = \rho(\mathcal{E}_n)\Delta\mathcal{E}, \tag{6.3.75}$$

where ΔN is the number of energy states within the energy interval $\Delta\mathcal{E}$ centered at $\mathcal{E}_n$. Now, we can compute the probability $\overline{P}_{ln}$ of the transition from the energy level $\mathcal{E}_l$ into energy interval $\Delta\mathcal{E}$ centered at $\mathcal{E}_n$:

$$\overline{P}_{ln} = \int_{\Delta\mathcal{E}} P_{ln}(\mathcal{E})\rho(\mathcal{E})d\mathcal{E}, \quad \text{where } \mathcal{E} = \mathcal{E}_l + \hbar w. \tag{6.3.76}$$

By using formula (6.3.74) in the last equation, we find

$$\overline{P}_{ln} = \frac{2\pi |W_{nl}|^2}{\hbar} \rho(\mathcal{E}_n)t. \tag{6.3.77}$$

We see that for large times this random transition occurs with constant rate

$$\overline{w}_{nl} = \frac{2\pi |W_{nl}|^2}{\hbar} \rho(\mathcal{E}_n). \tag{6.3.78}$$

The last formula as well as formula (6.3.74) are known as **Fermi's golden rule**. These formulas have many applications. In particular, they are extensively used in the semi-classical transport theory (discussed in the last section of Chapter 5) for specification of scattering rate $S(\mathbf{k}, \mathbf{k}')$.

Consider now the case when perturbations $\hat{V}(t)$ vary very slowly in time on the scale of $\frac{1}{\omega_{nl}}$. In this case, the expression under the integral sign in formula (6.3.58) consists of the product of slowly varying and quickly oscillating functions. For this reason, the value of this integral is quite small. This implies that the transitions between energy levels are highly unlikely. Such slow in time perturbations are called **adiabatic**, and they require special treatment which is presented below.

Consider the Hamiltonian $\hat{H}(t)$ of a quantum mechanical system that varies very slowly in time from its original form $\hat{H}(0)$ to its final form $\hat{H}(T)$. This slow variation of the Hamiltonian can be, for instance, caused by slow variations in time of applied magnetic field. It turns out that if a quantum mechanical system was originally in the n-th eigenstate of Hamiltonian $\hat{H}(0)$, it will remain in the n-th eigenstate of Hamiltonian $\hat{H}(t)$ as time progresses. The last statement is the essence of the so-called **adiabatic theorem**, which is due to the original work of **M. Born and V. Fock**.

Next, we shall present some justification of this theorem for the case when the energy spectra of Hamiltonians $\hat{H}(t)$ are discrete and non-degenerate.

Let $\Phi_n(t)$ and $\mathcal{E}_n(t)$ be instantaneous eigenstates and eigenvalues of the time-varying Hamiltonians, respectively. This means that

$$\hat{H}(t)\Phi_n(t) = \mathcal{E}_n(t)\Phi_n(t), \qquad (6.3.79)$$

and

$$\langle \Phi_k(t)|\Phi_n(t)\rangle = \delta_{kn}. \qquad (6.3.80)$$

Here and below, the dependence of $\Phi_k(t)$ on $\mathbf{r}$ is tacitly assumed. Suppose that at time $t = 0$ a quantum system under consideration was in the k-th eigenstate $\Phi_k(0)$. As time evolves, the wave function ψ of the system evolves as well in accordance with the time-dependent Schrödinger equation

$$i\hbar\frac{\partial \psi}{\partial t} = \hat{H}(t)\psi. \qquad (6.3.81)$$

It is our intention to demonstrate that at any instant of time t the wave function ψ coincides with the eigenstates $\Phi_k(t)$ up to some time-dependent phase factors. To do this, we expand $\psi(t)$ into series with respect to instantaneous eigenfunctions $\Phi_n(t)$:

$$\psi(t) = \sum_n c_n(t)\Phi_n(t)e^{i\theta_n(t)}, \qquad (6.3.82)$$

where $\theta_n(t)$ is the dynamic phase given by the formula

$$\theta_n(t) = -\frac{1}{\hbar} \int_0^t \mathcal{E}_n(\tau)d\tau. \tag{6.3.83}$$

This phase can be viewed as the dynamic generalization of the phase $\frac{\mathcal{E}_n}{\hbar}t$ of the stationary states.

By substituting the expansion (6.3.82) into equation (6.3.81), we derive

$$i\hbar \sum_n \left(\frac{dc_n(t)}{dt}\Phi_n(t) + c_n(t)\frac{\partial\Phi_n(t)}{\partial t} - \frac{i}{\hbar}c_n(t)\mathcal{E}_n(t)\Phi_n(t) \right)e^{i\theta_n(t)}$$

$$= \sum_n c_n(t)\hat{H}(t)\Phi_n(t)e^{i\theta_n(t)}. \tag{6.3.84}$$

In accordance with equation (6.3.79), the third and last terms in formula (6.3.84) cancel out. This leads to the equation

$$\sum_n \frac{dc_n(t)}{dt}\Phi_n(t)e^{i\theta_n(t)} = -\sum_n c_n(t)\frac{\partial\Phi_n(t)}{\partial t}e^{i\theta_n(t)}. \tag{6.3.85}$$

By using the orthonormality property (6.3.80) of instantaneous eigenfunctions $\Phi_n(t)$, from the last formula we derive

$$\frac{dc_m(t)}{dt} = -\sum_n c_n(t)\langle\Phi_m(t)|\frac{\partial\Phi_n(t)}{\partial t}\rangle e^{i[\theta_n(t)-\theta_m(t)]}. \tag{6.3.86}$$

Next, we want to evaluate the inner products $\langle\Phi_m(t)|\frac{\partial\Phi_n(t)}{\partial t}\rangle$. To this end, we shall differentiate with respect to time both sides of equation (6.3.79). This leads to

$$\frac{\partial\hat{H}(t)}{\partial t}\Phi_n(t) + \hat{H}(t)\frac{\partial\Phi_n(t)}{\partial t} = \frac{d\mathcal{E}_n(t)}{\partial t}\Phi_n(t) + \mathcal{E}_n(t)\frac{\partial\Phi_n(t)}{\partial t} \tag{6.3.87}$$

which can be further transformed as follows

$$\langle\Phi_m(t)|\frac{\partial\hat{H}(t)}{\partial t}\Phi_n(t)\rangle + \langle\Phi_m(t)|\hat{H}(t)\frac{\partial\Phi_n(t)}{\partial t}\rangle$$

$$= \langle\Phi_m(t)|\frac{d\mathcal{E}_n(t)}{dt}\Phi_n(t)\rangle + \mathcal{E}_n(t)\langle\Phi_m(t)|\frac{\partial\Phi_n(t)}{\partial t}\rangle. \tag{6.3.88}$$

Now, since operator $\hat{H}(t)$ is Hermitian, we can transform the second term in the last formula in the following way:

$$\langle\Phi_m(t)|\hat{H}(t)\frac{\partial\Phi_n(t)}{\partial t}\rangle = \langle\hat{H}(t)\Phi_m(t)|\frac{\partial\Phi_n(t)}{\partial t}\rangle = \mathcal{E}_m(t)\langle\Phi_m(t)|\frac{\partial\Phi_n(t)}{\partial t}\rangle.$$
$$\tag{6.3.89}$$

On the other hand, for the third term in formula (6.3.88) we have

$$\langle \Phi_m(t) | \frac{d\mathcal{E}_n(t)}{dt} \Phi_n(t) \rangle = \frac{d\mathcal{E}_n(t)}{dt} \delta_{mn}. \tag{6.3.90}$$

According to the last two formulas, equation (6.3.88) can be simplified for the case of $m \neq n$ as follows:

$$\langle \Phi_m(t) | \frac{\partial \hat{H}(t)}{\partial t} \Phi_n(t) \rangle = [\mathcal{E}_n(t) - \mathcal{E}_m(t)] \langle \Phi_m(t) | \frac{\partial \Phi_n(t)}{\partial t} \rangle, \tag{6.3.91}$$

which leads to

$$\langle \Phi_m(t) | \frac{\partial \Phi_n(t)}{\partial t} \rangle = \frac{\langle \Phi_m(t) | \frac{\partial \hat{H}(t)}{\partial t} \Phi_n(t) \rangle}{\mathcal{E}_n(t) - \mathcal{E}_m(t)}. \tag{6.3.92}$$

By using the last formula, equation (6.3.86) can be written as follows:

$$\frac{dc_m(t)}{dt} = -c_m(t) \langle \Phi_m(t) | \frac{\partial \Phi_m(t)}{\partial t} \rangle$$
$$- \sum_{n \neq m} c_n(t) \frac{\langle \Phi_m(t) | \frac{\partial \hat{H}(t)}{\partial t} \Phi_n(t) \rangle}{\mathcal{E}_n(t) - \mathcal{E}_m(t)} e^{i[\theta_m(t) - \theta_n(t)]}. \tag{6.3.93}$$

The last formula has been derived through exact mathematical transformations. Now, we shall make the **adiabatic approximation** by assuming that the time variations of Hamiltonian $\hat{H}(t)$ are so slow that

$$\frac{\partial \hat{H}(t)}{\partial t} \simeq 0. \tag{6.3.94}$$

This adiabatic approximation appreciably simplifies equation (6.3.93):

$$\frac{dc_m(t)}{dt} = -c_m(t) \langle \Phi_m(t) | \frac{\partial \Phi_m(t)}{\partial t} \rangle, \tag{6.3.95}$$

which is equivalent to

$$\frac{d}{dt}[\ln c_m(t)] = -\langle \Phi_m(t) | \frac{\partial \Phi_m(t)}{\partial t} \rangle. \tag{6.3.96}$$

By integrating this equation, we obtain

$$c_m(t) = c_m(0) e^{i\gamma_m(t)}, \tag{6.3.97}$$

where

$$\boxed{\gamma_m(t) = i \int_0^t \langle \Phi_m(\tau) | \frac{\partial \Phi_m(\tau)}{\partial \tau} \rangle d\tau.} \tag{6.3.98}$$

It was assumed at the beginning of our derivation that at time $t = 0$ the quantum mechanical system was in the eigenstate $\Phi_k(0)$. This implies according to formulas (6.3.82) and (6.3.83) the following initial conditions

$$c_m(0) = \delta_{mk}. \tag{6.3.99}$$

From formulas (6.3.97) and (6.3.99), it follows that

$$c_m(t) = \delta_{mk} e^{i\gamma_m(t)}. \tag{6.3.100}$$

Finally, from formulas (6.3.82) and (6.3.100), we conclude that

$$\boxed{\psi(t) = \Phi_k(t) e^{i\theta_k(t)} e^{i\gamma_k(t)}.} \tag{6.3.101}$$

Thus, it is established that, under adiabatic conditions, the quantum mechanical system evolves in time by remaining in the k-th eigenstate of the Hamiltonian $\hat{H}(t)$. This state acquires two phase factors: the dynamic phase factor $e^{i\theta_k(t)}$ and **geometric** phase factor $e^{i\gamma_k(t)}$. The geometric phase has a special significance in the case of cyclic adiabatic variations of Hamiltonians. In the latter case, it leads to the notion of the **Berry phase**, which in turn connects quantum mechanics with topology.

To introduce the notion of the Berry phase, consider the time evolving Hamiltonian $\hat{H}(\mathbf{R}(t))$. This Hamiltonian depends on time through time variations of some parameters represented by vector $\mathbf{R}(t)$. Then, the instantaneous eigenstates of this Hamiltonian are also functions of vector $\mathbf{R}(t)$:

$$\Phi_k(\mathbf{R}(t)). \tag{6.3.102}$$

This implies that

$$\frac{\partial \Phi_k(t)}{\partial t} = \frac{\partial \Phi_k(\mathbf{R}(t))}{\partial t} = \nabla_{\mathbf{R}} \Phi_k \cdot \frac{d\mathbf{R}}{dt}. \tag{6.3.103}$$

Now, the geometric phase $\gamma_k(t)$ can be written as follows

$$\gamma_k(t) = i \int_0^t \langle \Phi_k(\tau) | \frac{\partial \Phi_k(\tau)}{\partial \tau} \rangle d\tau = i \int_{\mathbf{R}(0)}^{\mathbf{R}(t)} \langle \Phi_k | \nabla_{\mathbf{R}} \Phi_k \rangle d\mathbf{R}, \tag{6.3.104}$$

where vectors $\mathbf{R}(0)$ and $\mathbf{R}(t)$ represent the initial and current values of parameters.

In the most interesting case of cyclic variations of the Hamiltonian in the geometric space of parameters $\mathbf{R}(t)$, we have

$$\mathbf{R}(T) = \mathbf{R}(0), \tag{6.3.105}$$

and

$$\boxed{\gamma_k(T) = i \oint_C \langle \Phi_k | \nabla_{\mathbf{R}} \Phi_k \rangle \cdot d\mathbf{R},} \tag{6.3.106}$$

where C is a closed loop in the parameter space. The phase $\gamma_k(T)$ is the **Berry phase**. It is clear from the last formula that $\gamma_k(T)$ depends only on geometry of C and does not depend on time. This is in contrast with the phase $\theta_k(t)$ defined by formula (6.3.83), and this explains why θ_k and γ_k are called dynamic and geometric phases, respectively. The phase $\gamma_k(T)$ was introduced by M. Berry who pointed out that this phase may have nonzero and measurable values.

By introducing the vector

$$\mathbf{A}_k(\mathbf{R}) = i\langle \Phi_k | \nabla_{\mathbf{R}} \Phi_k \rangle, \qquad (6.3.107)$$

the Berry phase can be written as

$$\gamma_k = \oint_C \mathbf{A}_k(\mathbf{R}) \cdot d\mathbf{R}. \qquad (6.3.108)$$

In the case of the three-dimensional parameter space, the vector

$$\mathbf{\Omega}_k = \nabla_{\mathbf{R}} \times \mathbf{A}_k(\mathbf{R}) \qquad (6.3.109)$$

can be introduced and, according to the Stokes theorem, the line integral in (6.3.108) can be transformed into the surface integral

$$\gamma_k = \int_S \mathbf{\Omega}_k \cdot d\mathbf{s}, \qquad (6.3.110)$$

where S is a surface with the boundary C. The most remarkable situation occurs when S is a closed surface. It can be shown that in this case

$$\frac{1}{2\pi} \oint_S \mathbf{\Omega}_k \cdot d\mathbf{s} = N, \qquad (6.3.111)$$

where integers N are called Chern numbers. In other words, the quantity in the left-hand side of formula (6.3.111) is quantized. This situation is mathematically similar to the famous Gauss-Bonnet theorem

$$\frac{1}{2\pi} \oint_S K ds = \chi(S), \qquad (6.3.112)$$

where K is the Gaussian curvature of S, while integer $\chi(S)$ is called the Euler number (Euler characteristic of S). This number is topological in nature and depends on the number of holes (or handles) of S. This number is invariant with respect to smooth deformations of S.

Similar language is being used with respect to formula (6.3.111). Here, $\mathbf{\Omega}_k$ is called Berry curvature, while Chern numbers are invariant with respect to smooth (in some sense, small) deformation of Hamiltonians. Formula (6.3.111) can be generalized to the multi-dimensional case. In this case, instead of vector $\mathbf{\Omega}_k$ the Berry curvature tensor is introduced. The notions of Berry curvature and Chern numbers as topological quantum numbers have deeply penetrated quantum mechanics. They are important, for instance, in the understanding of the integer quantum Hall effect. The detailed discussion of this matter is beyond the scope of this text.

6.4   Density Matrix. Optical Bloch Equation. Wigner Function and Quantum Transport

In our previous discussion, quantum mechanical states have been fully characterized by their wave functions. Such states are called **pure** states. However, in many applications of quantum mechanics, pure states cannot be clearly defined. Instead, it is only possible to specify **mixed** states which are statistical (probabilistic) mixtures of pure states. These mixed states can be characterized by the so-called **density matrix**, which is discussed in this section.

Before proceeding to this discussion of mixed states and the density matrix, we shall first discuss how average values of physical quantities in a pure state can be computed within the framework of the matrix form of quantum mechanics.

Consider a physical quantity g and its corresponding operator $\hat{g}$. In a pure state characterized by a wave function $\psi(\mathbf{r}, t)$ in coordinate representation the average value $\bar{g}(t)$ of g can be computed as follows:

$$\bar{g}(t) = \langle \psi | \hat{g} \psi \rangle. \tag{6.4.1}$$

Let ψ_n be eigenfunctions of operator $\hat{f}$ with discrete spectrum

$$\hat{f} \psi_n = f_n \psi_n, \tag{6.4.2}$$

$$\langle \psi_n, \psi_k \rangle = \delta_{kn}. \tag{6.4.3}$$

By using the orthonormal expansion

$$\psi(\mathbf{r}, t) = \sum_n a_n(t) \psi_n(\mathbf{r}), \tag{6.4.4}$$

we arrive at the $\hat{f}$-representation of $\psi(\mathbf{r}, t)$ by the infinite-dimensional vector consisting of expansion coefficients $a_n(t)$.

By substituting the last formula into equation (6.4.1), we obtain

$$\bar{g}(t) = \langle \sum_k a_k(t) \psi_k(\mathbf{r}) | \hat{g} \left(\sum_n a_n(t) \psi_n(\mathbf{r}) \right) \rangle, \tag{6.4.5}$$

which leads to

$$\bar{g}(t) = \sum_n \sum_k a_n(t) a_k^*(t) \langle \psi_k | \hat{g} | \psi_n \rangle. \tag{6.4.6}$$

By introducing the matrix

$$\hat{g} = \{ g_{kn} = \langle \psi_k | \hat{g} | \psi_n \rangle \}, \tag{6.4.7}$$

formula (6.4.6) can be written as follows

$$\bar{g}(t) = \sum_n \sum_k a_n(t) a_k^*(t) g_{kn}. \tag{6.4.8}$$

Next, we introduce the matrix

$$\hat{A}(t) = \{A_{nk}(t) = a_n(t) a_k^*(t)\} \tag{6.4.9}$$

and rewrite formula (6.4.8) in the form

$$\bar{g}(t) = \sum_n \sum_k A_{nk} g_{kn} = \sum_n (\hat{A}(t)\hat{g})_{nn}. \tag{6.4.10}$$

This implies that

$$\boxed{\bar{g}(t) = \text{Tr}(\hat{A}(t)\hat{g}),} \tag{6.4.11}$$

where the abbreviation "Tr" stands for the trace of the product of the matrices $\hat{A}(t)$ and $\hat{g}$, that is, for the sum of all diagonal elements of the matrix $\hat{A}(t)\hat{g}$.

The last formula has been derived for pure states. Next, we consider its generalization to the case of mixed states. As mentioned before, mixed states cannot be completely characterized by a single wave function. Instead, these states are characterized by specifying probabilities λ_m of being in different pure states $\psi^{(m)}(\mathbf{r}, t)$:

$$\lambda_m = \text{Probability}\{\psi(\mathbf{r}, t) = \psi^{(m)}(\mathbf{r}, t)\}. \tag{6.4.12}$$

By using expansions

$$\psi^{(m)}(\mathbf{r}, t) = \sum_n a_n^{(m)}(t) \psi_n(\mathbf{r}), \tag{6.4.13}$$

the average values $\bar{g}^{(m)}(t)$ of physical quantity g in the pure states $\psi^{(m)}(\mathbf{r}, t)$ can be computed as

$$\bar{g}^{(m)}(t) = \sum_n \sum_k a_n^{(m)}(t)(a_k^{(m)}(t))^* g_{kn}. \tag{6.4.14}$$

This implies that the average value $\tilde{\bar{g}}(t)$ of g in the mixed state specified by probabilities λ_m can be found by using the formula

$$\tilde{\bar{g}}(t) = \sum_m \lambda_m \bar{g}^{(m)}(t). \tag{6.4.15}$$

By combining the last two formulas, we arrive at the following expression

$$\tilde{\bar{g}}(t) = \sum_n \sum_k \left(\sum_m \lambda_m a_n^{(m)}(t)(a_k^{(m)}(t))^* \right) g_{kn}. \tag{6.4.16}$$

Now we introduce the density matrix $\hat{\rho}(t)$:

$$\hat{\rho}(t) = \{\rho_{nk}(t)\}, \tag{6.4.17}$$

whose matrix elements are defined by the formula

$$\boxed{\rho_{nk}(t) = \sum_m \lambda_m a_n^{(m)}(t)(a_k^{(m)}(t))^*.} \tag{6.4.18}$$

By using the density matrix, equation (6.4.16) can be written as follows

$$\tilde{\tilde{g}}(t) = \sum_n \sum_m \rho_{nk}(t)g_{kn}, \tag{6.4.19}$$

or

$$\tilde{\tilde{g}}(t) = \sum_n (\hat{\rho}(t)\hat{g})_{nn}. \tag{6.4.20}$$

The last formula can be written in the abbreviated form:

$$\boxed{\tilde{\tilde{g}}(t) = \mathrm{Tr}(\hat{\rho}(t)\hat{g}),} \tag{6.4.21}$$

which is the generalization of formula (6.4.11) to the case of mixed states.

It is apparent that formula (6.4.21) accounts for two kinds of uncertainties: for the fundamental uncertainties of measurements of physical quantity g in pure states $\psi^{(m)}(\mathbf{r}, t)$, as well as for the uncertainty of being in one of these states specified by probabilities λ_m. To compute $\tilde{\tilde{g}}(t)$ at various instants of time t by using formula (6.4.21), the density matrix $\hat{\rho}(t)$ should be known at those instants of time. The latter information can be attained by using the dynamic equation for the density matrix. This dynamic equation plays the same role for mixed states as the time-dependent Schrödinger equation for pure states.

The derivation of the dynamic equation for the density matrix proceeds as follows. From formula (6.4.18), it follows that

$$\frac{d\rho_{nk}(t)}{dt} = \sum_m \lambda_m \frac{da_n^{(m)}(t)}{dt}(a_k^{(m)}(t))^* + \sum_m \lambda_m a_n^{(m)}(t)\frac{d(a_k^{(m)}(t))^*}{dt}. \tag{6.4.22}$$

The pure states $\psi^{(m)}(\mathbf{r}, t)$ are solutions of the time-dependent Schrödinger equation

$$i\hbar\frac{\partial \psi^{(m)}}{\partial t} = \hat{H}\psi^{(m)}, \tag{6.4.23}$$

which can be written in the matrix form as follows (see formula (6.1.92)):

$$i\hbar\frac{da_n^{(m)}(t)}{dt} = \sum_j H_{nj}a_j^{(m)}(t). \tag{6.4.24}$$

Similarly,

$$-i\hbar\frac{d(a_k^{(m)}(t))^*}{dt} = \sum_j H_{kj}^*(a_j^{(m)}(t))^*. \tag{6.4.25}$$

Since the Hamiltonian is a Hermitian operator, we have

$$H_{kj}^* = H_{jk}, \tag{6.4.26}$$

and equation (6.4.25) can be written as

$$-i\hbar\frac{d(a_k^{(m)}(t))^*}{dt} = \sum_j H_{jk}(a_j^{(m)}(t))^*. \tag{6.4.27}$$

By substituting formulas (6.4.24) and (6.4.27) into equation (6.4.22), we derive

$$\frac{d\rho_{nk}(t)}{dt} = \frac{1}{i\hbar}\left(\sum_m \lambda_m \sum_j H_{nj}a_j^{(m)}(t)(a_k^{(m)}(t))^* \right.$$
$$\left. - \sum_m \lambda_m \sum_j H_{jk}a_n^{(m)}(t)(a_j^{(m)}(t))^* \right). \tag{6.4.28}$$

By changing the order of summations in the last formula, we find

$$\frac{d\rho_{nk}(t)}{dt} = \frac{1}{i\hbar}\left(\sum_j H_{nj}\left(\sum_m \lambda_m a_j^{(m)}(t)(a_k^{(m)}(t))^* \right) \right.$$
$$\left. - \sum_j H_{jk}\left(\sum_m \lambda_m a_n^{(m)}(t)(a_j^{(m)}(t))^* \right)\right). \tag{6.4.29}$$

In accordance with the relation (6.4.18), we have

$$\sum_m \lambda_m a_j^{(m)}(t)(a_k^{(m)}(t))^* = \rho_{jk}(t) \tag{6.4.30}$$

and

$$\sum_m \lambda_m a_n^{(m)}(t)(a_j^{(m)}(t))^* = \rho_{nj}(t). \tag{6.4.31}$$

Consequently, equation (6.4.29) can be written as follows

$$\frac{d\rho_{nk}(t)}{dt} = \frac{1}{i\hbar}\left(\sum_j H_{nj}\rho_{jk}(t) - \sum_j \rho_{nj}(t)H_{jk}\right), \tag{6.4.32}$$

which is tantamount to

$$\frac{d\rho_{nk}(t)}{dt} = \frac{1}{i\hbar}\left[(\hat{H}\hat{\rho}(t))_{nk} - (\hat{\rho}(t)\hat{H})_{nk}\right]. \tag{6.4.33}$$

The last formula implies that

$$i\hbar\frac{d\hat{\rho}}{dt} = \hat{H}\hat{\rho}(t) - \hat{\rho}(t)\hat{H}, \tag{6.4.34}$$

which can be also written as

$$\boxed{i\hbar\frac{d\hat{\rho}(t)}{dt} = \left[\hat{H},\hat{\rho}\right].} \tag{6.4.35}$$

This is the so-called **J. von Neumann equation**, which mathematically describes the time evolution of density matrix $\hat{\rho}$.

As an example, we consider this equation for two-level quantum mechanical systems described by the Hamiltonian

$$\hat{H} = \hat{H}^{(0)} + \hat{V}(t), \tag{6.4.36}$$

where

$$\hat{H}^{(0)}\psi_n = \mathcal{E}_n\psi_n, \quad (n = 1, 2). \tag{6.4.37}$$

We consider the matrix representation of the Hamiltonian $\hat{H}$ in the basis $\{\psi_n\}$, $(n = 1, 2)$. It is apparent that in this basis we have

$$H_{11}^{(0)} = \mathcal{E}_1, \quad H_{22}^{(0)} = \mathcal{E}_2, \quad H_{12}^{(0)} = H_{21}^{(0)} = 0. \tag{6.4.38}$$

Furthermore, we introduce the notations

$$V_{12}(t) = V_{21}^*(t) = \langle\psi_2|\hat{V}(t)|\psi_1\rangle \tag{6.4.39}$$

and assume (as is often the case) that

$$V_{11} = V_{22} = 0. \tag{6.4.40}$$

According to the last three formulas, the Hamiltonian $\hat{H}$ can be represented by the following matrix

$$\hat{H} = \begin{bmatrix} \mathcal{E}_1 & V_{12} \\ V_{21} & \mathcal{E}_2 \end{bmatrix}. \tag{6.4.41}$$

We shall also use the following matrix notation for the density operator

$$\hat{\rho}(t) = \begin{bmatrix} \rho_{11}(t) & \rho_{12}(t) \\ \rho_{21}(t) & \rho_{22}(t) \end{bmatrix}. \tag{6.4.42}$$

From the last two formulas, we find

$$\hat{H}\hat{\rho} = \begin{bmatrix} \mathcal{E}_1\rho_{11} + V_{12}\rho_{21} & \mathcal{E}_1\rho_{12} + V_{12}\rho_{22} \\ V_{21}\rho_{11} + \mathcal{E}_2\rho_{21} & V_{21}\rho_{12} + \mathcal{E}_2\rho_{22} \end{bmatrix}, \tag{6.4.43}$$

and

$$\hat{\rho}\hat{H} = \begin{bmatrix} \rho_{11}\mathcal{E}_1 + \rho_{12}V_{21} & \rho_{11}V_{12} + \rho_{12}\mathcal{E}_2 \\ \rho_{21}\mathcal{E}_1 + \rho_{22}V_{21} & \rho_{21}V_{12} + \rho_{22}\mathcal{E}_2 \end{bmatrix}. \tag{6.4.44}$$

By using the last two formulas, J. von Neumann equation (6.4.35) can be written as follows

$$\frac{d\hat{\rho}}{dt} = -\frac{i}{\hbar} \begin{bmatrix} V_{12}\rho_{21} - \rho_{12}V_{21} & -\hbar\omega_0\rho_{12} + V_{12}(\rho_{22} - \rho_{11}) \\ \hbar\omega_0\rho_{21} + V_{21}(\rho_{11} - \rho_{22}) & V_{21}\rho_{12} - \rho_{21}V_{12} \end{bmatrix}, \quad (6.4.45)$$

where ω_0 is defined by the relation

$$\mathcal{E}_2 - \mathcal{E}_1 = \hbar\omega_0. \quad (6.4.46)$$

The matrix equation (6.4.45) is equivalent to the following coupled differential equations

$$\frac{d\rho_{11}}{dt} = \frac{i}{\hbar}(\rho_{12}V_{21} - V_{12}\rho_{21}), \quad (6.4.47)$$

$$\frac{d\rho_{12}}{dt} = i\omega_0\rho_{12} + \frac{i}{\hbar}V_{12}(\rho_{11} - \rho_{22}), \quad (6.4.48)$$

$$\frac{d\rho_{21}}{dt} = -i\omega_0\rho_{21} + \frac{i}{\hbar}V_{21}(\rho_{22} - \rho_{11}), \quad (6.4.49)$$

$$\frac{d\rho_{22}}{dt} = -\frac{i}{\hbar}(V_{21}\rho_{12} - \rho_{21}V_{12}). \quad (6.4.50)$$

It is apparent from formulas (6.4.47) and (6.4.50) that

$$\frac{d\rho_{11}}{dt} = -\frac{d\rho_{22}}{dt}. \quad (6.4.51)$$

Consequently, we have three independent ordinary differential equations (6.4.47)–(6.4.49). It is remarkable that these three differential equations can be written in a very compact vector form. To arrive at this form, we introduce two vectors:

$$\mathbf{S} = \mathbf{e}_x(\rho_{21} + \rho_{12}) + \mathbf{e}_y i(\rho_{21} - \rho_{12}) + \mathbf{e}_z(\rho_{22} - \rho_{11}), \quad (6.4.52)$$

$$\mathcal{H} = \frac{1}{\hbar}[\mathbf{e}_x(V_{21} + V_{12}) + \mathbf{e}_y i(V_{21} - V_{12}) + \mathbf{e}_z \hbar\omega_0], \quad (6.4.53)$$

where, as before, $\mathbf{e}_x$, $\mathbf{e}_y$ and $\mathbf{e}_z$ are unit vectors along some x-, y- and z-axes. It is clear that Cartesian components of vectors $\mathbf{S}$ and $\mathcal{H}$ are real. Now, it can be demonstrated that differential equations (6.4.47)–(6.4.49) can be written as

$$\boxed{\frac{d\mathbf{S}}{dt} = -\mathbf{S} \times \mathcal{H}.} \quad (6.4.54)$$

Indeed, from formulas (6.4.48), (6.4.49) and (6.4.52), we find

$$\frac{dS_x}{dt} = \frac{d[\rho_{21} + \rho_{12}]}{dt} = -i\omega_0(\rho_{21} - \rho_{12}) + \frac{i}{\hbar}(\rho_{22} - \rho_{11})(V_{21} - V_{12}). \quad (6.4.55)$$

On the other hand,

$$-(\mathbf{S} \times \mathcal{H})_x = S_z \mathcal{H}_y - S_y \mathcal{H}_z, \tag{6.4.56}$$

which according to formulas (6.4.52), (6.4.53) and (6.4.55) leads to the equality

$$\frac{dS_x}{dt} = -(\mathbf{S} \times \mathcal{H})_x. \tag{6.4.57}$$

It is left as an exercise for the reader to verify the formulas

$$\frac{dS_y}{dt} = -(\mathbf{S} \times \mathcal{H})_y, \tag{6.4.58}$$

$$\frac{dS_z}{dt} = -(\mathbf{S} \times \mathcal{H})_z. \tag{6.4.59}$$

It is clear from equation (6.4.54) that

$$\mathbf{S} \cdot \frac{d\mathbf{S}}{dt} = 0, \tag{6.4.60}$$

and, consequently,

$$|\mathbf{S}| = S_0 = \text{const.} \tag{6.4.61}$$

The last equation implies that $|\mathbf{S}|$ is the constant of motion and that the time dynamics of $\mathbf{S}(t)$ occurs on the sphere (6.4.61). These dynamics are similar to the precessional dynamics of magnetic moment (or macrospin) under an applied magnetic field. This explains the adopted notations $\mathbf{S}$ and $\mathcal{H}$ in equation (6.4.54). In literature, this equation is called the optical Bloch equation and it is extensively used in quantum optics. Usually, a quantum mechanical system of interest is a part of a larger system. In other words, the system of interest is coupled to some reservoir or thermal bath. This coupling results in an additional (relaxation) term in the von Neumann equation (6.4.35) for the density matrix as well as in the optical Bloch equation (6.4.54) for two-level systems. As a result, the optical Bloch equation acquires the mathematical form which is identical to the Bloch equation for magnetization dynamics in paramagnetic materials. The latter equation is instrumental, for instance, in magnetic resonance imaging (MRI). A detailed discussion of all these issues is beyond the scope of this text.

Instead, we shall discuss how the density matrix can be used for the development of quantum mechanics in phase space. In classical mechanics momentum $\mathbf{p}$ and coordinate $\mathbf{r}$ of a particle are simultaneously measurable. For this reason, a joint probability distribution $f(\mathbf{r}, \mathbf{p})$ in phase space can be introduced and the Liouville equation for this distribution can be derived. This equation is at the very foundation of the classical transport theory. In quantum mechanics, $\mathbf{r}$ and $\mathbf{p}$ are not simultaneously measurable and, consequently, a joint probability distribution for these random quantities cannot be introduced. This, by the way, implies that quantum mechanical (fundamental) probabilities are, in some sense, mathematically different from classical probabilities of statistical mechanics. It turns out that in quantum mechanics quasiprobability distribution functions for $\mathbf{r}$ and $\mathbf{p}$ can be introduced and used for the development of quantum transport theory. This quasiprobability distribution is the so-called **Wigner function** which is also instrumental in the area of signal processing.

To introduce the Wigner function, we shall use the following expression for the density matrix in the coordinate representation

$$\hat{\rho}(\mathbf{r}, \mathbf{r}', t) = \sum_m \lambda_m \psi_m(\mathbf{r}, t) \psi_m^*(\mathbf{r}', t), \qquad (6.4.62)$$

where for the sake of notational simplicity we use symbols $\psi_m(\mathbf{r}, t)$ for the pure states instead of previous symbol $\psi^{(m)}(\mathbf{r}, t)$.

It is left as an exercise for the reader to demonstrate that the coordinate representation (6.4.62) of the density operator is equivalent to its matrix representation (6.4.17)–(6.4.18). It is apparent from the last formula that

$$\hat{\rho}(\mathbf{r}, \mathbf{r}, t) = \sum_m \lambda_m |\psi_m(\mathbf{r}, t)|^2 = P(\mathbf{r}, t), \qquad (6.4.63)$$

where $P(\mathbf{r}, t)$ has the meaning of probability density of $\mathbf{r}$ at time t.

Now, we introduce the so-called Wigner coordinates

$$\mathbf{x} = \frac{\mathbf{r} + \mathbf{r}'}{2}, \quad \mathbf{R} = \mathbf{r} - \mathbf{r}', \qquad (6.4.64)$$

and make the change of variables

$$\mathbf{r} = \mathbf{x} + \frac{\mathbf{R}}{2}, \quad \mathbf{r}' = \mathbf{x} - \frac{\mathbf{R}}{2} \qquad (6.4.65)$$

in $\hat{\rho}$. This leads to

$$\hat{\rho}(\mathbf{r}, \mathbf{r}', t) = \hat{\rho}\left(\mathbf{x} + \frac{\mathbf{R}}{2}, \mathbf{x} - \frac{\mathbf{R}}{2}, t\right) = \hat{\bar{\rho}}(\mathbf{x}, \mathbf{R}, t). \qquad (6.4.66)$$

The Wigner function is defined as

$$\boxed{w(\mathbf{x}, \mathbf{p}, t) = \frac{1}{(2\pi\hbar)^3} \int \hat{\bar{\rho}}(\mathbf{x}, \mathbf{R}, t) e^{-\frac{i}{\hbar}\mathbf{R}\cdot\mathbf{p}} d\mathbf{R}.} \qquad (6.4.67)$$

We intend to demonstrate that

$$P(\mathbf{r}, t) = \int w(\mathbf{x}, \mathbf{p}, t) d\mathbf{p}, \tag{6.4.68}$$

$$P(\mathbf{p}, t) = \int w(\mathbf{x}, \mathbf{p}, t) d\mathbf{x}, \tag{6.4.69}$$

where $P(\mathbf{p}, t)$ is the probability density of $\mathbf{p}$ at time t. Formulas (6.4.68) and (6.4.69) are similar to those which are valid for classical joint distribution functions. That is why the Wigner function $w(\mathbf{x}, \mathbf{p}, t)$ is called a quasiprobability distribution. The proof of formula (6.4.68) is straightforward and it goes as follows. From formula (6.4.67), we find

$$\int w(\mathbf{x}, \mathbf{p}, t) d\mathbf{p} = \int \hat{\rho}(\mathbf{x}, \mathbf{R}, t) \left(\frac{1}{(2\pi\hbar)^3} \int e^{-\frac{i}{\hbar}\mathbf{p}\cdot\mathbf{R}} d\mathbf{p} \right) d\mathbf{R}. \tag{6.4.70}$$

It is known that

$$\frac{1}{(2\pi\hbar)^3} \int e^{-\frac{i}{\hbar}\mathbf{p}\cdot\mathbf{R}} d\mathbf{p} = \delta(\mathbf{R}). \tag{6.4.71}$$

From the last two formulas, we infer that

$$\int w(\mathbf{x}, \mathbf{p}, t) d\mathbf{p} = \int \hat{\rho}(\mathbf{x}, \mathbf{R}, t) \delta(\mathbf{R}) d\mathbf{R} = \hat{\rho}(\mathbf{x}, 0, t). \tag{6.4.72}$$

It is also clear according to formulas (6.4.66), (6.4.65) and (6.4.63) that

$$\hat{\rho}(\mathbf{x}, 0, t) = \hat{\rho}(\mathbf{r}, \mathbf{r}, t) = P(\mathbf{r}, t) \tag{6.4.73}$$

and formula (6.4.68) is established.

Now, we proceed to the proof of formula (6.4.69) which is somewhat more involved. First, we introduce the momentum representations of pure states $\psi_m(\mathbf{r}, t)$:

$$c_m(\mathbf{p}, t) = \frac{1}{(2\pi\hbar)^{\frac{3}{2}}} \int \psi_m(\mathbf{r}, t) e^{-\frac{i}{\hbar}\mathbf{p}\cdot\mathbf{r}} d\mathbf{r}, \tag{6.4.74}$$

and

$$c_m^*(\mathbf{p}, t) = \frac{1}{(2\pi\hbar)^{\frac{3}{2}}} \int \psi_m^*(\mathbf{r}', t) e^{\frac{i}{\hbar}\mathbf{p}\cdot\mathbf{r}'} d\mathbf{r}'. \tag{6.4.75}$$

It is clear that the probability density $P(\mathbf{p}, t)$ is given by the formula

$$P(\mathbf{p}, t) = \sum_m \lambda_m |c_m(\mathbf{p}, t)|^2 = \sum_m \lambda_m c_m(\mathbf{p}, t) c_m^*(\mathbf{p}, t). \tag{6.4.76}$$

From the last three formulas, we derive:

$$P(\mathbf{p}, t) = \frac{1}{(2\pi\hbar)^3} \int \int \left(\sum_m \lambda_m \psi_m(\mathbf{r}, t)\psi_m^*(\mathbf{r}', t) \right) e^{-\frac{i}{\hbar}(\mathbf{r}-\mathbf{r}')\cdot\mathbf{P}} d\mathbf{r} d\mathbf{r}'.$$

$$(6.4.77)$$

In the last formula we shall make the following change of variables

$$(\mathbf{r}, \mathbf{r}') \Rightarrow (\mathbf{x}, \mathbf{R}) \tag{6.4.78}$$

with the Jacobian whose absolute value is equal to one. Consequently,

$$\begin{aligned}
P(\mathbf{p}, t) &= \frac{1}{(2\pi\hbar)^3} \int \left(\int \sum_m \lambda_m \psi_m(\mathbf{x} + \frac{\mathbf{R}}{2}, t)\psi(\mathbf{x} - \frac{\mathbf{R}}{2}, t) e^{-\frac{i}{\hbar}\mathbf{R}\cdot\mathbf{P}} d\mathbf{R} \right) d\mathbf{x} \\
&= \frac{1}{(2\pi\hbar)^3} \int \left(\int \hat{\rho}(\mathbf{x}, \mathbf{R}, t) e^{-\frac{i}{\hbar}\mathbf{R}\cdot\mathbf{P}} d\mathbf{R} \right) d\mathbf{x}.
\end{aligned}$$

$$(6.4.79)$$

By recalling the definition of the Wigner function (see (6.4.67)), from the last formula we find

$$P(\mathbf{p}, t) = \int w(\mathbf{x}, \mathbf{p}, t) d\mathbf{x}, \tag{6.4.80}$$

and the relation (6.4.69) is established.

Next, we shall derive the dynamic equation for the Wigner function. To this end, we shall differentiate formula (6.4.62) with respect to time

$$\frac{\partial \hat{\rho}(\mathbf{r}, \mathbf{r}', t)}{\partial t} = \sum_m \lambda_m \left(\frac{\partial \psi_m(\mathbf{r}, t)}{\partial t}\psi_m^*(\mathbf{r}', t) + \psi_m(\mathbf{r}, t)\frac{\partial \psi_m^*(\mathbf{r}', t)}{\partial t} \right). \tag{6.4.81}$$

According to the time-dependent Schrödinger equation, we have

$$\frac{\partial \psi_m(\mathbf{r}, t)}{\partial t} = \frac{1}{i\hbar}\hat{H}_\mathbf{r}\psi_m(\mathbf{r}, t), \tag{6.4.82}$$

and

$$\frac{\partial \psi_m^*(\mathbf{r}', t)}{\partial t} = -\frac{1}{i\hbar}\hat{H}_{\mathbf{r}'}\psi_m^*(\mathbf{r}, t), \tag{6.4.83}$$

where it is assumed that the Hamiltonian operator is given by the formula

$$\hat{H} = -\frac{\hbar^2}{2m}\nabla^2 + V(\mathbf{r}, t). \tag{6.4.84}$$

From the last four formulas, we derive

$$\boxed{i\hbar\frac{\partial \hat{\rho}}{\partial t} = -\frac{\hbar^2}{2m}\left[\nabla_\mathbf{r}^2\hat{\rho} - \nabla_{\mathbf{r}'}^2\hat{\rho}\right] + \left[\hat{V}(\mathbf{r}, t) - \hat{V}(\mathbf{r}', t)\right]\rho.} \tag{6.4.85}$$

This is the so-called **Heisenberg equation** for the density matrix. This equation is written in terms of $\mathbf{r}$ and $\mathbf{r}'$ variables. Now, we shall make the transition to $\mathbf{x}$ and $\mathbf{R}$ variables. From formula (6.4.66), we find:

$$\mathrm{grad}_{\mathbf{x}}\hat{\rho}\left(\mathbf{x}+\frac{\mathbf{R}}{2},\mathbf{x}-\frac{\mathbf{R}}{2},t\right) = \mathrm{grad}_{\mathbf{r}}\hat{\rho}+\mathrm{grad}_{\mathbf{r}'}\hat{\rho}. \qquad (6.4.86)$$

Furthermore,

$$\mathrm{div}_{\mathbf{R}}\mathrm{grad}_{\mathbf{x}}\hat{\rho} = \frac{1}{2}\mathrm{div}_{\mathbf{r}}(\mathrm{grad}_{\mathbf{r}}\rho) - \frac{1}{2}\mathrm{div}_{\mathbf{r}'}(\mathrm{grad}_{\mathbf{r}}\rho)$$
$$+ \frac{1}{2}\mathrm{div}_{\mathbf{r}}(\mathrm{grad}_{\mathbf{r}'}\hat{\rho}) - \frac{1}{2}\mathrm{div}_{\mathbf{r}'}(\mathrm{grad}_{\mathbf{r}'}\hat{\rho}). \qquad (6.4.87)$$

Due to the equality of mixed second-order partial derivatives, the second and third terms in the right-hand side of the last formula cancel out. Consequently,

$$\mathrm{div}_{\mathbf{R}}\mathrm{grad}_{\mathbf{x}}\hat{\rho} = \frac{1}{2}\nabla_{\mathbf{r}}^2\hat{\rho} - \frac{1}{2}\nabla_{\mathbf{r}'}^2\hat{\rho}. \qquad (6.4.88)$$

This implies that the Heisenberg equation (6.4.85) can be written as follows

$$i\hbar\frac{\partial\hat{\rho}}{\partial t} + \frac{\hbar^2}{m}\mathrm{div}_{\mathbf{R}}\mathrm{grad}_{\mathbf{x}}\hat{\rho} - \left(V(\mathbf{x}+\frac{\mathbf{R}}{2},t) - V(\mathbf{x}-\frac{\mathbf{R}}{2},t)\right)\hat{\rho} = 0. \qquad (6.4.89)$$

By performing the Fourier transform of the last equation and taking into account the definition of the Wigner function, we arrive at

$$i\hbar\frac{\partial w}{\partial t} + \frac{i\hbar}{m}\mathbf{p}\cdot\mathrm{grad}_{\mathbf{x}}w - \int\left(V(\mathbf{x}+\frac{\mathbf{R}}{2},t) - V(\mathbf{x}-\frac{\mathbf{R}}{2},t)\right)\hat{\rho}(\mathbf{x},\mathbf{R},t)e^{-\frac{i}{\hbar}\mathbf{p}\cdot\mathbf{R}}d\mathbf{R} = 0.$$
$$(6.4.90)$$

Next, we shall use the following Taylor expansions:

$$V(\mathbf{x}+\frac{\mathbf{R}}{2},t) = V(\mathbf{x},t) + \sum_{n=1}^{\infty}\frac{1}{n!}\left(\sum_{k=1}^{3}\frac{R_k}{2}\frac{\partial}{\partial x_k}\right)^n V(\mathbf{x},t), \qquad (6.4.91)$$

$$V(\mathbf{x}-\frac{\mathbf{R}}{2},t) = V(\mathbf{x},t) + \sum_{n=1}^{1}\frac{(-1)^n}{n!}\left(\sum_{k=1}^{3}\frac{R_k}{2}\frac{\partial}{\partial x_k}\right)^n V(\mathbf{x},t), \qquad (6.4.92)$$

where R_k and x_k are Cartesian components of vector $\mathbf{R}$ and $\mathbf{x}$, respectively.

From the last two formulas, we find that

$$V(\mathbf{x}+\frac{\mathbf{R}}{2},t) - V(\mathbf{x}-\frac{\mathbf{R}}{2},t) = \sum_{n=0}^{\infty}\frac{1}{4^n(2n+1)!}\left(\sum_{k=1}^{3}R_k\frac{\partial}{\partial x_k}\right)^{2n+1} V(\mathbf{x},t).$$
$$(6.4.93)$$

By substituting the last formula into equation (6.4.90) and taking into account the definition of the Wigner function, we derive

$$i\hbar\frac{\partial w}{\partial t} + \frac{i\hbar}{m}\mathbf{p}\cdot\text{grad}_\mathbf{x} w - \sum_{n=0}^{\infty}\frac{1}{4^n(2n+1)!}\left(\sum_{k=1}^{3}-i\hbar\frac{\partial}{\partial p_k}\frac{\partial}{\partial x_k}\right)^{2n+1}V(\mathbf{x})w = 0,$$

(6.4.94)

or

$$\boxed{\frac{\partial w}{\partial t} + \frac{\mathbf{p}}{m}\cdot\text{grad}_\mathbf{x} w + \sum_{n=0}^{\infty}\frac{(-1)^n\hbar^{2n}}{4^n(2n+1)!}\left(\sum_{k=1}^{3}\frac{\partial}{\partial p_k}\frac{\partial}{\partial x_k}\right)^{2n+1}V(\mathbf{x})w = 0.}$$

(6.4.95)

This is the quantum transport equation for the Wigner function. In the last two equations derivatives with respect to p_k are applied to $w(\mathbf{x},\mathbf{p},t)$, while derivatives with respect to x_k are applied to $V(\mathbf{x})$.

The solution to equation (6.4.95) can be sought in the form

$$w(\mathbf{x},\mathbf{p},t) = \sum_{n=0}^{\infty}\hbar^{2n}w_{2n}(\mathbf{x},\mathbf{p},t).$$

(6.4.96)

By substituting the last formula into equation (6.4.95) and equating to zero the term of zero-order with respect to $\hbar$, we find:

$$\boxed{\frac{\partial w_0}{\partial t} + \frac{\mathbf{p}}{m}\cdot\text{grad}_\mathbf{x} w_0 + \text{grad}_\mathbf{x}V(\mathbf{x})\cdot\text{grad}_\mathbf{p}w_0 = 0.}$$

(6.4.97)

This transport equation for w_0 has the mathematical form which is similar to the classical transport equation in statistical physics.

By substituting formula (6.4.96) into equation (6.4.95) and equating to zero the term of second-order with respect to $\hbar$, we obtain

$$\frac{\partial w_2}{\partial t} + \frac{\mathbf{p}}{m}\cdot\text{grad}_\mathbf{x} w_2 + \text{grad}_\mathbf{x}V(\mathbf{x})\cdot\text{grad}_\mathbf{p}w_2 = f_2,$$

(6.4.98)

where

$$\boxed{f_2 = -\frac{1}{24}\left(\sum_{k=1}^{3}\frac{\partial}{\partial p_k}\frac{\partial}{\partial x_k}\right)^3 V(\mathbf{x})w_0.}$$

(6.4.99)

Similar equations for w_{2n}, $(n > 1)$ can be derived.

Problems

(1) Prove formula (6.1.18).

(2) Prove the equality (6.1.29).

(3) Prove formulas (6.1.50) and (6.1.51).

(4) Prove formula (6.1.62).

(5) Derive formula (6.1.68) by using raising $\hat{a}^+$ and lowering $\hat{a}$ operators.

(6) Demonstrate the validity of formula (6.1.77).

(7) Prove that operators $\hat{L}_+$ and $\hat{L}_-$ are Hermitian adjoint.

(8) Derive formula (6.2.47) for the second-order correction for the eigenstates.

(9) Derive formula (6.2.49) for the third-order corrections for energy levels.

(10) Derive formulas (6.2.77)–(6.2.81).

(11) Derive the formula (6.2.88).

(12) Derive formulas (6.2.91) and (6.2.92).

(13) Derive formula (6.2.93).

(14) Prove the normal solvability conditions (6.2.102) in the case of degenerate spectrum.

(15) Derive formulas (6.3.42) and (6.3.43).

(16) Prove formula (6.3.71).

(17) Prove the validity of equations (6.4.58) and (6.4.59).

(18) Prove the equivalence of the coodinate representation (6.4.62) of the density operator to its matrix representation (6.4.17)–(6.4.18).

(19) Derive equations for w_{2n} for $n > 1$.

Chapter 7

Spin. Identical Particles. Second Quantization. Quantization of Electromagnetic Field

7.1 Spin

In all our previous discussion, it has been tacitly assumed that microscopic particles (electrons, for instance) can be fully characterized by only two intrinsic (internal) parameters: electric charge and mass. However, this is not the case. Experiments show that electrons (and other elementary particles) have another intrinsic parameter called spin. As we shall see later in this chapter, this internal parameter, spin, is of great significance because it controls the collective behavior of systems of identical particles and results in many important physical phenomena such as ferromagnetism, for instance. Furthermore, the use of spin properties of electrons has resulted in the emergence of a new area of applied science and technology called spintronics. It is currently believed that spintronics may eventually replace microelectronics at the nanoscale.

The experiments that first revealed the existence of electron spin were the observations of **doublets** in the energy spectrum of alkali metals as well as the observations of **double** splitting of beams of silver atoms (with $l = 0$) in Stern-Gerlach experiments. The beam splitting can only be attributed to the presence of angular momentum and associated with it magnetic moment. However, the **double** splitting precludes that the cause is the orbital angular momentum studied in Section 2.4 and used throughout our previous discussion. The existence of doublets in energy spectra led **W. Pauli** and (somewhat later) **G. Uhlenbeck and S. Goudsmit** to suggest that the electron has internal angular momentum termed spin. Since the electron is a structureless elementary particle, any interpretation of its spin as total angular momentum of its constituents is meaningless. This implies that the spin cannot be understood in classical terms. It must

239

be treated as an intrinsic electron parameter whose properties are similar to orbital angular momentum as suggested by Stern-Gerlach beam splitting experiments.

In quantum mechanics, every physical quantity is associated with a specific linear Hermitian operator. This brings the immediate question of how to find spin operators. We recall that the orbital angular momentum operator was defined in Section 2.4 as the linear Hermitian operator whose commutativity with the Hamiltonian for a closed system follows from the isotropicity of space, that is, from the rotational symmetry. On this basis, it was shown that the orbital angular momentum operators coincide (up to a multiplicative constant) with the operators of infinitesimally small rotations applied to spatial coordinates of wave functions. Since the spin is the intrinsic angular momentum of the electron itself, it is natural to relate spin operators to the infinitesimally small rotations of the electron itself. Indeed, the physical properties of a single (free) elementary particle must be invariant with respect to its rotations in space. This implies that the Hamiltonian operator of such a particle must be invariant with respect to such rotations and, consequently, must commute with operators of infinitesimally small rotations. This suggests that the operators of infinitesimally small rotations can be used to construct spin operators, and this is done below. In this way, the spin emerges as the internal parameter which reveals the symmetry properties of an elementary particle with respect to its rotations. In contrast, the orbital angular momentum reveals the symmetry properties of a closed system with respect to its spatial rotations.

It is known that the matrices of counter-clockwise rotations around x-, y- and z-axes of an arbitrarily chosen right-handed coordinate system are given by the following formulas, respectively,

$$\hat{R}_x(\theta_x) = \begin{bmatrix} 1 & 0 & 0 \\ 0 & \cos\theta_x & -\sin\theta_x \\ 0 & \sin\theta_x & \cos\theta_x \end{bmatrix}, \tag{7.1.1}$$

$$\hat{R}_y(\theta_y) = \begin{bmatrix} \cos\theta_y & 0 & \sin\theta_y \\ 0 & 1 & 0 \\ -\sin\theta_y & 0 & \cos\theta_y \end{bmatrix}, \tag{7.1.2}$$

$$\hat{R}_z(\theta_z) = \begin{bmatrix} \cos\theta_z & -\sin\theta_z & 0 \\ \sin\theta_z & \cos\theta_z & 0 \\ 0 & 0 & 1 \end{bmatrix}. \tag{7.1.3}$$

It is clear from the above formulas that the operators $\hat{I}_x(\delta\theta_x)$, $\hat{I}_y(\delta\theta_y)$ and $\hat{I}_z(\delta\theta_z)$ of infinitesimally small rotations around x-, y- and z-axes, respectively, are mathematically expressed as

$$\hat{I}_x(\delta\theta_x) = \delta\theta_x \hat{j}_x, \tag{7.1.4}$$

$$\hat{I}_y(\delta\theta_y) = \delta\theta_y \hat{j}_y, \tag{7.1.5}$$

$$\hat{I}_z(\delta\theta_z) = \delta\theta_z \hat{j}_z, \tag{7.1.6}$$

where the operators $\hat{j}_x$, $\hat{j}_y$ and $\hat{j}_z$ are given by the formulas

$$\hat{j}_x = \begin{bmatrix} 0 & 0 & 0 \\ 0 & 0 & -1 \\ 0 & 1 & 0 \end{bmatrix}, \tag{7.1.7}$$

$$\hat{j}_y = \begin{bmatrix} 0 & 0 & 1 \\ 0 & 0 & 0 \\ -1 & 0 & 0 \end{bmatrix}, \tag{7.1.8}$$

$$\hat{j}_z = \begin{bmatrix} 0 & -1 & 0 \\ 1 & 0 & 0 \\ 0 & 0 & 0 \end{bmatrix}. \tag{7.1.9}$$

Next, we introduce spin operators:

$$\hat{S}_x = i\hbar\hat{j}_x, \tag{7.1.10}$$

$$\hat{S}_y = i\hbar\hat{j}_y, \tag{7.1.11}$$

$$\hat{S}_z = i\hbar\hat{j}_z. \tag{7.1.12}$$

Here, the factor "i" is introduced to guarantee that operators $\hat{S}_x$, $\hat{S}_y$ and $\hat{S}_z$ are Hermitian, while the factor $\hbar$ is introduced to achieve the proper dimensions of the spin operators as well as to reflect properly the scale of values of spin.

By using formulas (7.1.7)–(7.1.12), it is easy to establish the following commutation relations

$$\hat{S}_x\hat{S}_y - \hat{S}_y\hat{S}_x = i\hbar\hat{S}_z, \tag{7.1.13}$$

$$\hat{S}_y\hat{S}_z - \hat{S}_z\hat{S}_y = i\hbar\hat{S}_x, \tag{7.1.14}$$

$$\hat{S}_z\hat{S}_x - \hat{S}_x\hat{S}_z = i\hbar\hat{S}_y. \tag{7.1.15}$$

These commutation relations are mathematically identical to the commutation relations for orbital angular momentum operators $\hat{L}_x$, $\hat{L}_y$ and $\hat{L}_z$ (see formulas (2.4.37)–(2.4.39)). By using this fact and by literally repeating the same reasoning as in Section 2.4, it can be established that the operator

$$\hat{\mathbf{S}}^2 = \hat{S}_x^2 + \hat{S}_y^2 + \hat{S}_z^2 \tag{7.1.16}$$

commutes with operator $\hat{S}_z$,

$$\hat{\mathbf{S}}^2 \hat{S}_z - \hat{S}_z \hat{\mathbf{S}}^2 = 0. \tag{7.1.17}$$

This implies that the physical quantities corresponding to operators $\hat{\mathbf{S}}^2$ and $\hat{S}_z$ are simultaneously measurable. Let us denote by $|sm\rangle$ the state in which these physical quantities have values $\hbar^2 s(s+1)$ and $\hbar m$, respectively:

$$\hat{\mathbf{S}}^2 |sm\rangle = \hbar^2 s(s+1)|sm\rangle, \tag{7.1.18}$$

$$\hat{S}_z |sm\rangle = \hbar m |sm\rangle. \tag{7.1.19}$$

By using the same line of reasoning as in Section 2.4, it can be demonstrated that for a given s, m may assume only the following values

$$-s, s+1, \dots s-1, s. \tag{7.1.20}$$

Since these sequential values are incremented by one, the difference $2s$ between the largest and the smallest possible values of s must be an integer or zero. This implies that s must be a multiple of $\frac{1}{2}$. Consequently, the possible values of s are:

$$s = 0, \frac{1}{2}, 1, \frac{3}{2}, \dots \tag{7.1.21}$$

Experiments show that for electrons

$$s = \frac{1}{2}, \tag{7.1.22}$$

and, according to (7.1.19) and (7.1.20) S_z has only two possible values

$$S_{z+} = \frac{\hbar}{2}, \quad S_{z-} = -\frac{\hbar}{2}. \tag{7.1.23}$$

Only this case of electrons with the spin $\frac{1}{2}$ is subsequently discussed. The spin of protons and neutrons is also $\frac{1}{2}$, and many facts discussed below are applicable to these particles as well.

It is clear from formulas (7.1.22) and (7.1.23) that for electrons there are only two distinct eigenstates $|sm\rangle$ corresponding to spin up ($\uparrow$) and spin down ($\downarrow$), respectively,

$$\left|\frac{1}{2}\frac{1}{2}\right\rangle \quad \text{and} \quad \left|\frac{1}{2}\left(-\frac{1}{2}\right)\right\rangle. \tag{7.1.24}$$

These eigenstates can be described by two-component vectors

$$\boldsymbol{\xi}_+ = \begin{pmatrix} 1 \\ 0 \end{pmatrix} \text{ and } \boldsymbol{\xi}_- = \begin{pmatrix} 0 \\ 1 \end{pmatrix}, \tag{7.1.25}$$

and the general spin-state of an electron is a linear combination of the above two eigenstates

$$\boldsymbol{\xi} = a_+ \boldsymbol{\xi}_+ + a_- \boldsymbol{\xi}_-, \tag{7.1.26}$$

where $|a_+|^2$ and $|a_-|^2$ are probabilities that the measurements of S_z in the state (7.1.26) produce the results $\frac{\hbar}{2}$ and $-\frac{\hbar}{2}$, respectively. Consequently,

$$|a_+|^2 + |a_-|^2 = 1. \tag{7.1.27}$$

It is also clear that equation (7.1.19) can be written for electrons as follows

$$\hat{S}_z \boldsymbol{\xi}_+ = \frac{\hbar}{2} \boldsymbol{\xi}_+, \tag{7.1.28}$$

$$\hat{S}_z \boldsymbol{\xi}_- = -\frac{\hbar}{2} \boldsymbol{\xi}_-. \tag{7.1.29}$$

Formula (7.1.25) and the last two equations imply that operator $\hat{S}_z$ can be represented by the two-dimensional matrix

$$\boxed{\hat{S}_z = \frac{\hbar}{2} \begin{pmatrix} 1 & 0 \\ 0 & -1 \end{pmatrix}.} \tag{7.1.30}$$

Next, we shall find two-dimensional matrix representations for operators $\hat{S}_x$ and $\hat{S}_y$. To this end, we shall use the "promotion" $\hat{S}_+$ and "demotion" $\hat{S}_-$ operators mathematically similar to operators $\hat{L}_+$ and $\hat{L}_-$ from Section 2.4:

$$\hat{S}_+ = \hat{S}_x + i\hat{S}_y, \tag{7.1.31}$$

$$\hat{S}_- = \hat{S}_x - i\hat{S}_y. \tag{7.1.32}$$

By using the same line of reasoning as in Section 2.4, it can be shown that

$$\hat{S}_+ \boldsymbol{\xi}_- = \hbar \boldsymbol{\xi}_+, \tag{7.1.33}$$

$$\hat{S}_+ \boldsymbol{\xi}_+ = \begin{pmatrix} 0 \\ 0 \end{pmatrix}, \tag{7.1.34}$$

$$\hat{S}_- \boldsymbol{\xi}_+ = \hbar \boldsymbol{\xi}_-, \tag{7.1.35}$$

$$\hat{S}_-\boldsymbol{\xi}_- = \begin{pmatrix} 0 \\ 0 \end{pmatrix}. \tag{7.1.36}$$

Let

$$\hat{S}_+ = \begin{pmatrix} a & b \\ c & d \end{pmatrix}. \tag{7.1.37}$$

Then from equation (7.1.33) we find

$$\hat{S}_+\boldsymbol{\xi}_- = \begin{pmatrix} a & b \\ c & d \end{pmatrix}\begin{pmatrix} 0 \\ 1 \end{pmatrix} = \begin{pmatrix} b \\ d \end{pmatrix} = \hbar\begin{pmatrix} 1 \\ 0 \end{pmatrix}. \tag{7.1.38}$$

Consequently,

$$b = \hbar, \quad d = 0. \tag{7.1.39}$$

Similarly, from equation (7.1.34), we derive

$$\hat{S}_+\boldsymbol{\xi}_+ = \begin{pmatrix} a & b \\ c & d \end{pmatrix}\begin{pmatrix} 1 \\ 0 \end{pmatrix} = \begin{pmatrix} a \\ c \end{pmatrix} = \hbar\begin{pmatrix} 0 \\ 0 \end{pmatrix}. \tag{7.1.40}$$

Consequently,

$$a = 0, \quad c = 0. \tag{7.1.41}$$

Thus, we have found that

$$\hat{S}_+ = \hbar\begin{pmatrix} 0 & 1 \\ 0 & 0 \end{pmatrix}. \tag{7.1.42}$$

By using the same line of reasoning, it can be shown that

$$\hat{S}_- = \hbar\begin{pmatrix} 0 & 0 \\ 1 & 0 \end{pmatrix}. \tag{7.1.43}$$

From formulas (7.1.31) and (7.1.32) follows that

$$\hat{S}_x = \frac{1}{2}(\hat{S}_+ + \hat{S}_-), \tag{7.1.44}$$

$$\hat{S}_y = -\frac{i}{2}(\hat{S}_+ - \hat{S}_-). \tag{7.1.45}$$

From the last two formulas as well as formulas (7.1.44) and (7.1.45) we conclude that

$$\boxed{\hat{S}_x = \frac{\hbar}{2}\begin{pmatrix} 0 & 1 \\ 1 & 0 \end{pmatrix},} \tag{7.1.46}$$

$$\boxed{\hat{S}_y = \frac{\hbar}{2}\begin{pmatrix} 0 & -i \\ i & 0 \end{pmatrix}.} \tag{7.1.47}$$

It is clear that formulas (7.1.30), (7.1.46) and (7.1.47) can be written as follows

$$\hat{S}_x = \frac{\hbar}{2}\hat{\sigma}_x, \quad \hat{S}_y = \frac{\hbar}{2}\hat{\sigma}_y, \quad \hat{S}_z = \frac{\hbar}{2}\hat{\sigma}_z, \tag{7.1.48}$$

where $\hat{\sigma}_x$, $\hat{\sigma}_y$ and $\hat{\sigma}_z$ are the famous **Pauli matrices**

$$\hat{\sigma}_x = \begin{pmatrix} 0 & 1 \\ 1 & 0 \end{pmatrix}, \tag{7.1.49}$$

$$\hat{\sigma}_y = \begin{pmatrix} 0 & -i \\ i & 0 \end{pmatrix}, \tag{7.1.50}$$

$$\hat{\sigma}_z = \begin{pmatrix} 1 & 0 \\ 0 & -1 \end{pmatrix}, \tag{7.1.51}$$

which satisfy the commutation relations

$$[\hat{\sigma}_x, \hat{\sigma}_y] = 2i\hat{\sigma}_z, \quad [\hat{\sigma}_y, \hat{\sigma}_z] = 2i\hat{\sigma}_x, \quad [\hat{\sigma}_z, \hat{\sigma}_x] = 2i\hat{\sigma}_y. \tag{7.1.52}$$

It can be shown (and left as an exercise for the reader) that the eigenvalues of $\hat{S}_x$ and $\hat{S}_y$ are $\frac{\hbar}{2}$ and $-\frac{\hbar}{2}$, while the corresponding eigenvectors are given by the formulas

$$\boldsymbol{\xi}_+^{(x)} = \begin{pmatrix} \frac{1}{\sqrt{2}} \\ \frac{1}{\sqrt{2}} \end{pmatrix}, \quad \boldsymbol{\xi}_-^{(x)} = \begin{pmatrix} \frac{1}{\sqrt{2}} \\ -\frac{1}{\sqrt{2}} \end{pmatrix}, \tag{7.1.53}$$

$$\boldsymbol{\xi}_+^{(y)} = i\begin{pmatrix} \frac{1}{\sqrt{2}} \\ \frac{1}{\sqrt{2}} \end{pmatrix}, \quad \boldsymbol{\xi}_-^{(y)} = i\begin{pmatrix} \frac{1}{\sqrt{2}} \\ -\frac{1}{\sqrt{2}} \end{pmatrix}. \tag{7.1.54}$$

An attentive reader may have noticed that in the case of $s = \frac{1}{2}$ there are two matrix representations of spin operators $\hat{S}_x$, $\hat{S}_y$ and $\hat{S}_z$: by three-dimensional matrices (see formulas (7.1.7)—(7.1.12)) and by two-dimensional matrices (see formulas (7.1.30), (7.1.46), and (7.1.47)). Matrices (7.1.7), (7.1.8) and (7.1.9) belong to the class of three-dimensional orthogonal matrices with determinants equal to $+1$. On the other hand, Pauli matrices (7.1.49), (7.1.50) and (7.1.51) belong to the class of two-dimensional Hermitian matrices with determinants equal to -1. The fact that the same spin operators can be represented by the matrices from these two different classes suggests that there exists some connection between the above two classes of matrices. This connection is studied in the group theory (namely, in the theory of groups SO(3) and SU(2)).

Next, as an example, consider a system of two electrons with spins $\hat{\mathbf{S}}_1$ and $\hat{\mathbf{S}}_2$. We want to find out what the total spin $\hat{\mathbf{S}}$ of this system is,

$$\hat{\mathbf{S}} = \hat{\mathbf{S}}_1 + \hat{\mathbf{S}}_2. \tag{7.1.55}$$

where $\hat{\mathbf{S}}_1 = \mathbf{e}_x S_{1x} + \mathbf{e}_y S_{1y} + \mathbf{e}_z S_{1z}$ and $\hat{\mathbf{S}}_2 = \mathbf{e}_x S_{2x} + \mathbf{e}_y S_{2y} + \mathbf{e}_z S_{2z}$.

Consider first

$$\hat{S}_z = \hat{S}_{1z} + \hat{S}_{2z}. \tag{7.1.56}$$

It is clear that as far as $\hat{S}_z$ is concerned there are four distinct states: ↑↑ which is denoted as $|++\rangle$, ↑↓ which is denoted as $|+-\rangle$, ↓↑ which is denoted as $|-+\rangle$ and ↓↓ denoted as $|--\rangle$. In terms of two-dimensional vectors (spinors) $\boldsymbol{\xi}_+$ and $\boldsymbol{\xi}_-$ (see formula (7.1.25)) these states can be represented as follows

$$|++\rangle = \begin{pmatrix} 1 \\ 0 \end{pmatrix}_1 \begin{pmatrix} 1 \\ 0 \end{pmatrix}_2 \Rightarrow s_z = 1, \tag{7.1.57}$$

$$|+-\rangle = \begin{pmatrix} 1 \\ 0 \end{pmatrix}_1 \begin{pmatrix} 0 \\ 1 \end{pmatrix}_2 \Rightarrow s_z = 0, \tag{7.1.58}$$

$$|-+\rangle = \begin{pmatrix} 0 \\ 1 \end{pmatrix}_1 \begin{pmatrix} 1 \\ 0 \end{pmatrix}_2 \Rightarrow s_z = 0, \tag{7.1.59}$$

$$|--\rangle = \begin{pmatrix} 0 \\ 1 \end{pmatrix}_1 \begin{pmatrix} 0 \\ 1 \end{pmatrix}_2 \Rightarrow s_z = -1. \tag{7.1.60}$$

In the last four formulas subscripts "1" and "2" for two-dimensional vectors indicate that they represent the first and second spins, respectively, while the total values of s_z immediately follow from formula (7.1.56).

Next, we demonstrate that as far as the total spin s is concerned there are three states (triplet) with $s = 1$ and one state (singlet) with $s = 0$:

triplet

$$\left. \begin{array}{l} |1,1\rangle = |++\rangle \\ |1,0\rangle = \frac{1}{\sqrt{2}}\left(|+-\rangle + |-+\rangle\right) \\ |1,-1\rangle = |--\rangle \end{array} \right\} \Rightarrow s = 1, \tag{7.1.61}$$

singlet

$$|0,0\rangle = \frac{1}{\sqrt{2}}\left(|+-\rangle - |-+\rangle\right) \Rightarrow s = 0. \tag{7.1.62}$$

The demonstration is performed by using the formula

$$\hat{\mathbf{S}}^2 = (\hat{\mathbf{S}}_1 + \hat{\mathbf{S}}_2)^2 = \hat{\mathbf{S}}_1^2 + \hat{\mathbf{S}}_2^2 + 2(\hat{S}_{1x}\hat{S}_{2x} + \hat{S}_{1y}\hat{S}_{2y} + \hat{S}_{1z}\hat{S}_{2z}). \tag{7.1.63}$$

It is done below for the state $|++\rangle$. It is apparent that

$$\hat{S}_1^2|++\rangle = \frac{3}{4}\hbar^2|++\rangle, \tag{7.1.64}$$

$$\hat{S}_2^2|++\rangle = \frac{3}{4}\hbar^2|++\rangle. \tag{7.1.65}$$

Furthermore, by using formulas (7.1.46), we find:

$$\hat{S}_{1x}\hat{S}_{2x}|++\rangle = \hat{S}_{1x}\begin{pmatrix}1\\0\end{pmatrix}_1 \hat{S}_{2x}\begin{pmatrix}1\\0\end{pmatrix}_2 = \frac{\hbar^2}{4}\begin{pmatrix}0\\1\end{pmatrix}_1\begin{pmatrix}0\\1\end{pmatrix}_2 = \frac{\hbar^2}{4}|--\rangle. \tag{7.1.66}$$

Similarly,

$$\hat{S}_{1y}\hat{S}_{2y}|++\rangle = \hat{S}_{1y}\begin{pmatrix}1\\0\end{pmatrix}_1 \hat{S}_{2y}\begin{pmatrix}1\\0\end{pmatrix}_2 = \frac{(-i)^2\hbar^2}{4}\begin{pmatrix}0\\1\end{pmatrix}_1\begin{pmatrix}0\\1\end{pmatrix}_2 = -\frac{\hbar^2}{4}|--\rangle, \tag{7.1.67}$$

and

$$\hat{S}_{1x}\hat{S}_{2x}|++\rangle + \hat{S}_{1y}\hat{S}_{2y}|++\rangle = 0. \tag{7.1.68}$$

Finally,

$$\hat{S}_{1z}\hat{S}_{2z}|++\rangle = \frac{\hbar^2}{4}\begin{pmatrix}1\\0\end{pmatrix}_1\begin{pmatrix}1\\0\end{pmatrix}_2 = \frac{\hbar^2}{4}|++\rangle. \tag{7.1.69}$$

From formulas (7.1.63), (7.1.64), (7.1.65), (7.1.68) and (7.1.69) follows that

$$\hat{S}^2|++\rangle = 2\hbar^2|++\rangle. \tag{7.1.70}$$

On the other hand,

$$\hat{S}^2|++\rangle = \hbar^2 s(s+1)|++\rangle. \tag{7.1.71}$$

By comparing the last two formulas, we find

$$s = 1, \tag{7.1.72}$$

and, consequently the state $|++\rangle$ is the state $|11\rangle$. It is left as an exercise for the reader to finish the proof of formula (7.1.61) and prove the formula (7.1.62).

The discussed example is a particular case of a more general problem of addition of two different spins or two different orbital angular momenta. Namely, if we combine two spins with $s = s_1$ and $s = s_2$, then there are states with the following values of the total spin

$$s_1 + s_2, \ s_1 + s_2 - 1, \ s_1 + s_2 - 2, \ \ldots |s_1 - s_2|. \tag{7.1.73}$$

The combined states $|sm\rangle$ with total spin s and $s_z = m$ are superpositions of various composites $|s_1 m_1\rangle|s_2 m_2\rangle$ of the individual states. These superpositions are given by the formula

$$|sm\rangle = \sum_{m_1 + m_2 = m} C^{s_1, s_2, s}_{m_1, m_2, m} |s_1 m_1\rangle|s_2 m_2\rangle, \qquad (7.1.74)$$

where $C^{s_1, s_2, s}_{m_1, m_2, m}$ are known as **Clebsch-Gordan coefficients**.

Similarly, in the case of addition of two orbital angular momenta l_1 and l_2 there are states with the following values of the total angular momentum l:

$$l = l_1 + l_2, \, l_1 + l_2 - 1, \, l_1 + l_2 - 2, \, \ldots |l_1 - l_2|. \qquad (7.1.75)$$

There is a formula for the wave functions of states with above values of l which is similar to formula (7.1.74). General formulas for Clebsh-Gordan coefficients can be derived using the group theory.

Now, after the above digression, we shall proceed to the further discussion of electron spin. It is known that the orbital angular momentum is related to the magnetic moment by the formula

$$\mathbf{M} = \gamma \mathbf{L}, \qquad (7.1.76)$$

where γ is the gyromagnetic ratio given by the formula

$$\gamma = \frac{e}{2m}. \qquad (7.1.77)$$

It turns out that electrons have magnetic moment as well which is related to their spin by the formula

$$\mathbf{M}_e = \gamma_e \mathbf{S}, \qquad (7.1.78)$$

where γ_e is the gyromagnetic ratio of electrons which is equal to

$$\gamma_e = \frac{e}{m} = 2\gamma. \qquad (7.1.79)$$

The last formula can be derived from **the Dirac relativistic theory** of the electron. This formula is also in agreement with measurements performed by using the **Einstein-de Haas experiment**. The difference between γ and γ_e can be viewed as another indication of different physical nature of $\mathbf{L}$ and $\mathbf{S}$.

Next, we shall discuss how the Schrödinger equation of a spinless particle must be modified to account for electron spin. First, we note that the wave function of an electron must be dependent on the z-component of electron spin

$$\psi(\mathbf{r}, s_z, t). \qquad (7.1.80)$$

Since s_z may assume only two values $\frac{1}{2}$ and $-\frac{1}{2}$, the wave function (7.1.80) can be mathematically viewed as a two-component (vectorial) wave function

$$\boldsymbol{\psi}(\mathbf{r}, t) = \begin{pmatrix} \psi^+(\mathbf{r}, t) \\ \psi^-(\mathbf{r}, t) \end{pmatrix}, \tag{7.1.81}$$

where $\psi^+(\mathbf{r}, t)$ and $\psi^-(\mathbf{r}, t)$ can be viewed as spin up and spin down wave functions, respectively. It is apparent that the wave function $\boldsymbol{\psi}(\mathbf{r}, t)$ can be written as follows

$$\boldsymbol{\psi}(\mathbf{r}, t) = \psi^+(\mathbf{r}, t)\boldsymbol{\xi}_+ + \psi^-(\mathbf{r}, t)\boldsymbol{\xi}_-, \tag{7.1.82}$$

where $\boldsymbol{\xi}_+$ and $\boldsymbol{\xi}_-$ are defined by formula (7.1.25).

It is clear that the wave equation for an electron with spin should be written with respect to the wave function $\boldsymbol{\psi}(\mathbf{r}, t)$. It is also clear that the previous Hamiltonian

$$\hat{H}_0 = \frac{(\hat{\mathbf{p}} + e\mathbf{A})^2}{2m} - e\varphi \tag{7.1.83}$$

should be modified to account for the energy of electron magnetic moment in magnetic field. The classical expression for this energy is given by formula

$$\mathcal{E} = -\mathbf{M}_e \cdot \mathbf{B}. \tag{7.1.84}$$

It is natural to modify the Hamiltonian (7.1.83) by adding the term which is obtained from formula (7.1.84) by replacing $\mathbf{M}_e$ with the operator $\gamma_e \hat{\mathbf{S}} = \gamma_e \frac{\hbar}{2} \hat{\boldsymbol{\sigma}}$. Thus,

$$\boxed{\hat{H} = \frac{(\hat{\mathbf{p}} + e\mathbf{A})^2}{2m} - e\varphi - \frac{\gamma_e \hbar}{2} \mathbf{B} \cdot \hat{\boldsymbol{\sigma}}.} \tag{7.1.85}$$

In the last formula $\mathbf{B} \cdot \hat{\boldsymbol{\sigma}}$ has the following meaning

$$\mathbf{B} \cdot \hat{\boldsymbol{\sigma}} = B_x \hat{\sigma}_x + B_y \hat{\sigma}_y + B_z \hat{\sigma}_z, \tag{7.1.86}$$

where $\hat{\sigma}_x$, $\hat{\sigma}_y$ and $\hat{\sigma}_z$ are Pauli matrices given by formulas (7.1.49)–(7.1.51).

Now, the time-dependent wave equation for an electron with spin can be written as follows

$$\boxed{i\hbar \frac{\partial}{\partial t} \begin{pmatrix} \psi_+ \\ \psi_- \end{pmatrix} = \left[\frac{(\hat{\mathbf{p}} + e\mathbf{A})^2}{2m} - e\varphi \right] \begin{pmatrix} \psi_+ \\ \psi_- \end{pmatrix} - \frac{e\hbar}{2m} \mathbf{B} \cdot \hat{\boldsymbol{\sigma}} \begin{pmatrix} \psi_+ \\ \psi_- \end{pmatrix}.} \tag{7.1.87}$$

This is the so-called **Pauli equation**. It is worthwhile to mention that the Hamiltonian (7.1.85) can be written in another (equivalent and more compact) form

$$\hat{H} = \frac{\hat{\boldsymbol{\sigma}} \cdot (\hat{\mathbf{p}} + e\mathbf{A})^2}{2m} - e\varphi. \tag{7.1.88}$$

This form of the Hamiltonian is convenient in order to prove that the Pauli equation is invariant with respect to local change of phase of the wave function (see Section 2.5).

Next, we consider the problem of an electron in uniform magnetic field $\mathbf{B}_0$ which was discussed in Section 4.4 without taking into account electron spin. As before, it is assumed that the magnetic field direction is along the z-axis:

$$\mathbf{B}_0 = \mathbf{e}_z B_0. \tag{7.1.89}$$

It is also assumed that there is no applied electric field. Consequently,

$$\varphi = 0. \tag{7.1.90}$$

Now, the Pauli equation (7.1.87) can be written as

$$i\hbar \frac{\partial}{\partial t} \begin{pmatrix} \psi_+ \\ \psi_- \end{pmatrix} = \frac{(\hat{\mathbf{p}} + e\mathbf{A})^2}{2m} \begin{pmatrix} \psi_+ \\ \psi_- \end{pmatrix} - \frac{e\hbar}{2m} B_0 \hat{\sigma}_z \begin{pmatrix} \psi_+ \\ \psi_- \end{pmatrix}. \tag{7.1.91}$$

By recalling that

$$\omega_c = \frac{eB_0}{m} \tag{7.1.92}$$

is the cyclotron frequency, from equation (7.1.91), we derive the following two equations for the "spin up" and "spin down" energy spectra, respectively,

$$\frac{(\hat{\mathbf{p}} + e\mathbf{A})^2}{2m} \Phi^+ = \left(\mathcal{E}^+ + \frac{\hbar\omega_c}{2} \right) \Phi^+, \tag{7.1.93}$$

$$\frac{(\hat{\mathbf{p}} + e\mathbf{A})^2}{2m} \Phi^- = \left(\mathcal{E}^- - \frac{\hbar\omega_c}{2} \right) \Phi^-. \tag{7.1.94}$$

The last two equations are mathematically identical to the eigenvalue equation for Landau energy levels studied in Section 4.4. Consequently, by using formula (4.4.34), we find

$$\mathcal{E}_k^+ = \mathcal{E}_k - \frac{\hbar\omega_c}{2} = \hbar\omega_c k + \frac{p_z^2}{2m}, \quad (k = 0, 1, 2, \dots), \tag{7.1.95}$$

$$\mathcal{E}_k^- = \mathcal{E}_k + \frac{\hbar\omega_c}{2} = \hbar\omega_c(k + 1) + \frac{p_z^2}{2m}, \quad (k = 0, 1, 2, \dots). \tag{7.1.96}$$

The discrete components of spin up and spin down energy spectra are illustrated by Figure 7.1. Thus, it is clear that the effect of electron spin leads to the splitting of Landau energy levels and to the formation of spin up and spin down energy spectra. Somewhat similar physical phenomena occur in

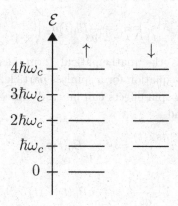

Fig. 7.1

ferromagnetic conductors where spin up and spin down energy sub-bands appear.

Next, consider the case of time-varying magnetic field

$$\mathbf{B}_0(t) = \mathbf{e}_z B_0(t). \tag{7.1.97}$$

The time-dependent Pauli equation (7.1.87) can be written in this case as follows

$$i\hbar \frac{\partial}{\partial t} \begin{pmatrix} \psi_+ \\ \psi_- \end{pmatrix} = \hat{H}_0 \begin{pmatrix} \psi_+ \\ \psi_- \end{pmatrix} - \frac{e\hbar}{2m} B_0(t) \hat{\sigma}_z \begin{pmatrix} \psi_+ \\ \psi_- \end{pmatrix}, \tag{7.1.98}$$

where $\hat{H}_0$ is the Hamiltonian specified by formula (7.1.83).

We shall look for the solution of equation (7.1.98) in the form

$$\begin{pmatrix} \psi_+(\mathbf{r}, t) \\ \psi_-(\mathbf{r}, t) \end{pmatrix} = \varphi(\mathbf{r}, t) \begin{pmatrix} s_+(t) \\ s_-(t) \end{pmatrix}. \tag{7.1.99}$$

By substituting the last formula into equation (7.1.98), we find

$$i\hbar \frac{\partial \varphi(\mathbf{r}, t)}{\partial t} \begin{pmatrix} s_+(t) \\ s_-(t) \end{pmatrix} + i\hbar \varphi(\mathbf{r}, t) \frac{d}{dt} \begin{pmatrix} s_+(t) \\ s_-(t) \end{pmatrix}$$

$$= (\hat{H}_0 \varphi(\mathbf{r}, t)) \begin{pmatrix} s_+(t) \\ s_-(t) \end{pmatrix} - \varphi(\mathbf{r}, t) \frac{e\hbar}{2m} B_0(t) \hat{\sigma}_z \begin{pmatrix} s_+(t) \\ s_-(t) \end{pmatrix}. \tag{7.1.100}$$

The last equation can be split into two equations:

$$i\hbar \frac{\partial \varphi(\mathbf{r}, t)}{\partial t} = \hat{H}_0 \varphi(\mathbf{r}, t), \tag{7.1.101}$$

$$i\frac{d}{dt}\begin{pmatrix} s_+(t) \\ s_-(t) \end{pmatrix} = -\frac{e}{2m}B_0(t)\hat{\sigma}_z\begin{pmatrix} s_+(t) \\ s_-(t) \end{pmatrix}. \qquad (7.1.102)$$

Equation (7.1.101) has the mathematical form identical to the time-dependent Schrödinger equation for a spinless particle, while the last equation which accounts for spin effects can be easily integrated. Indeed, from the last equation we find

$$i\frac{ds_+(t)}{dt} = -\frac{e}{2m}B_0(t)s_+(t), \qquad (7.1.103)$$

$$i\frac{ds_-(t)}{dt} = \frac{e}{2m}B_0(t)s_-(t). \qquad (7.1.104)$$

The solution of the last two equations can be written, respectively, as follows

$$s_+(t) = s_+(0)e^{i\frac{e}{2m}\int_0^t B_0(\tau)d\tau}, \qquad (7.1.105)$$

$$s_-(t) = s_-(0)e^{-i\frac{e}{2m}\int_0^t B_0(\tau)d\tau}. \qquad (7.1.106)$$

Thus, the dynamics of $\varphi(\mathbf{r}, t)$ and $\begin{pmatrix} s_+(t) \\ s_-(t) \end{pmatrix}$ are completely decoupled.

Finally, we conclude this section by deriving the differential equation for the average value of spin. To this end, we shall write the Hamiltonian in formula (7.1.85) in the form

$$\hat{H} = \hat{H}_0 - \gamma_e\mathbf{B}(t) \cdot \hat{\mathbf{S}} \qquad (7.1.107)$$

and recall formula (2.2.41) from Section 2.2. According to that formula

$$\frac{d\hat{\mathbf{S}}}{dt} = \frac{i}{\hbar}\left[\hat{\mathbf{S}}, \hat{H}\right]. \qquad (7.1.108)$$

Since the Hamiltonian $\hat{H}_0$ does not depend on $\hat{\mathbf{S}}$ (see formula (7.1.83)), we find that

$$\left[\hat{\mathbf{S}}, \hat{H}_0\right] = 0. \qquad (7.1.109)$$

Consequently, equation (7.1.108) can be written as follows

$$\frac{d\hat{\mathbf{S}}}{dt} = -\frac{i\gamma_e}{\hbar}\left[\hat{\mathbf{S}}, \mathbf{B}(t)\hat{\mathbf{S}}\right]. \qquad (7.1.110)$$

It is clear that

$$\left[\hat{S}_x, \mathbf{B}(t)\hat{\mathbf{S}}\right] = \left[\hat{S}_x, B_x(t)\hat{S}_x + B_y(t)\hat{S}_y + B_z(t)\hat{S}_z\right]. \qquad (7.1.111)$$

By recalling the commutation relations (7.1.13)–(7.1.15), from the last formula we find

$$[\hat{S}_x, \mathbf{B}(t)\hat{\mathbf{S}}] = i\hbar \left(B_y(t)\hat{S}_z - B_z(t)\hat{S}_y \right), \tag{7.1.112}$$

or ,

$$\left[\hat{S}_x, \mathbf{B}(t)\hat{\mathbf{S}}\right] = -i\hbar \left(\hat{\mathbf{S}} \times \mathbf{B}(t) \right)_x. \tag{7.1.113}$$

In the same way, it can be established that

$$[\hat{S}_y, \mathbf{B}(t)\hat{\mathbf{S}}] = -i\hbar \left(\hat{\mathbf{S}} \times \mathbf{B}(t) \right)_y, \tag{7.1.114}$$

$$[\hat{S}_z, \mathbf{B}(t)\hat{\mathbf{S}}] = -i\hbar \left(\hat{\mathbf{S}} \times \mathbf{B}(t) \right)_z. \tag{7.1.115}$$

The last three formulas imply that

$$\left[\hat{\mathbf{S}}, \mathbf{B}(t)\hat{\mathbf{S}}\right] = -i\hbar \left(\hat{\mathbf{S}} \times \mathbf{B}(t) \right). \tag{7.1.116}$$

By substituting the last formula into equation (7.1.110), we obtain

$$\boxed{\frac{d\hat{\mathbf{S}}}{dt} = -\gamma_e(\hat{\mathbf{S}} \times \mathbf{B}(t)).} \tag{7.1.117}$$

By taking the quantum mechanical average of both sides of the last equation and by interpreting the quantum mechanical average of $\gamma_e\hat{\mathbf{S}}$ as the average magnetic moment $\mathbf{M}$ (see equation (7.1.78)), from the last formula we derive

$$\boxed{\frac{d\mathbf{M}}{dt} = -\gamma_e \left(\mathbf{M} \times \mathbf{B}(t) \right).} \tag{7.1.118}$$

It is apparent from the last equation that

$$\mathbf{M} \cdot \frac{d\mathbf{M}}{dt} = 0. \tag{7.1.119}$$

This implies that

$$|\mathbf{M}(t)| = M_0 = \text{const}, \tag{7.1.120}$$

which means that the tip of $\mathbf{M}(t)$ moves along the sphere defined by equation (7.1.120). In a particular case of uniform and constant in time magnetic field $\mathbf{B}_0 = \mathbf{e}_z B_0$, equation (7.1.118) can be written as follows

$$\frac{d\mathbf{M}}{dt} = -\gamma_e B_0(\mathbf{M} \times \mathbf{e}_z). \tag{7.1.121}$$

It is easy to show that the solution of the last equation has the form

$$M_x(t) = A\sin(\omega t + \varphi), \tag{7.1.122}$$

$$M_y(t) = A\cos(\omega t + \varphi), \tag{7.1.123}$$

$$M_z = \text{const}, \tag{7.1.124}$$

and

$$\omega = \gamma_e B_0. \tag{7.1.125}$$

The last formulas suggest that the dynamics of $\mathbf{M}(t)$ is the so-called **Larmor precessions** of $\mathbf{M}(t)$ around the z-axis (around the applied magnetic field) with constant frequency ω called the **Larmor frequency** (see Figure 7.2).

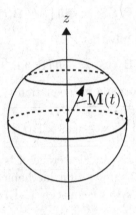

Fig. 7.2

7.2 Identical Particles. Exchange Interaction

In nature, there are identical particles which have the same intrinsic physical properties such as mass, electric charge and spin. Examples of such identical particles are electrons, protons and photons. (Photons will be discussed in the last section of this chapter.) In classical physics, identical particles can be distinguished because they move along distinct trajectories. Indeed, if identical particles are enumerated at some instant of time in accordance with their measured spatial locations, then each of them can be followed according to their motion along distinct trajectories and consequently, they can be in principle distinguished from one another at subsequent instants of time despite their identical intrinsic physical properties.

In quantum mechanics, the notion of motion along a trajectory is not valid. For this reason, if a spatial location of one of the identical particles is measured at some instant of time, then it is not possible in principle to identify if the same particle was localized by position measurements performed at subsequent instants of time. In this way, the identical particles lose their distinguishability in quantum mechanics, and we arrive at **the principle of indistinguishability of identical particles**. This principle implies that the numeration of identical particles is arbitrary, and that wave functions of ensembles of identical particles should be invariant (up to a phase factor) with respect to the permutation (changing) of numeration of identical particles.

Consider the wave function of N identical particles

$$\psi(\eta_1, \eta_2, \ldots \eta_k, \ldots \eta_j, \ldots \eta_N, t), \tag{7.2.1}$$

where for the sake of brevity symbols η_k are used for the notation of spatial coordinates and spins of particles. Now, we introduce the operator $\hat{P}_{kj}$ of permutation (exchange) of the order of two particles

$$\hat{P}_{kj}\psi(\eta_1, \eta_2, \ldots \eta_k, \ldots \eta_j, \ldots \eta_N, t) = \psi(\eta_1, \eta_2, \ldots \eta_j, \ldots \eta_k, \ldots \eta_N, t). \tag{7.2.2}$$

It is apparent that the description of a system of N identical particles by the wave function $\psi(\eta_1, \eta_2, \ldots \eta_j, \ldots \eta_k, \ldots \eta_N, t)$ is equivalent to the description of the same system of particles by the original wave function (7.2.1). This implies that

$$\psi(\eta_1, \eta_2, \ldots \eta_j, \ldots \eta_k, \ldots \eta_N, t) = e^{i\alpha}\psi(\eta_1, \eta_2, \ldots \eta_k, \ldots \eta_j, \ldots \eta_N, t). \tag{7.2.3}$$

It is clear from the last two formulas that

$$\hat{P}_{kj}\psi(\eta_1, \eta_2, \ldots \eta_k, \ldots \eta_j, \ldots \eta_N, t) = e^{i\alpha}\psi(\eta_1, \eta_2, \ldots \eta_k, \ldots \eta_j, \ldots \eta_N, t). \tag{7.2.4}$$

This means that the operator of permutation of two identical particles can be viewed as the operator of multiplication by $e^{i\alpha}$.

It is also clear from formula (7.2.2) that by applying the operator of permutation of two identical particles two consecutive times we obtain the original wave function, i.e., the wave function with the original numeration of elementary particles:

$$\hat{P}_{kj}^2\psi(\eta_1, \eta_2, \ldots \eta_k, \ldots \eta_j, \ldots \eta_N, t) = \psi(\eta_1, \eta_2, \ldots \eta_k, \ldots \eta_j, \ldots \eta_N, t). \tag{7.2.5}$$

However, according to formula (7.2.4), we have

$$\hat{P}_{kj}^2\psi = \hat{P}_{kj}\left(\hat{P}_{kj}\psi\right) = e^{i\alpha}\hat{P}_{kj}\psi = e^{i2\alpha}\psi. \tag{7.2.6}$$

By comparing formulas (7.2.5) and (7.2.6), we conclude that

$$e^{i2\alpha} = 1, \tag{7.2.7}$$

and

$$e^{i\alpha} = \pm 1. \tag{7.2.8}$$

Thus, it can be concluded that there may exist two distinct types of identical particles: the identical particles whose wave functions $\psi^{(s)}$ are symmetric with respect to permutations of any two identical particles

$$\hat{P}_{kj}\psi^{(s)} = \psi^{(s)}, \tag{7.2.9}$$

and the identical particles whose wave functions $\psi^{(a)}$ are antisymmetric with respect to permutations of any two identical particles

$$\hat{P}_{kj}\psi^{(a)} = -\psi^{(a)}. \tag{7.2.10}$$

Next, we demonstrate that the Hamiltonian of a system of identical particles commutes with the two particle permutation operators. Indeed, according to the time-dependent Schrödinger equation we have

$$i\hbar\frac{\partial\psi}{\partial t} = \hat{H}\psi, \tag{7.2.11}$$

as well as

$$i\hbar\frac{\partial}{\partial t}\left(\hat{P}_{kj}\psi\right) = \hat{H}\left(\hat{P}_{kj}\psi\right). \tag{7.2.12}$$

However, by using equation (7.2.11), we obtain

$$i\hbar\frac{\partial}{\partial t}\left(\hat{P}_{kj}\psi\right) = \hat{P}_{kj}\left(i\hbar\frac{\partial\psi}{\partial t}\right) = \hat{P}_{kj}\left(\hat{H}\psi\right). \tag{7.2.13}$$

From the last two formulas, we find

$$\hat{H}\left(\hat{P}_{kj}\psi\right) = \hat{P}_{kj}\left(\hat{H}\psi\right), \tag{7.2.14}$$

which implies that

$$\left[\hat{H}, \hat{P}_{kj}\right] = 0. \tag{7.2.15}$$

By using the commutation relation (7.2.15), we shall demonstrate that the symmetry type of the wave function with respect to two particle permutation is preserved in time. Indeed, let $\psi^{(s)}$ be symmetric at some instant of time t. Then, from formulas (7.2.9) and (7.2.14), it follows that

$$\hat{P}_{kj}\left(\hat{H}\psi^{(s)}\right) = \hat{H}\psi^{(s)} \tag{7.2.16}$$

which means that $\hat{H}\psi^{(s)}$ is symmetric. Then, from the Schrödinger equation

$$i\hbar\frac{\partial\psi^{(s)}}{\partial t} = \hat{H}\psi^{(s)} \qquad (7.2.17)$$

we find that

$$d_t\psi^{(s)} = -\frac{i}{\hbar}\left(\hat{H}\psi^{(s)}\right)dt, \qquad (7.2.18)$$

where $d_t\psi^{(s)}$ is the infinitesimally small increment of $\psi^{(s)}$ during the time interval dt. From the last formula and equation (7.2.16), we conclude that

$$\hat{P}_{kj}\left(d_t\psi^{(s)}\right) = d_t\psi^{(s)}, \qquad (7.2.19)$$

which means that $d_t\psi^{(s)}$ is symmetric. This, in turn, implies that the symmetry of $\psi^{(s)}$ is preserved in time. By using the same line of reasoning, it can be established that the antisymmetry of the wave function $\psi^{(a)}$ is also preserved in time.

Next, we shall discuss how the symmetric and antisymmetric wave functions can be mathematically constructed in the case of noninteracting identical particles. The Hamiltonian for these particles can be written as follows

$$\hat{H} = \sum_{i=1}^{N}\left[-\frac{\hbar^2}{2m}\nabla_i^2 + U(\eta_i)\right] = \sum_{i=1}^{N}\hat{H}_i, \qquad (7.2.20)$$

where $\hat{H}_i$ are the identical Hamiltonians of individual particles. Let Φ_n be the stationary states (eigenstates) of these particles

$$\hat{H}_i\Phi_n = \mathcal{E}_n\Phi_n. \qquad (7.2.21)$$

Consider first the case of two identical particles. Then, the symmetric stationary states of these two particles can be written as follows

$$\Phi^{(s)}(\eta_1, \eta_2) = \frac{1}{\sqrt{2}}\left[\Phi_{n_1}(\eta_1)\Phi_{n_2}(\eta_2) + \Phi_{n_1}(\eta_2)\Phi_{n_2}(\eta_1)\right]. \qquad (7.2.22)$$

It is apparent that this wave function is symmetric with respect to the permutation of indices 1 and 2. The factor $\frac{1}{\sqrt{2}}$ appears in the last formula to keep $\Phi^{(s)}(\eta_1, \eta_2)$ normalized.

The last formula is valid when $n_1 \neq n_2$. In the case $n_1 = n_2 = n$, it must be replaced by

$$\Phi^{(s)}(\eta_1, \eta_2) = \Phi_n(\eta_1)\Phi_n(\eta_2). \qquad (7.2.23)$$

It is easy to see that the antisymmetric wave function of two identical particles has the form

$$\Phi^{(a)}(\eta_1, \eta_2) = \frac{1}{\sqrt{2}} \left[\Phi_{n_1}(\eta_1)\Phi_{n_2}(\eta_2) - \Phi_{n_1}(\eta_2)\Phi_{n_2}(\eta_1) \right]. \qquad (7.2.24)$$

Formulas (7.2.22) and (7.2.24) can be generalized to the case of arbitrary number N of identical particles. Indeed, in the case of identical noninteracting elementary particles with symmetric wave functions, these wave functions can be written as follows

$$\Phi^{(s)}(\eta_1, \eta_2, \ldots \eta_n) = \left[\frac{N_1! N_2! \ldots}{N!} \right]^{\frac{1}{2}} \sum \Phi_{n_1}(\eta_1)\Phi_{n_2}(\eta_2) \cdots \Phi_{n_N}(\eta_N),$$

$$(7.2.25)$$

where the summation is taken over all distinct (different) permutations of different indices $n_1, n_2, \ldots n_N$ and N_i is the number of how many of those indices have the same value i. It is clear that

$$N = \sum_i N_i. \qquad (7.2.26)$$

It is also clear that the number of terms in the sum in formula (7.2.25) is equal to

$$\frac{N!}{N_1! N_2! \cdots}. \qquad (7.2.27)$$

Taking into account the orthogonality of eigenfunctions Φ_{n_k} it is easy to realize that the factor before the sum in formula (7.2.25) is the normalization coefficient.

Formula (7.2.24) can be generalized to the case of N identical particles as follows

$$\Phi^{(a)}(\eta_1, \eta_2, \ldots \eta_N) = \frac{1}{\sqrt{N!}} \begin{vmatrix} \Phi_{n_1}(\eta_1) & \Phi_{n_1}(\eta_2) & \cdots & \Phi_{n_1}(\eta_N) \\ \Phi_{n_2}(\eta_1) & \Phi_{n_2}(\eta_2) & \cdots & \Phi_{n_2}(\eta_N) \\ \cdots\cdots\cdots\cdots\cdots\cdots\cdots\cdots\cdots\cdots \\ \Phi_{n_N}(\eta_1) & \Phi_{n_N}(\eta_2) & \cdots & \Phi_{n_N}(\eta_N) \end{vmatrix}. \qquad (7.2.28)$$

The determinant in the last formula is known as the **Slater determinant**. It is apparent from the last formula that the application of operator $\hat{P}_{kj}$ results in the interchanging of columns k and j in the above determinant, which changes the sign of the determinant. This means that the wave function represented by this determinant is antisymmetric. It is also clear that the factor $\frac{1}{\sqrt{N!}}$ is the normalization coefficient.

It is clear from the presented discussion that there may exist two distinct types of identical particles: identical particles called **fermions** whose wave functions are antisymmetric, and identical particles called **bosons** whose wave functions are symmetric. It turns out that all particles whose spins are half odd integers are fermions, while all particles whose spins are integers are bosons. Thus, electrons, protons and neutrons which have spin $\frac{1}{2}$, are fermions, while photons have spin 1 and they are bosons. Within the framework of the nonrelativistic quantum mechanics, the above connection between spins of identical particles and the type of symmetry of their wave functions is postulated and accepted as an experimental fact. In the relativistic quantum theory, the above connection between spins of identical particles and the symmetries of their wave functions is theoretically demonstrated. It seems somewhat enigmatic that relativity is the basis for this connection and the question can be asked why it is not possible to establish this connection without invoking relativity.

It is clear from the presented discussion that spins of elementary particles control their collective behavior. This controllability is especially apparent in the case of fermions. Indeed, it follows from formula (7.2.24) as well as formula (7.2.28) that fermions cannot be in the same states, because otherwise their wave functions would vanish. This is known as the **Pauli exclusion principle**. In contrast, in the case of bosons there may be wave functions with an arbitrarily large number of identical particles in the same states (see discussion of formula (7.2.25)). This phenomenon is known as **Bose-Einstein condensation**. The Pauli exclusion principle is instrumental in explaining the periodic table of chemical elements, the so-called **Mendeleev** table. The Pauli exclusion principle is also very helpful in the understanding of energy band structures in solids and their classification as insulators, conductors and semiconductors. For instance, a crystal in which each energy band has either all its states occupied by electrons or all empty is an insulator. The reason is that electrons in such a crystal cannot respond to an applied electric field because their transitions to occupied energy states are prohibited by the Pauli principle. This is not the case for conductors where there are energy bands with appreciable numbers of occupied and unfilled states, and the electron transitions to unoccupied states within these bands are not prohibited by the Pauli principle.

The distinction between fermions and bosons manifests itself in different statistics for these two kinds of identical particles. To derive the formulas for these statistics, we shall treat ensembles of noninteracting bosons or fermions as grand canonical ensembles. For such ensembles, the probability

$P(N_n)$ of N_n particles being in a state with energy $\mathcal{E}_n$ is given by the formula

$$P(N_n) = \frac{1}{\mathcal{Z}} \exp\left[-N_n(\mathcal{E}_n - \mu)/kT\right]. \tag{7.2.29}$$

Here μ is the chemical potential, k is Boltzmann's constant, while $\mathcal{Z}$ is the partition function which can be viewed as a normalization factor:

$$\mathcal{Z} = \sum_n \exp\left[-N_n(\mathcal{E}_n - \mu)/kT\right]. \tag{7.2.30}$$

We shall first consider ensembles of bosons. In this case there is no restriction on the number of boson particles in a state with energy $\mathcal{E}_n$. Consequently, the average (mean) value $\overline{N}_n$ of the number of particles being in the above state can be computed as follows

$$\overline{N}_n = \frac{\sum_{N_n=0}^{\infty} N_n P(N_n)}{\sum_{N_n=0}^{\infty} P(N_n)}. \tag{7.2.31}$$

Now, by using equation (7.2.29), the last formula can be transformed as follows

$$\overline{N}_n = \frac{\sum_{N_n=0}^{\infty} N_n \exp\left[-N_n(\mathcal{E}_n - \mu)/kT\right]}{\sum_{N_n=0}^{\infty} \exp\left[-N_n(\mathcal{E}_n - \mu)/kT\right]}. \tag{7.2.32}$$

To simplify the last expression, we introduce the notation

$$x = \exp\left[-(\mathcal{E}_n - \mu)/kT\right] < 1 \tag{7.2.33}$$

and write formula (7.2.32) in the form

$$\overline{N}_n = \frac{\sum_{N_n=0}^{\infty} N_n x^{N_n}}{\sum_{N_n=0}^{\infty} x^{N_n}}. \tag{7.2.34}$$

It is clear that

$$\sum_{N_n=0}^{\infty} x^{N_n} = \frac{1}{1-x}. \tag{7.2.35}$$

By treating x in the last formula as a variable and by differentiating both sides of the last formula with respect to x, we find

$$\sum_{N_n=0}^{\infty} N_n x^{N_n-1} = \frac{1}{(1-x)^2}. \tag{7.2.36}$$

Consequently,

$$\sum_{N_n=0}^{\infty} N_n x^{N_n} = \frac{x}{(1-x)^2}. \tag{7.2.37}$$

By substituting formulas (7.2.35) and (7.2.37) into equation (7.2.34), we derive

$$\overline{N}_n = \frac{x}{1-x} = \frac{1}{\frac{1}{x}-1}. \qquad (7.2.38)$$

Now, by recalling formula (7.2.33) we obtain

$$\boxed{\overline{N}_n = \frac{1}{\exp\left[(\mathcal{E}_n - \mu)/kT\right]-1}.} \qquad (7.2.39)$$

This is the **Bose-Einstein distribution**.

Next, we shall discuss the statistics for ensembles of fermions. According to the Pauli exclusion principle, there are only two possible values of N_n:

$$N_n = 0 \ \text{ and } \ N_n = 1, \qquad (7.2.40)$$

and formula (7.2.31) implies that

$$\overline{N}_n = \frac{P(1)}{P(0) + P(1)}. \qquad (7.2.41)$$

By using equation (7.2.29) in the last formula, we arrive at

$$\overline{N}_n = \frac{\exp\left[-(\mathcal{E}_n - \mu)/kT\right]}{1 + \exp\left[-(\mathcal{E}_n - \mu)/kT\right]}, \qquad (7.2.42)$$

or

$$\boxed{\overline{N}_n = \frac{1}{\exp\left[(\mathcal{E}_n - \mu)/kT\right]+1}.} \qquad (7.2.43)$$

This is the **Fermi-Dirac distribution**. It is clear from the last formula that for any state $\mathcal{E}_n$

$$\overline{N}_n < 1 \qquad (7.2.44)$$

as implied by the Pauli exclusion principle.

In the limiting case of $T \to 0$, from formula (7.2.43) we obtain

$$\overline{N}_n = \begin{cases} 1, & \text{if } \mathcal{E}_n < \mathcal{E}_f, \\ 0, & \text{if } \mathcal{E}_n > \mathcal{E}_f, \end{cases} \qquad (7.2.45)$$

where the Fermi energy (Fermi level) $\mathcal{E}_f$ is introduced as follows

$$\mathcal{E}_f = \mu \ \text{ at } \ T = 0. \qquad (7.2.46)$$

Distribution (7.2.45) is illustrated by Figure 7.3 below. The Fermi-Dirac distribution for $T > 0$ is illustrated by Figure 7.4. In the case of sufficiently

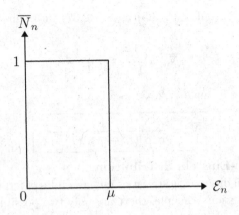

Fig. 7.3

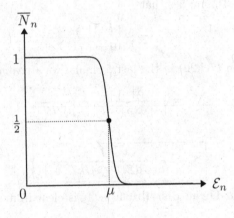

Fig. 7.4

high energies $\mathcal{E}_n$

$$\exp\left[(\mathcal{E}_n - \mu)/kT\right] \gg 1, \tag{7.2.47}$$

and both Bose-Einstein and Fermi-Dirac distributions are reduced to the Boltzmann distribution

$$\overline{N}_n = \exp\left[-(\mathcal{E}_n - \mu)/kT\right]. \tag{7.2.48}$$

We have discussed so far two types of identical particles: fermions and bosons, which have different wave function symmetries and different statistics. It turns out that in the case of two-dimensional systems, there are particles (quasiparticles, to be precise) called "anyons" whose wave function symmetries are different from those observed for fermions and bosons. As a result, their statistical properties are described by so-called "fractional" quantum statistics which are different from Fermi-Dirac and Bose-Einstein statistics. The study of anyons and fractional statistics has been a very active area of research during the last twenty five years. This study has been driven by promising applications of anyons in the field of fractional quantum Hall effect and high temperature superconductivity. The detailed discussion of this matter is clearly beyond the scope of this text.

Previously, we considered noninteracting identical particles. Now, we shall turn to the discussion of interacting identical particles, and we shall demonstrate that spin-dependent symmetries of wave functions result in a new physical phenomenon which is called the **exchange interaction**. This interaction is purely quantum mechanical in origin. We shall first limit our discussion to the simplest case of two electrically interacting electrons described by the following Hamiltonian

$$\hat{H} = -\frac{\hbar^2}{2m}(\nabla_1^2 + \nabla_2^2) + U(\mathbf{r}_1) + U(\mathbf{r}_2) + V(|\mathbf{r}_1 - \mathbf{r}_2|), \qquad (7.2.49)$$

where the last term $V(|\mathbf{r}_1 - \mathbf{r}_2|)$ describes the interaction. Suppose that we know the eigenstates $\varphi_k(\mathbf{r})$ of the single particle Hamiltonian

$$\hat{H}_0 = -\frac{\hbar^2}{2m}\nabla^2 + U(\mathbf{r}). \qquad (7.2.50)$$

Then, there are two distinct coordinate forms for the wave function of two noninteracting electrons. The first form corresponds to the "singlet" spin state of two electrons. This state (see formula (7.1.62)) is antisymmetric with respect to the permutation of electron spins. Since the overall wave function of two electrons must be antisymmetric with respect to the simultaneous permutation of spins and spatial coordinates, we conclude the coordinate form of the wave function in the singlet state is symmetric with respect to the electron coordinate permutations. This means that

$$\varphi^{(0)}(\mathbf{r}_1, \mathbf{r}_2) = \frac{1}{\sqrt{2}}\left[\varphi_1(\mathbf{r}_1)\varphi_2(\mathbf{r}_2) + \varphi_1(\mathbf{r}_2)\varphi_2(\mathbf{r}_1)\right], \qquad (7.2.51)$$

where the superscript "(0)" reflects the total zero spin value of the two electrons. The second coordinate form of the wave function of two identical electrons corresponds to three "triplet" spin states of two electrons. These states are symmetric with respect to the permutation of electron spins (see formula (7.1.61)). Consequently,

$$\varphi^{(1)}(\mathbf{r}_1, \mathbf{r}_2) = \frac{1}{\sqrt{2}}\left[\varphi_1(\mathbf{r}_1)\varphi_2(\mathbf{r}_2) - \varphi_1(\mathbf{r}_2)\varphi_2(\mathbf{r}_1)\right], \qquad (7.2.52)$$

where superscript (1) reflects the total spin value of two spins in the "triplet" state.

Now, we shall use the time-independent perturbation theory to find how the electron interaction affects the energy spectrum of two electrons. According to formula (6.2.30), we find that the first-order corrections for the energy levels of singlet and triplet states can be computed by using the following formulas, respectively,

$$\tilde{\mathcal{E}}^{(0)} = \langle \varphi^{(0)}(\mathbf{r}_1, \mathbf{r}_2) | V(|\mathbf{r}_1 - \mathbf{r}_2|) | \varphi^{(0)}(\mathbf{r}_1, \mathbf{r}_2) \rangle, \tag{7.2.53}$$

$$\tilde{\mathcal{E}}^{(1)} = \langle \varphi^{(1)}(\mathbf{r}_1, \mathbf{r}_2) | V(|\mathbf{r}_1 - \mathbf{r}_2|) | \varphi^{(1)}(\mathbf{r}_1, \mathbf{r}_2) \rangle. \tag{7.2.54}$$

By combining formulas (7.2.51) and (7.2.53) as well as formulas (7.2.52) and (7.2.54), we find after simple transformations that

$$\tilde{\mathcal{E}}^{(0)} = A + J, \tag{7.2.55}$$

$$\tilde{\mathcal{E}}^{(1)} = A - J, \tag{7.2.56}$$

where

$$A = \int V(|\mathbf{r}_1 - \mathbf{r}_2|) |\varphi_1(\mathbf{r}_1)|^2 |\varphi_2(\mathbf{r}_2)|^2 \, d\mathbf{r}_1 d\mathbf{r}_2, \tag{7.2.57}$$

$$J = \int V(|\mathbf{r}_1 - \mathbf{r}_2|) \varphi_1(\mathbf{r}_1) \varphi_2(\mathbf{r}_2) \varphi_1^*(\mathbf{r}_2) \varphi_2^*(\mathbf{r}_1) \, d\mathbf{r}_1 d\mathbf{r}_2. \tag{7.2.58}$$

The Hamiltonian (7.2.49) of two interacting electrons does not depend on their spins. However, according to formulas (7.2.55) and (7.2.56), the energy of these electrons depends on their total spin. It looks as if there is some underlying interaction that results in this dependence. This interaction is called the **exchange interaction**. There is no real physical force behind this interaction; it is rather a consequence of symmetry of wave functions of identical particles. It is clear from formulas (7.2.55) and (7.2.56) that the "strength" of the exchange interaction is characterized by the integral J, which is called the exchange integral. The value of this integral is determined by the overlap of the wave functions φ_1 and φ_2.

Since the difference in energy levels (7.2.55) and (7.2.56) is of spin origin, it is desirable to introduce the so-called exchange Hamiltonian operator which depends only on electron spins and whose eigenvalues are $\pm J$. In this way the exchange interaction can be directly related to spins. Such a Hamiltonian operator was first introduced **by P. Dirac** and it has the form

$$\boxed{\hat{H}_{ex} = -\frac{J}{2}\left(1 + \frac{4}{\hbar^2}\hat{\mathbf{S}}_1 \cdot \hat{\mathbf{S}}_2\right).} \tag{7.2.59}$$

It can be shown that J and $-J$ are eigenvalues of $\hat{H}_{ex}$ corresponding to singlet and triplet states, respectively. Namely,

$$\hat{H}_{ex}|\text{singlet}\rangle = J|\text{singlet}\rangle, \tag{7.2.60}$$

and

$$\hat{H}_{ex}|\text{triplet}\rangle = -J|\text{triplet}\rangle, \tag{7.2.61}$$

where $|\text{triplet}\rangle$ stands for any of three states in formula (7.1.61). Indeed, by using the same line of reasoning as in the derivation of formulas (7.1.66)–(7.1.69) it can be shown that

$$\hat{\mathbf{S}}_1 \cdot \hat{\mathbf{S}}_2 |+-\rangle = \frac{\hbar^2}{2}|-+\rangle - \frac{\hbar^2}{4}|+-\rangle \tag{7.2.62}$$

and

$$\hat{\mathbf{S}}_1 \cdot \hat{\mathbf{S}}_2 |-+\rangle = \frac{\hbar^2}{2}|+-\rangle - \frac{\hbar^2}{4}|-+\rangle. \tag{7.2.63}$$

Consequently, from formulas (7.2.59), (7.2.62) and (7.2.63) we find

$$\begin{aligned}
\hat{H}_{ex}|\text{singlet}\rangle &= \hat{H}_{ex}\left[\frac{1}{\sqrt{2}}(|+-\rangle - |-+\rangle)\right] \\
&= J\left[\frac{1}{\sqrt{2}}(|+-\rangle - |-+\rangle)\right] = J|\text{singlet}\rangle.
\end{aligned} \tag{7.2.64}$$

Formula (7.2.61) can be established in a similar way.

The Hamiltonian (7.2.59) was further generalized by W. Heisenberg to the case of many electrons. The Heisenberg exchange Hamiltonian is given by the formula

$$\boxed{\hat{H}_{ex} = -\sum_{i,j} J_{ij}\hat{\mathbf{S}}_i \cdot \hat{\mathbf{S}}_j,} \tag{7.2.65}$$

and it is extensively used in the quantum mechanical theory of ferromagnetism.

In ferromagnets (iron, nickel, cobalt) electron spins are spontaneously aligned along some direction as a result of their exchange interaction. Such an alignment occurs because it is energetically favorable under certain conditions. This alignment of electron spins results in the alignment of their magnetic moments which leads on the macroscopic level to the emergence of spontaneous magnetization $\mathbf{M}$ of constant magnitude

$$|\mathbf{M}| = M_s = const. \tag{7.2.66}$$

Although the exchange interaction keeps spins (and their magnetic moments) aligned, it does not align them in a specific direction. The direction of alignment is controlled by magnetocrystalline anisotropy and applied magnetic field. The primary source of magnetocrystalline anisotropy is the spin-orbit interaction. The spin-orbit interaction occurs because the electric field of a crystal lattice manifests itself as the magnetic field in the reference frame of moving (orbiting) electrons and, consequently, interacts with spin magnetic moments. It is clear that this interaction reflects the anisotropic structure of the crystal lattice through the structure of the electric field of this lattice. The direction of $\mathbf{M}$ is determined from the minimum condition for free micromagnetic energy. In a simple case of spatially uniform magnetization, which typically occurs at the nanoscale, the micromagnetic energy $w(\mathbf{M})$ has two terms:

$$w(\mathbf{M}) = w_{an}(\mathbf{M}) - \mu_0 \mathbf{M} \cdot \mathbf{H}, \qquad (7.2.67)$$

where the first term in the right-hand side of formula (7.2.67) is due to magnetic anisotropy, while the last term is the energy of magnetization $\mathbf{M}$ in the applied field $\mathbf{H}$.

The condition of minimum of $w_m(\mathbf{M})$ subject to the constraint (7.2.66) can be written as follows

$$\frac{\partial w(\mathbf{M})}{\partial \mathbf{M}} \cdot \delta \mathbf{M} = 0, \qquad (7.2.68)$$

where

$$\delta \mathbf{M} = \mathbf{M} \times \delta \mathbf{a}. \qquad (7.2.69)$$

From the last two formulas we derive

$$\delta \mathbf{a} \cdot \left(\mathbf{M} \times \frac{\partial w(\mathbf{M})}{\partial \mathbf{M}} \right) = 0, \qquad (7.2.70)$$

which leads to

$$\mathbf{M} \times \frac{\partial w(\mathbf{M})}{\partial \mathbf{M}} = 0. \qquad (7.2.71)$$

Now, we introduce the effective magnetic field

$$\mathbf{H}_{eff} = -\frac{1}{\mu_0} \frac{\partial w(\mathbf{M})}{\partial \mathbf{M}} \qquad (7.2.72)$$

and write equation (7.2.71) in the form

$$\boxed{\mathbf{M} \times \mathbf{H}_{eff} = 0.} \qquad (7.2.73)$$

The last equation is a simple version of the **W. F. Brown equation** which is extensively used in micromagnetics to compute equilibrium configurations of magnetization in ferromagnetic objects.

It is clear from formulas (7.2.67) and (7.2.72) that the effective magnetic field has two components:

$$\mathbf{H}_{eff} = \mathbf{H}_{an} + \mathbf{H}, \tag{7.2.74}$$

where

$$\mathbf{H}_{an} = -\frac{1}{\mu_0}\frac{\partial w_{an}(\boldsymbol{\mu})}{\partial \mathbf{M}}. \tag{7.2.75}$$

The Brown equation (7.2.73) determines the equilibrium alignment of magnetization $\mathbf{M}$. It turns out that by using formulas (7.2.66) and (7.2.73) the dynamic equation for $\mathbf{M}(t)$ can be derived. Indeed, from the constraint (7.2.66) we find

$$\frac{d\mathbf{M}(t)}{dt} = \boldsymbol{v}(\mathbf{M}, t), \tag{7.2.76}$$

where vector $\boldsymbol{v}(\mathbf{M}, t)$ is tangential to the sphere $|\mathbf{M}(t)| = const$. Consequently at any instant of time this vector can be decomposed in terms of orthogonal basis vectors $\mathbf{M} \times \mathbf{H}_{eff}$ and $\mathbf{M} \times (\mathbf{M} \times \mathbf{H}_{eff})$:

$$\boxed{\frac{d\mathbf{M}}{dt} = -\gamma(\mathbf{M} \times \mathbf{H}_{eff}) - \frac{\alpha\gamma}{M_s}(\mathbf{M} \times (\mathbf{M} \times \mathbf{H}_{eff})).} \tag{7.2.77}$$

This decomposition is consistent with the equilibrium condition (7.2.73). Indeed, at equilibrium $\left(\frac{d\mathbf{M}}{dt} = 0\right)$, the last equation is reduced to

$$\gamma(\mathbf{M} \times \mathbf{H}_{eff}) + \frac{\alpha\gamma}{M_s}(\mathbf{M} \times (\mathbf{M} \times \mathbf{H}_{eff})) = 0, \tag{7.2.78}$$

which is only possible if equation (7.2.73) is valid.

Equation (7.2.77) is the celebrated **Landau-Lifshitz** equation. There are two terms in the right-hand side of this equation. The first term describes the precession of magnetization, while the second term is due to the damping phenomena caused by the thermal bath. This damping results in relaxation of magnetization to equilibrium. Without damping (that is, when the damping constant α is equal to zero), the Landau-Lifshitz equation (7.2.77) is reduced to

$$\frac{d\mathbf{M}}{dt} = -\gamma(\mathbf{M} \times \mathbf{H}_{eff}), \tag{7.2.79}$$

which is mathematically similar to equation (7.1.118). This equation describes periodic oscillations of magnetization.

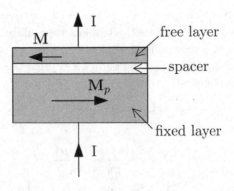

Fig. 7.5

Next, we shall turn to the brief discussion of spin-torque nano-oscillators, which is a very active area of research in spintronics. A typical three-layer structure of such oscillators is shown in Figure 7.5. This structure consists of two ferromagnetic layers separated by a nonmagnetic spacer. The magnetization of one of the magnetic layers is fixed due to large anisotropy and volume or as a result of pinning by additional under-layers. Another magnetic layer is called the "free" layer, because it is not exchanged-coupled to the "fixed" layer. This can be achieved in accordance with the RKKY (Ruderman, Kittel, Kasuya and Yosida) theory which predicts that the exchange coupling between these ferromagnetic layers separated by a nonmagnetic spacer should oscillate between ferromagnetic and antiferromagnetic as a function of the distance between the layers. Consequently, the spacer thickness can be chosen to guarantee the absence of appreciable exchange coupling between ferromagnetic layers. If a dc conducting or (tunneling) current is injected through the spacer, the electrons of this current get polarized. The latter means that they predominantly have the same spin (and magnetic moment) polarization as the magnetization in the "fixed" layer. The injection of a spin-polarized current into the "free" layer results in an additional torque term in the Landau-Lifshitz equation, called the **spin-torque** or **Slonczewski** term:

$$\frac{d\mathbf{M}}{dt} = -\gamma(\mathbf{M} \times \mathbf{H}_{eff}) - \frac{\alpha\gamma}{M_s}(\mathbf{M} \times (\mathbf{M} \times \mathbf{H}_{eff})) + \frac{\beta}{M_s}\frac{\mathbf{M} \times (\mathbf{M} \times \mathbf{e}_p)}{1 + c_p\mathbf{M} \cdot \mathbf{e}_p}.$$
$$(7.2.80)$$

Here $\mathbf{e}_p$ is the unit vector parallel to $\mathbf{M}_p$ (see Figure 7.5) and β is proportional to the injected current. It turns out that along some precessional trajectories defined by equation (7.2.79) the cancellation of the damping and spin-torque terms in the last equation may occur. This implies that the magnetization of the "free" layer will periodically precess along this trajectory. The resistance (often called the magnetoresistance) of the three-layer structure depends on the direction of magnetization in the "free" layer with respect to the direction of magnetization in the "fixed" layer. Consequently, this resistance oscillates during the precession of the magnetization in the "free" layer. This leads to the appearance of ac voltage across the three-layer structure caused by injection of dc current through it. The precessional trajectory along which the compensation of the damping and spin-torque terms occur depends on the value I of the injected spin-polarized current. This implies that the frequency of the ac voltage across the three-layer structure can be controlled by dc current. It is worthwhile to mention that the described effect is dual to the ac **Josephson effect** (discussed in Section 3.3) when the applied dc voltage results in ac current and the frequency of this current is controlled by the value of the applied dc voltage.

It is interesting to mention that there is also the spin-diode effect when ac current flow through the three-layer structure results in the appearance of the dc component of the voltage across this structure.

It is apparent that spin-torque nano-oscillators can be viewed as dc-to-ac power converters, while the spin-torque diode effect can be viewed as ac-to-dc power conversion, both occurring at the nano-scale. This suggests that the proper cascading of these dc-to-ac and ac-to-dc power converters is very promising for the development of nano-scale dc-to-dc power converters which will be needed as dc power supplies for nanospintronics devices. It is remarkable that, unlike conventional switch-mode dc-to-dc power converters, this spintronics-based power conversion can be accomplished without any switching.

In the conclusion of this section, we shall briefly mention another very active area of research in spintronics which deals with magnetic tunnel junctions and spin-dependent tunneling. In such junctions, two ferromagnetic conducting layers are separated by a very thin (usually a few nanometers) insulator, typically aluminum oxide (Al_2O_3) or crystalline magnesium oxide (MgO). If a bias voltage is applied between two ferromagnetic conducting layers then the tunneling current appears. It turns out that the tunneling current is strongly dependent on the relative orientation of magnetizations in the two ferromagnetic layers which can be controlled by an applied magnetic field. This means that such magnetic junctions can be switched by applied magnetic fields between the two states with very high and very low values of magnetoresistance. Such magnetic tunnel junctions have already found applications in read heads of modern hard disk drives and in non-volatile magnetic random access memories (MRAMs).

7.3 Second Quantization. (Representation of Occupation Numbers)

In the case of a large number of identical particles, the coordinate representation of their wave functions is not very informative. In this case, the representation of occupation numbers is used because it is physically more relevant. In this representation, one is concerned with probabilities of certain numbers of identical particles being in specific distinct quantum states. Such numbers are called occupation numbers. The second quantization is the mathematical formalism of the occupation-number representation. The second quantization technique was developed by **P. Dirac for bosons** and by **P. Jordan and E. Wigner for fermions**. This technique has since permeated various research areas of quantum physics, and it is especially instrumental in quantum field theory and solid state physics.

In this section, we shall present only the very basic facts related to second quantization. We start with the discussion of the case of boson particles. The wave function of N noninteracting boson particles is symmetric, and it is given by formula (7.2.25). In this formula, $N_1, N_2, \ldots$ are the occupation numbers, that is, the numbers of particles in states described by wave functions $\Phi_1(\eta), \Phi_2(\eta), \ldots$, respectively. These occupation numbers completely define the wave function $\Phi^{(s)}$ in the above-mentioned formula. For this reason, it is convenient to write this formula as follows

$$\Phi^{(s)}_{N_1,N_2,\ldots} = \left[\frac{N_1!N_2!\cdots}{N!}\right]^{\frac{1}{2}} \sum \Phi_{n_1}(\eta_1)\Phi_{n_2}(\eta_2)\cdots\Phi_{n_N}(\eta_N). \quad (7.3.1)$$

It is apparent that functions $\Phi^{(s)}_{N_1,N_2\ldots}$ for different occupation numbers are orthonormal. This means that

$$\langle \Phi^{(s)}_{N_1,N_2,\ldots} | \Phi^{(s)}_{N_1',N_2',\ldots} \rangle = \delta_{N_1,N_1'}\delta_{N_2,N_2'}\cdots \quad (7.3.2)$$

Any wave function $\psi(\eta_1, \eta_2, \ldots \eta_N, t)$ which is symmetric with respect to permutations of η_k for any two particles can be expanded into the following series

$$\psi(\eta_1, \ldots \eta_N, t) = \sum_{N_k} c(N_1, N_2, \ldots, t)\Phi^{(s)}_{N_1 N_2 \ldots}, \quad (7.3.3)$$

where the summation is performed over all possible values of $N_1, N_2, \ldots N_k, \ldots$.

It is clear that

$$|c(N_1, N_2, \ldots, t)|^2 \tag{7.3.4}$$

has the meaning of probability that the state $\Phi_1(\eta)$ is occupied by N_1 particles, the state $\Phi_2(\eta)$ is occupied by N_2 particles and so on. In other words, $c(N_1, N_2, \ldots, t)$ can be viewed as the occupation-number representation of the state described by the wave function $\psi(\eta_1, \ldots \eta_N, t)$. Our goal is to transform the time-dependent Schrödinger equation

$$i\hbar \frac{\partial \psi}{\partial t} = \hat{H}\psi \tag{7.3.5}$$

written in coordinate representation into the time-dependent Schrödinger equation

$$i\hbar \frac{\partial c(N_1, N_2, \ldots, t)}{\partial t} = \hat{H} c(N_1, N_2, \ldots, t) \tag{7.3.6}$$

written in the occupation-number representation.

This goal can be achieved by using annihilation $\hat{a}_k$ and creation $\hat{a}_k^+$ operators defined by the following formulas, respectively,

$$\hat{a}_k c(N_1, N_2, \ldots, N_k, \ldots) = \sqrt{N_k} c(N_1, N_2, \ldots N_k - 1, \ldots), \tag{7.3.7}$$

$$\hat{a}_i^+ c(N_1, N_2, \ldots N_i, \ldots) = \sqrt{N_i + 1} c(N_1, N_2, \ldots N_i + 1, \ldots). \tag{7.3.8}$$

It can be verified (and this is left as an exercise for the reader) that the following commutation relations are valid for operators $\hat{a}_k$ and $\hat{a}_i^+$:

$$\hat{a}_k \hat{a}_i - \hat{a}_i \hat{a}_k = 0, \tag{7.3.9}$$

$$\hat{a}_k^+ \hat{a}_i^+ - \hat{a}_i^+ \hat{a}_k^+ = 0, \tag{7.3.10}$$

$$\hat{a}_i^+ \hat{a}_i = N_i \hat{I}, \tag{7.3.11}$$

$$\hat{a}_i \hat{a}_i^+ = (N_i + 1)\hat{I}, \tag{7.3.12}$$

$$\hat{a}_k \hat{a}_i^+ - \hat{a}_i^+ \hat{a}_k = \delta_{ki} \hat{I}, \tag{7.3.13}$$

where, as before, δ_{ki} is the Kronecker delta, while $\hat{I}$ is the identity operator. The Hamiltonian in equation (7.3.6) can be expressed in terms of annihilation and creation operators. Let us do this for the important case when the Hamiltonian has the following form

$$\hat{H} = \sum_{j=1}^{N} \hat{H}(\eta_j) + \frac{1}{2} \sum_{j \neq p}^{N} \hat{V}(\eta_j, \eta_p), \tag{7.3.14}$$

where operator $\hat{H}(\eta_j)$ acts on coordinates and spin of particle j, while operator $\hat{V}(\eta_j, \eta_p)$ acts on coordinates and spins of particles j and p. Since the particles are indistinguishable, operators $\hat{H}(\eta_j)$ are the same for all particles, while operators $\hat{V}(\eta_j, \eta_p)$ are the same for any pair of particles.

The above Hamiltonian can be written as follows

$$\hat{H} = \hat{H}^{(1)} + \hat{H}^{(2)}, \qquad (7.3.15)$$

where

$$\hat{H}^{(1)} = \sum_{j=1}^{N} \hat{H}(\eta_j) \qquad (7.3.16)$$

is the so-called "single-particle" component of $\hat{H}$, while

$$\hat{H}^{(2)} = \frac{1}{2} \sum_{j \neq p}^{N} \hat{V}(\eta_j, \eta_p) \qquad (7.3.17)$$

is the "two-particle" component of $\hat{H}$.

It turns out (and it is demonstrated below) that in the occupation-number representation the Hamiltonian (7.3.7) is given by the formula

$$\boxed{\hat{H} = \sum_{i,k} H_{ik} \hat{a}_i^+ \hat{a}_k + \frac{1}{2} \sum_{m,n} \sum_{i,k} V_{mn,ik} \hat{a}_i^+ \hat{a}_k^+ \hat{a}_m \hat{a}_n} \qquad (7.3.18)$$

where

$$H_{ik} = \langle \Phi_k | \hat{H}(\eta_j) | \hat{\Phi}_i \rangle, \qquad (7.3.19)$$

and

$$V_{mn,ik} = \langle \Phi_k \hat{\Phi}_i | \hat{V}(\eta_j, \eta_p) | \Phi_m \Phi_n \rangle. \qquad (7.3.20)$$

In the particular case when functions $\Phi_{n_k}(\eta_j)$ in formula (7.3.1) are the eigenfunctions of the Hamiltonian $\hat{H}(\eta_j)$, formula (7.3.18) can be simplified. Indeed, in this case

$$H_{ik} = \mathcal{E}_i \delta_{ik} \qquad (7.3.21)$$

with $\mathcal{E}_i$ being the eigenvalue of $\hat{H}(\eta_j)$. This leads, according to relation (7.3.11), to the following simplification of formula (7.3.18):

$$\boxed{\hat{H} = \left(\sum_i \mathcal{E}_i N_i \right) \hat{I} + \frac{1}{2} \sum_{m,n} \sum_{i,k} V_{mn,ik} \hat{a}_i^+ \hat{a}_k^+ \hat{a}_m \hat{a}_n.} \qquad (7.3.22)$$

Now, we shall outline the derivation of formula (7.3.18). This derivation is based on the following relations:

$$\Phi^{(s)}_{N_1,\ldots N_k\ldots} = \sum_k \Phi_k(\eta_j)\sqrt{\frac{N_k}{N}}\,\Phi^{(s)}_{N_1,\ldots N_k-1,\ldots}, \qquad (7.3.23)$$

$$\Phi^{(s)}_{N_1,\ldots N_i,\ldots N_k,\ldots} = \sum_{i,k} \Phi_i(\eta_j)\Phi_k(\eta_p)\sqrt{\frac{N_k N_i}{N(N-1)}}\,\Phi^{(s)}_{N_1,\ldots N_k-1,\ldots N_i-1\ldots}.$$

$$(7.3.24)$$

The validity of formula (7.3.23) can be justified by using the following reasoning. For any k, all terms in formula (7.3.1) which contain $\Phi_k(\eta_j)$ can be identified and grouped together. Then, these terms can be represented as the product of $\Phi_k(\eta_j)$ and the function which is different from $\Phi^{(s)}_{N_1,\ldots N_k-1,\ldots}$ by the excessive factor $\sqrt{\frac{N_k}{N}}$. By performing this procedure for all k, we arrive at formula (7.3.23). Formula (7.3.24) can be justified by applying formula (7.3.23) to $\Phi^{(s)}_{N_1,\ldots N_k-1,\ldots}$ in (7.3.23).

Next, we proceed to the transformation of equation (7.3.5) into equation (7.3.6). To this end, by using the expansion (7.3.3) we shall write equation (7.3.5) as follows

$$i\hbar\sum_{N_k'} \frac{\partial c(N_1',N_2',\ldots,t)}{\partial t}\,\hat{\Phi}^{(s)}_{N_1',N_2',\ldots} = \sum_{N_k'} c(N_1',N_2',\ldots,t)\hat{H}\Phi^{(s)}_{N_1',N_2',\ldots}.$$

$$(7.3.25)$$

By using the orthogonality condition (7.3.2), from the last equation we derive:

$$i\hbar\frac{\partial c(N_1,N_2,\ldots,t)}{\partial t} = \sum_{N_k'} c(N_1',N_2',\ldots,t)\langle\Phi^{(s)}_{N_1,N_2,\ldots}|\hat{H}|\Phi^{(s)}_{N_1',N_2',\ldots}\rangle.$$

$$(7.3.26)$$

By using the relation (7.3.15), the last formula can be represented in the form

$$i\hbar\frac{\partial c(N_1,N_2,\ldots,t)}{\partial t}$$
$$= \sum_{N_k'} c(N_1',N_2',\ldots,t)\langle\hat{H}^{(1)}\Phi^{(s)}_{N_1,N_2\ldots}|\Phi^{(s)}_{N_1',N_2',\ldots}\rangle$$
$$+ \sum_{N_k'} c(N_1',N_2',\ldots,t)\langle\hat{H}^{(2)}\Phi^{(s)}_{N_1,N_2\ldots}|\Phi^{(s)}_{N_1',N_2',\ldots}\rangle. \qquad (7.3.27)$$

Next, by using formula (7.3.16), we find that

$$\langle \hat{H}^{(1)} \Phi^{(s)}_{N_1,N_2,\ldots} | \Phi^{(s)}_{N'_1,N'_2,\ldots} \rangle = \sum_{j=1}^{N} \langle \hat{H}(\eta_j) \Phi^{(s)}_{N_1,N_2,\ldots} | \Phi^{(s)}_{N'_1,N'_2} \rangle. \qquad (7.3.28)$$

By using relation (7.3.23), the last equation can be further transformed as follows:

$$\langle \hat{H}^{(1)} \Phi^{(s)}_{N_1,N_2,\ldots} | \Phi^{(s)}_{N'_1,N'_2,\ldots} \rangle$$

$$= \sum_{i,k} \frac{\sqrt{N_k N'_i}}{N} \left(\sum_{j=1}^{N} \langle \hat{H}(\eta_j) \Phi_k | \Phi_i \rangle \right) \langle \Phi^{(s)}_{N_1,\ldots N_k-1,\ldots} | \Phi^{(s)}_{N'_1,\ldots N'_i-1,\ldots} \rangle. \qquad (7.3.29)$$

It is apparent that

$$\langle \hat{H}(\eta_j) \Phi_k | \Phi_i \rangle \qquad (7.3.30)$$

is the same for any j and it is equal to H_{ik} (see formula (7.3.19)). Consequently,

$$\sum_{j=1}^{N} \langle \hat{H}(\eta_j) \Phi_k | \Phi_i \rangle = N H_{ik}. \qquad (7.3.31)$$

Furthermore, according to the orthogonality condition (7.3.2), we find

$$\langle \Phi^{(s)}_{N_1,\ldots N_k-1,\ldots} | \Phi^{(s)}_{N'_1,\ldots N'_i-1,\ldots} \rangle = \delta_{N_1,N'_1} \cdots \delta_{N_k-1,N'_k} \cdots \delta_{N_i,N'_i-1} \cdots \qquad (7.3.32)$$

By using the last two formulas in equation (7.3.29), we obtain:

$$\langle \hat{H}^{(1)} \Phi^{(s)}_{N_1,N_2,\ldots} | \Phi^{(s)}_{N'_1,N'_2,\ldots} \rangle$$

$$= \sum_{i,k} H_{ik} \sqrt{N_k N'_i} \delta_{N_1,N'_1} \cdots \delta_{N_k-1,N'_k} \cdots \delta_{N_i,N'_i-1} \cdots. \qquad (7.3.33)$$

By substituting the last formula in the first term in the right-hand side of equation (7.3.27), we find:

$$\sum_{N'_k} c(N'_1, N'_2, \ldots, t) \langle \hat{H}^{(1)} \Phi^{(s)}_{N_1,N_2,\ldots} | \Phi^{(s)}_{N'_1,N'_2,\ldots} \rangle$$

$$= \sum_{i,k} H_{ik} \sqrt{N_k (N_i + 1)} c(N_1, \ldots N_k - 1, \ldots N_i + 1, \ldots). \qquad (7.3.34)$$

Now, by using the definition of annihilation and creation operators (see (7.3.7) and (7.3.8)), the last equation can be transformed as follows

$$\sum_{N'_k} c(N'_1, N'_2, \ldots, t) \langle \hat{H}^{(1)} \Phi^{(s)}_{N_1,N_2,\ldots} | \Phi^{(s)}_{N'_1,N'_2,\ldots} \rangle$$

$$= \sum_{i,k} H_{ik} \hat{a}_i^+ \hat{a}_k c(N_1, N_2, \ldots, t). \qquad (7.3.35)$$

By using formulas (7.3.17) and (7.3.24) and by repeating almost the same line of reasoning as before, it can be established that

$$\langle \hat{H}^{(2)} \Phi^{(s)}_{N_1,N_2,\ldots} | \Phi^{(s)}_{N'_1,N'_2,\ldots} \rangle$$

$$= \frac{1}{2} \sum_{m,n} \sum_{i,k} V_{mn,ik} \sqrt{N_m N_n N'_i N'_k} \cdot \tag{7.3.36}$$

$$\delta_{N_1,N'_1} \cdots \delta_{N_m-1,N'_m} \cdots \delta_{N_n-1,N'_n} \cdots \delta_{N_i,N'_i-1} \cdots \delta_{N_k,N'_k-1} \cdots .$$

By substituting the above formula in the last term of equation (7.3.27), we find that

$$\sum_{N'_k} c(N'_1, N'_2, \ldots, t) \langle \hat{H}^{(2)} \Phi^{(s)}_{N_1,N_2,\ldots} | \Phi^{(s)}_{N'_1,N'_2,\ldots} \rangle$$

$$= \frac{1}{2} \sum_{m,n} \sum_{i,k} V_{mn,ik} \sqrt{N_m N_n (N_i+1)(N_k+1)} \cdot \tag{7.3.37}$$

$$c(N_1, \ldots N_m - 1, \ldots N_n - 1, \ldots N_i + 1, \ldots N_k + 1 \ldots).$$

By recalling relations (7.3.7) and (7.3.8), formula (7.3.37) can be further transformed as follows

$$\sum_{N'_k} c(N'_1, N'_2, \ldots, t) \langle \hat{H}^{(2)} \Phi^{(s)}_{N_1,N_2,\ldots} | \Phi^{(s)}_{N'_1,N'_2,\ldots} \rangle$$

$$= \frac{1}{2} \sum_{m,n} \sum_{i,k} V_{mn,ik} \hat{a}^+_i \hat{a}^+_k \hat{a}_m \hat{a}_n c(N_1, N_2, \ldots, t). \tag{7.3.38}$$

By combining formulas (7.3.27), (7.3.35) and (7.3.38), we finally arrive at the time-dependent Schrödinger equation in the occupation-number representation

$$\boxed{\begin{aligned} i\hbar \frac{\partial c(N_1, N_2, \ldots, t)}{\partial t} &= \sum_{i,k} H_{ik} \hat{a}^+_i \hat{a}_k c(N_1, N_2, \ldots, t) + \\ \frac{1}{2} \sum_{m,n} \sum_{i,k} V_{mn,ik} \hat{a}^+_i \hat{a}^+_k \hat{a}_m \hat{a}_n & c(N_1, N_2, \ldots, t). \end{aligned}} \tag{7.3.39}$$

This also proves the validity of formula (7.3.18). This formula can be extended to more complicated Hamiltonians that contain terms describing multi-particle interactions.

In our derivations, we have used inner products. This implies that the states of identical particles are elements of proper Hilbert space. However, the nature of this Hilbert space was not explicitly specified. It is clear that this Hilbert space must consist of functions of correct permutation symmetry. Furthermore, it turns out that the proper Hilbert space has the structure of direct sum of Hilbert subspaces representing respectively one particle states, two particle states, etc., for which functions $\Phi^{(s)}_{N_1,N_2\ldots}$ can be used as basis functions. The annihilation and creation operators map one of such subspaces into another. This Hilbert space is called the **Fock space**, and it was introduced by Russian mathematical physicist V. Fock. The Fock space is very instrumental for the rigorous mathematical study of the second quantization technique.

.The Hamiltonian (7.3.18) can be also represented in another mathematical form. To do this, consider a wave function $\psi(\eta)$ of a single boson particle and expand it into series with respect to eigenfunctions $\Phi_n(\eta)$:

$$\psi(\eta) = \sum_k a_k \Phi_k(\eta). \qquad (7.3.40)$$

Now, we shall treat the expansion coefficients a_n not as numbers but as annihilation and creation operators. Namely, we introduce the following wave function operators:

$$\hat{\psi}(\eta) = \sum_k \hat{a}_k \Phi_k(\eta), \qquad (7.3.41)$$

$$\hat{\psi}^+(\eta) = \sum_i \hat{a}_i^+ \Phi_i^*(\eta), \qquad (7.3.42)$$

where $\Phi_i^*(\eta)$ is the complex conjugate of $\Phi_i(\eta)$. Then, it can be shown that the Hamiltonian (7.3.18) can be represented in the following form

$$\hat{H} = \int \hat{\psi}^+(\eta)\hat{H}(\eta)\hat{\psi}(\eta)\,d\eta$$
$$+ \frac{1}{2}\int\int \hat{\psi}^+(\eta)\hat{\psi}^+(\eta')\hat{V}(\eta,\eta')\hat{\psi}(\eta)\hat{\psi}(\eta')\,d\eta d\eta'. \qquad (7.3.43)$$

Indeed, by using formulas (7.3.41) and (7.3.42), we derive

$$\int \hat{\psi}^+(\eta)\hat{H}(\eta)\hat{\psi}(\eta)\,d\eta = \sum_{i,k} \hat{a}_i^+ \hat{a}_k \int \Phi_i^*(\eta)\hat{H}(\eta)\Phi_k(\eta)d\eta$$
$$= \sum_{i,k} H_{ik}\hat{a}_i^+ \hat{a}_k. \qquad (7.3.44)$$

Similarly, it is easy to see that

$$\frac{1}{2}\int\int \hat{\psi}^+(\eta)\hat{\psi}^+(\eta')\hat{V}(\eta,\eta')\hat{\psi}(\eta)\hat{\psi}(\eta')\,d\eta d\eta'$$
$$= \frac{1}{2}\sum_{m,n}\sum_{i,k} V_{mn,ik}\hat{a}_i^+ \hat{a}_k^+ \hat{a}_m \hat{a}_n. \qquad (7.3.45)$$

The transition from formula (7.3.40) to formulas (7.3.41) and (7.3.42) was historically one of the reasons why the occupation-number representation technique was called the second quantization technique. The logic behind this naming is that the wave function description of microscopic systems can be regarded as the first quantization procedure, while the replacement of wave functions by operators can be viewed as second quantization.

Finally, it is worthwhile to note that the following commutation relation is valid for wave function operators

$$\hat{\psi}(\eta)\hat{\psi}^+(\eta') - \hat{\psi}^+(\eta')\hat{\psi}(\eta) = \delta(\eta - \eta')\hat{I}. \tag{7.3.46}$$

This commutation relation can be proven by using the identity

$$\sum_i \Phi_i^*(\eta)\Phi_i(\eta') = \delta(\eta' - \eta), \tag{7.3.47}$$

which can be viewed as the expansion of Dirac function $\delta(\eta' - \eta)$ into series with respect to functions $\Phi_i(\eta')$. The details of proving formulas (7.3.46) and (7.3.47) are left as an exercise for the reader.

The mathematically similar formalism has been also developed for identical fermion particles. According to the Pauli exclusion principle, occupation numbers N_n of quantum states for such particles have only two distinct values

$$N_n = \begin{cases} 1_n \\ 0_n \end{cases}. \tag{7.3.48}$$

This leads to the following modifications in the definitions of annihilation and creation operators:

$$\hat{a}_n^+ c(N_1, \ldots 0_n, \ldots, t) = \pm c(N_1, \ldots 1_n, \ldots, t), \tag{7.3.49}$$

$$\hat{a}_n^+ c(N_1, \ldots 1_n, \ldots, t) = 0, \tag{7.3.50}$$

$$\hat{a}_n c(N_1, \ldots 1_n, \ldots, t) = \pm c(N_1, \ldots 0_n, \ldots, t), \tag{7.3.51}$$

$$\hat{a}_n c(N_1, \ldots 0_n, \ldots, t) = 0. \tag{7.3.52}$$

In formulas (7.3.49) and (7.3.51) sign "+" is taken if an even number of states preceding the state $|n\rangle$ are occupied, that is, if

$$\sum_{i<n} N_i \text{ is even}, \tag{7.3.53}$$

and sign "$-$" is taken if

$$\sum_{i<n} N_i \text{ is odd.} \tag{7.3.54}$$

Formulas (7.3.49) and (7.3.50) can be combined and written as the following one formula

$$\hat{a}_n^+ c(N_1, \ldots N_n, \ldots, t) = \pm(1 - N_n)c(N_1, \ldots N_n + 1, \ldots, t). \tag{7.3.55}$$

Similarly, formulas (7.3.51) and (7.3.52) can be combined as follows:

$$\hat{a}_n c(N_1, \ldots N_n, \ldots, t) = \pm N_n c(N_1, \ldots, N_n - 1, \ldots, t). \tag{7.3.56}$$

By using the above formulas, it can be demonstrated that

$$\hat{a}_n^+ \hat{a}_n = N_n \hat{I}, \tag{7.3.57}$$

$$\hat{a}_n \hat{a}_n^+ = (1 - N_n)\hat{I}, \tag{7.3.58}$$

and

$$\hat{a}_n \hat{a}_m^+ + \hat{a}_m^+ \hat{a}_n = \delta_{mn} \hat{I}, \tag{7.3.59}$$

$$\hat{a}_n \hat{a}_m + \hat{a}_m \hat{a}_n = 0, \tag{7.3.60}$$

$$\hat{a}_n^+ \hat{a}^+ + \hat{a}_m^+ \hat{a}_n^+ = 0. \tag{7.3.61}$$

It is apparent from the above formulas that operators $\hat{a}_n$ and $\hat{a}_m$ as well as $\hat{a}_n^+$ and $\hat{a}_m^+$ are anticommutative, which is in contrast with the case of boson particles when the corresponding operators are commutative. This is the consequence of the above "sign" rule. Furthermore, in the case of bosons, operators $\hat{a}_n$ and $\hat{a}_m$ are totally independent; each of these operators acts only on variables N_n and N_m, respectively, and the result of their action does not depend on values of other occupation numbers. This is not the case for fermions, where the result of action of $\hat{a}_n$ (or $\hat{a}_n^+$) depends not only on the value of N_n but on the value of sum $\sum_{i<n} N_i$ as well. This is the reason why the actions of annihilation and creation operators cannot be viewed as totally independent.

Despite the above-mentioned difference, formula (7.3.18) for the Hamiltonian can be derived for the case of identical fermions as well. The derivation is more involved than in the case of bosons, and it is omitted here. Instead, as an example, we consider the problem of transition probabilities in ensembles of identical boson particles. We shall use the Schrödinger equation (7.3.6) with the Hamiltonian (7.3.22):

$$i\hbar \frac{\partial c(N_1, N_2, \ldots, t)}{dt} = \mathcal{E}c(N_1, N_2, \ldots, t)$$

$$+ \frac{1}{2} \sum_{m,n} \sum_{i,k} V_{mn,ik} \hat{a}_m^+ \hat{a}_n^+ \hat{a}_i \hat{a}_k c(N_1, N_2, \ldots, t), \tag{7.3.62}$$

where

$$\mathcal{E} = \sum_i \mathcal{E}_i N_i. \tag{7.3.63}$$

Equation (7.3.62) can be somewhat simplified by introducing new function $b(N_1, N_2, \ldots, t)$:

$$c(N_1, N_2, \ldots, t) = e^{-\frac{i}{\hbar}\mathcal{E}t} b(N_1, N_2, \ldots, t). \tag{7.3.64}$$

It is clear that

$$|c(N_1, N_2, \ldots, t)|^2 = |b(N_1, N_2, \ldots, t)|^2. \tag{7.3.65}$$

It is also apparent that due to the definition of annihilation and creation operators we have

$$\hat{a}_m^+ \hat{a}_n^+ \hat{a}_i \hat{a}_k c(N_1, N_2, \ldots, t)$$
$$= e^{-\frac{i}{\hbar}\mathcal{E}t} e^{-\frac{i}{\hbar}(\mathcal{E}_m + \mathcal{E}_n - \mathcal{E}_i - \mathcal{E}_k)t} \hat{a}_n^+ \hat{a}_m^+ \hat{a}_i \hat{a}_k b(N_1, N_2, \ldots, t). \tag{7.3.66}$$

By substituting formulas (7.3.64) and (7.3.66) into equation (7.3.62), we derive

$$i\hbar \frac{\partial b(N_1, N_2, \ldots, t)}{\partial t}$$
$$= \frac{1}{2} \sum_{m,n} \sum_{i,k} e^{-\frac{i}{\hbar}(\mathcal{E}_m + \mathcal{E}_n - \mathcal{E}_i - \mathcal{E}_k)} V_{mn,ik} \hat{a}_m^+ \hat{a}_n^+ \hat{a}_i \hat{a}_k b(N_1, N_2, \ldots, t). \tag{7.3.67}$$

It is evident that, since $b(N_1, N_2, \ldots, t)$ is only different from $c(N_1, N_2, \ldots, t)$ by the time-dependent factor $e^{-\frac{i}{\hbar}\mathcal{E}t}$, the rules of actions of operators $\hat{a}_m^+, \hat{a}_n^+, \hat{a}_i$ and $\hat{a}_k$ on $b(N_1, N_2, \ldots, t)$ are the same as the rules of actions of these operators on $c(N_1, N_2, \ldots, t)$.

Now, consider the following initial conditions at $t = 0$

$$N_k = N_k^{(0)}, \quad (k = 1, 2, \ldots). \tag{7.3.68}$$

This implies that

$$b\left(N_1^{(0)}, N_2^{(0)}, \ldots, 0\right) = 1, \tag{7.3.69}$$

while

$$b(N_1, N_2, \ldots, 0) = 0 \tag{7.3.70}$$

for all other values of N_k.

Next, we assume that the interaction term $\hat{H}^{(2)}$ in the Hamiltonian, and, consequently, $V_{mn,ik}$ are small. Then, we can use the time-dependent perturbation theory for the solution of equation (7.3.67). As discussed in Section 6.3, the first-order approximation $b^{(1)}(N_1, N_2, \ldots, t)$ can be obtained by replacing $b(N_1, N_2, \ldots, t)$ in the right-hand side of equation (7.3.67) by its initial conditions (7.3.69) and (7.3.70). This leads to the equations

$$i\hbar \frac{d}{dt} b^{(1)} \left(N_1^{(0)}, \ldots N_m^{(0)} + 1, \ldots N_n^{(0)} + 1, \ldots N_i^{(0)} - 1, \ldots, N_k^{(0)} - 1, \ldots, t \right)$$

$$= e^{-\frac{i}{\hbar}(\mathcal{E}_m + \mathcal{E}_n - \mathcal{E}_i - \mathcal{E}_k)} \sqrt{\left(N_n^{(0)} + 1 \right) \left(N_m^{(0)} + 1 \right) N_i^{(0)} N_k^{(0)}} V_{mn,ik},$$

$$(7.3.71)$$

valid for various m, n, i and k.

By integrating the last equations and computing the probabilities of transitions per unit time

$$P_{mn,ik} = \frac{d}{dt} \left| b^{(1)} \right|^2, \qquad (7.3.72)$$

the following formula can be derived in the same manner as in Section 6.3:

$$P_{mn,ik} = \frac{2\pi}{\hbar} \left(N_m^{(0)} + 1 \right) \left(N_n^{(0)} + 1 \right) N_i^{(0)} N_k^{(0)} |V_{mn,ik}|^2 \, \delta(\mathcal{E}_m + \mathcal{E}_n - \mathcal{E}_i - \mathcal{E}_k).$$

$$(7.3.73)$$

In the case of fermion ensembles, the last equation is modified in the manner suggested by formulas (7.3.55) and (7.3.56). This leads to the following result:

$$P_{mn,ik} = \frac{2\pi}{\hbar} \left(1 - N_m^{(0)} \right) \left(1 - N_n^{(0)} \right) N_i^{(0)} N_k^{(0)} |V_{mn,ik}|^2 \, \delta(\mathcal{E}_m + \mathcal{E}_n - \mathcal{E}_i - \mathcal{E}_k).$$

$$(7.3.74)$$

It is clear from the last two formulas that in the case of ensembles of identical particles, the probabilities of transitions depend not only on the number of particles $N_i^{(0)}$ and $N_k^{(0)}$ in the initial state but on the occupation of the final state as well. For bosons (see formula (7.3.73)), these probabilities are enhanced by the occupation of the final state leading to the phenomena of the **Bose-Einstein condensation**. In the case of fermions (see formula (7.3.74)), the occupation of the final state is prohibitive to the transition, which is in accordance with the Pauli exclusion principle.

7.4 Quantization of Electromagnetic Field

In this section, we shall discuss the quantization of free (sourceless) electromagnetic field. Historically, the quantum nature of light and light absorption was first postulated to explain the frequency spectrum of black-body radiation as well as the photoelectric effect and Compton effect. With the advent of modern quantum mechanics, various theoretical techniques for electromagnetic field quantization have been developed. These are very sophisticated techniques, especially if a relativistic (Lorentz) invariant theory must be constructed. In this section, we shall discuss the simplest electromagnetic field quantization technique when the issue of Lorentz invariance is left unattended. The central idea of this quantization technique is to mathematically represent radiation (free) electromagnetic field as a set of harmonic oscillators and then quantize these oscillators in the same way as discussed in Section 4.2. The further fundamental step is to represent energy levels of these oscillators in terms of elementary particles called photons. In this way, photons appear as elementary energy excitations of electromagnetic field. As a result, the corpuscular nature of electromagnetic field is revealed.

The idea that elementary particles can be viewed as energy excitations of corresponding fields permeates modern quantum field theory. According to this idea, electrons and positrons are also treated as elementary excitations (bundles of energy) of the electron field (also called Dirac field because it is the quantized version of the relativistic Dirac equation for electrons). As a result of interaction, bundles (quanta) of energy may be transferred from electron field to electromagnetic field, and this may result in electron-positron pair annihilations and photon creations. When quanta of energy are transferred from electromagnetic field to electron field, this results in electron-positron pair creations and photon annihilations. This discussion suggests that quantum fields are fundamental, while elementary particles are excitations (quanta of energy) of these fields.

After the above digression, we return to the discussion of electromagnetic field quantization, and we first discuss how the radiation field can be mathematically modeled as a set of harmonic oscillators. We start with the basic equations for free (sourceless) electromagnetic fields, which can be written as follows

$$\nabla \times \mathbf{E} = -\frac{\partial \mathbf{B}}{\partial t}, \tag{7.4.1}$$

$$\nabla \times \mathbf{B} = \frac{1}{c^2}\frac{\partial \mathbf{E}}{\partial t}, \tag{7.4.2}$$

$$\nabla \cdot \mathbf{B} = 0, \tag{7.4.3}$$

$$\nabla \cdot \mathbf{E} = 0, \tag{7.4.4}$$

where, as usual,

$$c = \frac{1}{\sqrt{\mu_0 \varepsilon_0}}. \tag{7.4.5}$$

We shall next introduce magnetic vector potential

$$\mathbf{B} = \nabla \times \mathbf{A}, \tag{7.4.6}$$

and, to achieve its uniqueness, we shall impose the Coulomb gauge

$$\nabla \cdot \mathbf{A} = 0. \tag{7.4.7}$$

From equations (7.4.1) and (7.4.6), we derive

$$\nabla \times \mathbf{E} = -\nabla \times \left(\frac{\partial \mathbf{A}}{\partial t} \right), \tag{7.4.8}$$

and

$$\mathbf{E} = -\frac{\partial \mathbf{A}}{\partial t}. \tag{7.4.9}$$

By substituting formulas (7.4.6) and (7.4.9) into equation (7.4.2), we find

$$\nabla \times \nabla \times \mathbf{A} = -\frac{1}{c^2} \frac{\partial^2 \mathbf{A}}{\partial t^2}. \tag{7.4.10}$$

Since

$$\nabla \times \nabla \times \mathbf{A} = \nabla(\nabla \cdot \mathbf{A}) - \nabla^2 \mathbf{A}, \tag{7.4.11}$$

from the last two formulas and equation (7.4.7), we derive

$$\nabla^2 \mathbf{A} - \frac{1}{c^2} \frac{\partial^2 \mathbf{A}}{\partial t^2} = 0. \tag{7.4.12}$$

It should be kept in mind that the magnetic vector potential $\mathbf{A}$ simultaneously satisfies two equations (7.4.12) and (7.4.7).

Next, we consider plane wave solutions of these equations. To avoid some mathematical difficulties related to continuous spectrum of these solutions, we consider only solutions that satisfy the following periodicity condition

$$\mathbf{A}(\mathbf{r}) = \mathbf{A}(\mathbf{r} + \mathbf{e}_x L) = \mathbf{A}(\mathbf{r} + \mathbf{e}_y L) = \mathbf{A}(\mathbf{r} + \mathbf{e}_z L), \tag{7.4.13}$$

where, as before, $\mathbf{e}_x$, $\mathbf{e}_y$ and $\mathbf{e}_z$ are unit vectors directed along x, y and z-axes, while L is some sufficiently large length. In other words, one may consider the free electromagnetic field confined to a cube with edge L and satisfying periodic boundary conditions on its faces. This treatment is usually justified by the fact that L does not appear in the final results.

We look for the solution of equations (7.4.12) and (7.4.7) in the form

$$\mathbf{A}(\mathbf{r}, t) = b(t)\mathbf{U}(\mathbf{r}). \tag{7.4.14}$$

By substituting the last formula into equation (7.4.12), after simple transformations we obtain

$$\frac{\nabla^2 \mathbf{U}(\mathbf{r})}{\mathbf{U}(\mathbf{r})} = \frac{1}{c^2} \frac{\frac{d^2 b(t)}{dt^2}}{b(t)} = -k^2, \qquad (7.4.15)$$

where k is some number which does not depend on $\mathbf{r}$ and t, and the minus sign in (7.4.15) is taken to guarantee periodic in space and time solutions.

From the last equation, we find

$$\nabla^2 \mathbf{U}(\mathbf{r}) + k^2 \mathbf{U}(\mathbf{r}) = 0. \qquad (7.4.16)$$

It is easy to check that plane wave solutions of the last equation can be written in the form

$$\mathbf{U}_\alpha(\mathbf{r}) = \frac{1}{\sqrt{V}} \mathbf{e}_\alpha e^{i\mathbf{k}\cdot\mathbf{r}}, \qquad (7.4.17)$$

where $V = L^3$, $\mathbf{e}_\alpha$ is a unit vector and

$$|\mathbf{k}|^2 = k^2. \qquad (7.4.18)$$

Furthermore, from formulas (7.4.14) and (7.4.17) we find that equation (7.4.7) is satisfied if

$$\mathbf{e}_\alpha \cdot \mathbf{k} = 0, \quad (\alpha = 1, 2). \qquad (7.4.19)$$

There are only two linearly independent unit vectors $\mathbf{e}_\alpha$ that satisfy the above orthogonality condition. This orthogonality condition implies that the plane waves (7.4.19) are transverse, because $\mathbf{U}(\mathbf{r})$ is orthogonal to the propagation direction $\mathbf{k}$. The latter is the reason why the Coulomb gauge (7.4.7) is also known as the transverse gauge.

It is clear from formulas (7.4.14) and (7.4.17) that the periodicity condition (7.4.13) is satisfied if the wave vector $\mathbf{k}$ coincides with one of the vectors given by the formula

$$\mathbf{k}_\lambda = \mathbf{e}_x \frac{2\pi\lambda_x}{L} + \mathbf{e}_y \frac{2\pi\lambda_y}{L} + \mathbf{e}_z \frac{2\pi\lambda_z}{L}, \qquad (7.4.20)$$

where λ_x, λ_y and λ_z are integers and symbol λ stands for a set of these three integers

$$\lambda = (\lambda_x, \lambda_y, \lambda_z). \qquad (7.4.21)$$

By taking into account the restriction (7.4.20), the plane wave solutions (7.4.17) of equation (7.4.16) can be written as follows

$$\mathbf{U}_{\lambda\alpha}(\mathbf{r}) = \frac{1}{\sqrt{V}} \mathbf{e}_\alpha e^{i\mathbf{k}_\lambda \cdot \mathbf{r}}, \quad (\alpha = 1, 2). \qquad (7.4.22)$$

Thus, $\mathbf{U}_{\lambda\alpha}(\mathbf{r})$ is defined by four parameters $\lambda_x, \lambda_y, \lambda_z$ and α. It is evident that parameter α defines the polarization of the plane wave $\mathbf{U}_{\lambda\alpha}(\mathbf{r})$.

In our subsequent discussion, we shall also use the functions

$$\mathbf{U}_{\lambda\alpha}^*(\mathbf{r}) = \frac{1}{\sqrt{V}}\mathbf{e}_\alpha e^{-i\mathbf{k}_\lambda \cdot \mathbf{r}}, \tag{7.4.23}$$

which are as well solutions of equation (7.4.16). By using the equality

$$\int_V e^{i\mathbf{k}_\lambda \cdot \mathbf{r}} e^{-i\mathbf{k}_{\lambda'} \cdot \mathbf{r}}\, dV = V\delta_{\lambda\lambda'}, \tag{7.4.24}$$

the following orthogonality conditions can be established

$$\int_V \mathbf{U}_{\lambda\alpha} \cdot \mathbf{U}_{\lambda'\alpha'}\, dV = 0, \tag{7.4.25}$$

$$\int_V \mathbf{U}_{\lambda\alpha}^* \cdot \mathbf{U}_{\lambda'\alpha'}^*\, dV = 0, \tag{7.4.26}$$

$$\int_V \mathbf{U}_{\lambda\alpha}^* \cdot \mathbf{U}_{\lambda'\alpha'}\, dV = \delta_{\lambda\lambda'}\delta_{\alpha\alpha'}. \tag{7.4.27}$$

Now, we turn again to formula (7.4.15) and, taking into account restrictions (7.4.20), we obtain the following equation

$$\frac{d^2 b_\lambda(t)}{dt^2} + \omega_\lambda^2 b_\lambda(t) = 0, \tag{7.4.28}$$

where

$$\omega_\lambda = c\,|\mathbf{k}_\lambda|. \tag{7.4.29}$$

There are two linearly independent solutions of equation (7.4.28):

$$b_\lambda(t) = b_\lambda e^{-i\omega_\lambda t}, \quad b_\lambda^*(t) = b_\lambda^* e^{i\omega_\lambda t}. \tag{7.4.30}$$

By using formulas (7.4.14), (7.4.22), (7.4.23) and (7.4.30), we obtain the following plane wave expansion for magnetic vector potential

$$\mathbf{A}(\mathbf{r},t) = \sum \left[b_{\lambda\alpha}(t)\mathbf{U}_{\lambda\alpha}(\mathbf{r}) + b_{\lambda\alpha}^*(t)\mathbf{U}_{\lambda\alpha}^*(\mathbf{r}) \right], \tag{7.4.31}$$

where the summation is performed over all possible λ and α.

From the last formula and equations (7.4.6) and (7.4.9) we derive the following expressions for $\mathbf{B}(t)$ and $\mathbf{E}(t)$, respectively,

$$\mathbf{B}(t) = i\sum \left[b_{\lambda\alpha}(t)(\mathbf{k}_\lambda \times \mathbf{U}_{\lambda\alpha}(\mathbf{r})) + b_{\lambda\alpha}^*(t)(\mathbf{k}_\lambda \times \mathbf{U}_{\lambda\alpha}^*(\mathbf{r})) \right], \tag{7.4.32}$$

$$\mathbf{E} = -\sum \left[\dot{b}_{\lambda\alpha}(t)\mathbf{U}_{\lambda\alpha}(\mathbf{r}) + \dot{b}_{\lambda\alpha}^*(t)\mathbf{U}_{\lambda\alpha}^*(\mathbf{r}) \right], \tag{7.4.33}$$

where $\dot{b}_{\lambda\alpha}(r)$ and $\dot{b}^*_{\lambda\alpha}(t)$ are time-derivatives of $b_{\lambda\alpha}(t)$ and $b^*_{\lambda\alpha}(t)$, respectively.

Now, we shall use the following expression for energy of electromagnetic field

$$\mathcal{E} = \frac{\varepsilon_0}{2} \int_V \left(|\mathbf{E}(t)|^2 + c^2 |\mathbf{B}(t)|^2 \right) dV. \tag{7.4.34}$$

By substituting formulas (7.4.32) and (7.4.33) into equation (7.4.34) and taking into account the vector identities

$$\mathbf{k}_\lambda \times \mathbf{e}_1 = |\mathbf{k}_\lambda| \, \mathbf{e}_2, \tag{7.4.35}$$

$$\mathbf{k}_\lambda \times \mathbf{e}_2 = - |\mathbf{k}_\lambda| \, \mathbf{e}_1, \tag{7.4.36}$$

as well as the orthogonality conditions (7.4.25), (7.4.26) and (7.4.27), the following expression for the energy can be derived

$$\mathcal{E} = \sum \left[\dot{b}_{\lambda\alpha}(t)\dot{b}^*_{\lambda\alpha}(t) + \omega_\lambda^2 b_{\lambda\alpha}(t)b^*_{\lambda\alpha}(t) \right]. \tag{7.4.37}$$

The derivation is left as an exercise for the reader.

By recalling formula (7.4.30), the last equation can be further simplified as follows

$$\mathcal{E} = 2 \sum \omega_\lambda^2 b_{\lambda\alpha}(t)b^*_{\lambda\alpha}(t). \tag{7.4.38}$$

We shall further transform the last expression by introducing new variables

$$q_{\lambda\alpha} = b_{\lambda\alpha} + b^*_{\lambda\alpha}, \tag{7.4.39}$$

$$p_{\lambda\alpha} = -i\omega_\lambda(b_{\lambda\alpha} - b^*_{\lambda\alpha}). \tag{7.4.40}$$

It is apparent that equation (7.4.28) can be written in terms of $q_{\lambda\alpha}$ as follows

$$\frac{d^2 q_{\lambda\alpha}}{dt^2} + \omega_\lambda^2 q_{\lambda\alpha} = 0. \tag{7.4.41}$$

It is also clear that formula (7.4.38) can be written in the form

$$H = \mathcal{E} = \frac{1}{2} \sum \left[p_{\lambda\alpha}^2 + \omega_\lambda^2 q_{\lambda\alpha}^2 \right]. \tag{7.4.42}$$

It is easy to recognize that for any λ and α equation (7.4.41) has the mathematical form of the differential equation for the harmonic oscillator. Furthermore, the expression (7.4.42) for the energy of free (transverse) electromagnetic field has the mathematical form of the sum of energy of harmonic oscillators corresponding to various λ and α. Thus, it can be concluded that the free electromagnetic field can be mathematically represented as a set of noninteracting (not coupled) harmonic oscillators.

The letter H is used in formula (7.4.42) because $p_{\lambda\alpha}$ and $q_{\lambda\alpha}$ can be treated as canonical momenta and coordinates and H as the Hamiltonian function. Indeed, from formulas (7.4.30) and (7.4.39) we find

$$\dot{q}_{\lambda\alpha} = p_{\lambda\alpha}, \tag{7.4.43}$$

$$\dot{p}_{\lambda\alpha} = -\omega_\lambda^2 q_{\lambda\alpha}, \tag{7.4.44}$$

which by using formula (7.4.42) can be written as the following Hamiltonian equations

$$\dot{q}_{\lambda\alpha} = \frac{\partial H}{\partial p_{\lambda\alpha}}, \tag{7.4.45}$$

$$\dot{p}_{\lambda\alpha} = -\frac{\partial H}{\partial q_{\lambda\alpha}}. \tag{7.4.46}$$

Thus, the free electromagnetic field can be mathematically treated as the infinite-dimensional Hamiltonian systems. Consequently, it can be quantized as the Hamiltonian systems. In doing so, we replace the Hamiltonian function by the Hamiltonian operator. The expression for this Hamiltonian operator is obtained by replacing canonical momenta and coordinates by the corresponding operators which satisfy the same commutation relations as discussed in Section 2.3. Accordingly, we have

$$\hat{H}_{ef} = \sum \hat{H}_{\lambda\alpha} = \frac{1}{2} \sum (\hat{p}_{\lambda\alpha}^2 + \omega_\lambda^2 \hat{q}_{\lambda\alpha}^2), \tag{7.4.47}$$

where

$$[\hat{p}_{\lambda\alpha}, \hat{q}_{\lambda'\alpha'}] = -i\hbar\delta_{\lambda\alpha,\lambda'\alpha'}\hat{I}, \tag{7.4.48}$$

$$[\hat{p}_{\lambda\alpha}, \hat{p}_{\lambda'\alpha'}] = 0, \tag{7.4.49}$$

$$[\hat{q}_{\lambda\alpha}, \hat{q}_{\lambda'\alpha'}] = 0. \tag{7.4.50}$$

Then, by using the same line of reasoning as in the discussion of the harmonic oscillator in Section 4.2, we introduce operators

$$\hat{a}_{\lambda\alpha}^+ = \sqrt{\frac{1}{2\hbar\omega_\lambda}}(\omega\hat{q}_{\lambda\alpha} - i\hat{p}_{\lambda\alpha}), \tag{7.4.51}$$

$$\hat{a}_{\lambda\alpha} = \sqrt{\frac{1}{2\hbar\omega_\lambda}}(\omega\hat{q}_{\lambda\alpha} + i\hat{p}_{\lambda\alpha}), \tag{7.4.52}$$

and, by using formulas (7.4.48)–(7.4.50), we establish the commutation relations

$$[\hat{a}_{\lambda\alpha}, \hat{a}^+_{\lambda'\alpha'}] = \delta_{\lambda\alpha,\lambda'\alpha'}\hat{I}, \tag{7.4.53}$$

$$[\hat{a}_{\lambda\alpha}, \hat{a}_{\lambda'\alpha'}] = 0, \tag{7.4.54}$$

$$[\hat{a}^+_{\lambda\alpha}, \hat{a}^+_{\lambda'\alpha'}] = 0. \tag{7.4.55}$$

Furthermore, as in Section 4.2, we can establish the following formula for the Hamiltonian (7.4.47)

$$\boxed{\hat{H}_{ef} = \sum \hat{H}_{\lambda\alpha} = \sum \hbar\omega_\lambda \left(\hat{a}^+_{\lambda\alpha}\hat{a}_{\lambda\alpha} + \frac{1}{2} \right)} \tag{7.4.56}$$

and the validity of the following relations for eigenvalues $\mathcal{E}_{n_{\lambda\alpha}}$ and eigenfunctions $|n_{\lambda\alpha}\rangle$ of operators $\hat{H}_\lambda$:

$$\hat{H}_{\lambda\alpha}|n_{\lambda\alpha}\rangle = \mathcal{E}_{n_{\lambda\alpha}}|n_{\lambda\alpha}\rangle, \tag{7.4.57}$$

$$\mathcal{E}_{n_{\lambda\alpha}} = \hbar\omega_\lambda \left(n_{\lambda\alpha} + \frac{1}{2} \right), \tag{7.4.58}$$

where $n_{\lambda\alpha}$ is a positive integer. The last three formulas suggest that the energy of the electromagnetic field is quantized as the energy of harmonic oscillators. **The next fundamental step** is to represent the discrete energy levels of electromagnetic fields in terms of elementary particles called photons. Namely, we postulate that each photon of frequency ω_λ and polarization α has the energy

$$\boxed{\mathcal{E}^{ph}_{\lambda\alpha} = \hbar\omega_\lambda,} \tag{7.4.59}$$

and, consequently, the energy level $\mathcal{E}_{n_{\lambda\alpha}}$ of the electromagnetic field can be viewed as a state $|n_{\lambda\alpha}\rangle$ where there are $n_{\lambda\alpha}$ photons of frequency ω_λ and polarization α. In the state $|0_\lambda\rangle$ with energy $\mathcal{E}^{(\lambda)}_0 = \frac{\hbar\omega_\lambda}{2}$ there are no photons of frequency ω_λ. The photons of the same frequency and polarization can be treated as identical elementary particles. It is clear from the commutation relations (7.4.53)–(7.4.55) that these identical particles can be viewed as bosons. According to formulas (7.4.56)–(7.4.58), a general state of quantized electromagnetic field may contain $n_{\lambda_1\alpha}$ photons of frequency $\omega_{\lambda_1}, n_{\lambda_2,\alpha}$ photons of frequency $\omega_{\lambda_2}, \ldots, n_{\lambda_r\alpha}$ photons of frequency ω_{λ_r}. The following notation

$$|n_{\lambda_1\alpha}, n_{\lambda_2\alpha}, \ldots n_{\lambda_r\alpha}\rangle \tag{7.4.60}$$

can be used for this state.

The energy of the electromagnetic field in the above state due to the presence of these photons is given by the formula

$$\mathcal{E} = \sum_{j,\alpha} n_{\lambda_j \alpha} \hbar \omega_{\lambda j}. \tag{7.4.61}$$

Next, we turn to the discussion of momentum of quantized electromagnetic field. The momentum of classical electromagnetic field is defined by the equation

$$\mathbf{P}_{ef} = \varepsilon_0 \int_V (\mathbf{E} \times \mathbf{B})\, dV. \tag{7.4.62}$$

By using formulas (7.4.32), (7.4.33), (7.4.35), (7.4.36) as well as the orthogonality conditions (7.4.25)–(7.4.27), the following expression can be derived for this momentum

$$\mathbf{P}_{ef} = \sum \mathbf{P}_{\lambda\alpha} = 2 \sum \omega_\lambda b_{\lambda\alpha} b_{\lambda\alpha}^* \mathbf{k}_\lambda. \tag{7.4.63}$$

In terms of $p_{\lambda\alpha}$ and $q_{\lambda\alpha}$ variables, the electromagnetic momentum can be written as follows

$$\mathbf{P}_{ef} = \frac{1}{2} \sum \frac{\mathbf{k}_\lambda}{\omega_\lambda} \left[p_{\lambda\alpha}^2 + \omega_\lambda^2 q_{\lambda\alpha}^2 \right]. \tag{7.4.64}$$

By replacing $p_{\lambda\alpha}$ and $q_{\lambda\alpha}$ by operators and then by using formulas (7.4.51) and (7.4.52), we arrive at the following expression for the operator of electromagnetic field momentum

$$\widehat{\mathbf{P}}_{ef} = \sum \widehat{\mathbf{P}}_{\lambda\alpha} = \frac{1}{2} \sum \mathbf{k}_\lambda \hbar \left(\hat{a}_{\lambda\alpha}^+ \hat{a}_{\lambda\alpha} + \frac{1}{2} \right). \tag{7.4.65}$$

It is clear that operators $\hat{H}_{\lambda\alpha}$ and $\widehat{\mathbf{P}}_{\lambda\alpha}$ commute and have the same eigenfunctions $|n_{\lambda\alpha}\rangle$. It is also apparent that

$$\widehat{\mathbf{P}}_\lambda |n_{\lambda\alpha}\rangle = \mathbf{k}_\lambda \hbar \left(n_{\lambda\alpha} + \frac{1}{2} \right) |n_{\lambda\alpha}\rangle. \tag{7.4.66}$$

This suggests that each photon of frequency ω_λ has the momentum

$$\boxed{\mathbf{P}_{\lambda\alpha}^{ph} = \hbar \mathbf{k}_\lambda.} \tag{7.4.67}$$

From the last formula and equation (7.4.29) we find

$$\left| \mathbf{P}_{\lambda\alpha}^{ph} \right| = \frac{\hbar \omega_\lambda}{c}, \tag{7.4.68}$$

which according to the relation (7.4.59) leads to

$$\mathcal{E}_{\lambda\alpha}^{ph} = \left| \mathbf{P}_{\lambda\alpha}^{ph} \right| c. \tag{7.4.69}$$

By recalling the relativistic energy-momentum relation

$$\mathcal{E}^2 = (pc)^2 + (m_0 c^2)^2, \tag{7.4.70}$$

from formula (7.4.69) we conclude that the photon is massless

$$m_0 = 0. \tag{7.4.71}$$

The next question that can be asked is related to photon spin. The issue of photon spin is somewhat complicated and will not be discussed here in detail. We just mention that the spin of the photon is related to its polarization and it is equal to one. In particular, if a light beam is circularly polarized, then each of its photons carries a spin angular momentum equal to $\pm\hbar$ with positive sign for "left" and negative sign for "right" circular polarizations.

In summary, a photon is an elementary electrically neutral massless particle moving with the speed of light whose energy and momentum are given by formulas (7.4.59) and (7.4.67), respectively, and whose spin is equal to one.

In quantizing free electromagnetic field, we replace canonical momenta $p_{\lambda\alpha}$ and coordinates $q_{\lambda\alpha}$ by operators. These momenta and coordinates are related to $b_{\lambda\alpha}$ and $b_{\lambda\alpha}^*$ by formulas (7.4.39) and (7.4.40). This implies that as a result of quantization $b_{\lambda\alpha}$ and $b_{\lambda\alpha}^*$ become operators $\hat{b}_{\lambda\alpha}$ and $\hat{b}_{\lambda\alpha}^+$, respectively. It is easy to see from formulas (7.4.39), (7.4.40), (7.4.51) and (7.4.52) that the following relations are valid

$$\hat{b}_{\lambda\alpha} = \sqrt{\frac{\hbar}{2\omega_\lambda}}\hat{a}_{\lambda\alpha}, \quad \hat{b}_{\lambda\alpha}^+ = \sqrt{\frac{\hbar}{2\omega_\lambda}}\hat{a}_{\lambda\alpha}^+. \tag{7.4.72}$$

Now, by using formulas (7.4.22), (7.4.23), (7.4.30) and (7.4.31) and by replacing $b_{\lambda\alpha}$ and $b_{\lambda\alpha}^*$ by operators (7.4.72), we arrive at the following expression for the operator of the magnetic vector potential

$$\hat{\mathbf{A}}(\mathbf{r}, t) = \sum \left[\hat{a}_{\lambda\alpha}\mathbf{A}_{\lambda\alpha}(\mathbf{r}, t) + \hat{a}_{\lambda\alpha}^+\mathbf{A}_{\lambda\alpha}^*(\mathbf{r}, t) \right], \tag{7.4.73}$$

where

$$\mathbf{A}_{\lambda\alpha}(\mathbf{r}, t) = \sqrt{\frac{\hbar}{2V\omega_\lambda}}\mathbf{e}_\alpha e^{i(\mathbf{k}_\lambda \cdot \mathbf{r} - \omega_\lambda t)}. \tag{7.4.74}$$

By using the above expression for the operator of the magnetic vector potential, we shall briefly discuss the problem of interaction of a quantum mechanical system (called "atom") with quantized electromagnetic field. For such a system, the Hamiltonian has three distinct parts

$$\hat{H} = \hat{H}_0 + \hat{H}_{ef} + \hat{V}, \tag{7.4.75}$$

where $\hat{H}_0$ and $\hat{H}_{ef}$ are the Hamiltonians for the quantum mechanical system and the quantized electromagnetic field (see formula (7.4.56)), respectively, while the operator $\hat{V}$ describes the interaction between them. For this operator, we shall use the formula (see (2.5.30)):

$$\hat{V} = -\frac{e}{m}\hat{\mathbf{p}} \cdot \hat{\mathbf{A}} = -\frac{e}{m}\hat{\mathbf{p}} \cdot \sum \left[\hat{a}_{\lambda\alpha}\mathbf{A}_{\lambda\alpha}(\mathbf{r},t) + \hat{a}_{\lambda\alpha}^{+}\mathbf{A}_{\lambda\alpha}^{*}(\mathbf{r},t) \right]. \quad (7.4.76)$$

In the absence of interaction the stationary states of the overall system are described by products of eigenfunctions of operators $\hat{H}_0$ and $\hat{H}_{ef}$. We shall use the notation $|\nu\rangle|n_{\lambda\alpha}\rangle$ for such stationary states, where ν is the symbol for a complete set of parameters that completely characterize the stationary state of $\hat{H}_0$, while $n_{\lambda\alpha}$ in the number of photons with specified wave vector $\mathbf{k}_\lambda$ and polarization α. We shall use the first-order time dependent perturbation theory to evaluate the probability of transitions caused by interaction operator $\hat{V}$. By taking into account the relations

$$\langle n_{\lambda,\alpha} + 1 | \hat{a}_{\lambda\alpha}^{+} | n_{\lambda\alpha} \rangle = \sqrt{n_{\lambda,\alpha} + 1}, \quad (7.4.77)$$

$$\langle n_{\lambda,\alpha} - 1 | \hat{a}_{\lambda\alpha} | n_{\lambda\alpha} \rangle = \sqrt{n_{\lambda,\alpha}}, \quad (7.4.78)$$

we conclude that the matrix elements of $\hat{V}$ in the first-order time dependent perturbation theory are not equal to zero only for transition between the states with number of photons with specific λ and α different by one. The probability of such transitions is proportional to the square of the magnitude of such matrix elements. Consequently, according to the last two formulas, we have

$$\text{Prob}(|\nu\rangle|n_{\lambda,\alpha}\rangle \to |\nu'\rangle|n_{\lambda\alpha} + 1\rangle) \sim n_{\lambda\alpha} + 1, \quad (7.4.79)$$

$$\text{Prob}(|\nu\rangle|n_{\lambda,\alpha}\rangle \to |\nu''\rangle|n_{\lambda\alpha} - 1\rangle) \sim n_{\lambda\alpha}. \quad (7.4.80)$$

It is clear that the probability in formula (7.4.79) is the probability of photon emission, while the probability in formula (7.4.80) is the probability of absorption. It is clear that emission dominates absorption. This is because of spontaneous emission which according to formula (7.4.79) may occur even in the absence of photons, that is, when $n_{\lambda\alpha} = 0$. Formula (7.4.79) also accounts for stimulated emission by demonstrating that the probability of emission is increased in the presence of photons.

We conclude this section with the brief discussion of one controversial issue related to the quantization of electromagnetic field. This is the issue of infinite energy of the ground (or vacuum) state, that is, the state in which there are no photons of any kind. Indeed, according to formula (7.4.56), we find the following expression for the energy $\mathcal{E}_0$ of the ground state

$$\mathcal{E}_0 = \sum \frac{\hbar\omega_\lambda}{2} = \infty. \tag{7.4.81}$$

This energy is infinite, because the summation is performed over the infinite number of all possible photon modes of increasing frequency. This infinity does not affect many predictions of the theory in which not the total energy but rather energy differences of different quantum states are of interest. Furthermore, the quantization procedure itself can be mathematically modified to dispense completely with the notion of the vacuum state. Indeed, we quantized electromagnetic field by replacing canonical momenta and coordinates by operators in the following expression for energy of a single harmonic oscillator

$$H_{\lambda\alpha} = \frac{1}{2}(p_{\lambda\alpha}^2 + \omega_\lambda^2 q_{\lambda\alpha}^2). \tag{7.4.82}$$

However, this quantization can be done differently by using the equivalent mathematical formula

$$H_{\lambda\alpha} = \frac{1}{2}(\omega q_{\lambda\alpha} - ip_{\lambda\alpha})(\omega q_{\lambda\alpha} + ip_{\lambda\alpha}) \tag{7.4.83}$$

and by replacing $q_{\lambda\alpha}$ and $p_{\lambda\alpha}$ by operators

$$\hat{H}_{\lambda\alpha} = \frac{1}{2}(\omega \hat{q}_{\lambda\alpha} - i\hat{p}_{\lambda\alpha})(\omega \hat{q}_{\lambda\alpha} + i\hat{p}_{\lambda\alpha}), \tag{7.4.84}$$

which according to formulas (7.4.51) and (7.4.52) leads to the following expressions for the Hamiltonians:

$$\hat{H}_{\lambda\alpha} = \hbar\omega \hat{a}_{\lambda\alpha}^+ \hat{a}_{\lambda\alpha}, \tag{7.4.85}$$

and

$$\hat{H}_{ef} = \sum \hat{H}_{\lambda\alpha} = \sum \hbar\omega \hat{a}_{\lambda\alpha}^+ \hat{a}_{\lambda\alpha} \tag{7.4.86}$$

with no ground state of infinite energy.

Nevertheless, the notion of the dynamic (fluctuating) vacuum state is physically very meaningful. There is evidence that the fluctuations of this state are responsible for the Casimir force between two uncharged conductive plates in a vacuum separated by a few nanometers. Furthermore, since energy is equivalent to mass and mass affects gravitation, the energy of the vacuum states may have far-reaching gravitational effects which influence, for instance, the expansion of the universe. There are also speculations (scientific conjectures) that the vacuum energy of the electromagnetic field is related to dark matter. The further discussion of these issues is well beyond the scope of this text.

Problems

(1) Explain why the double splitting of electron beams in Stern-Gerlach experiments cannot be attributed to orbital angular momentum.

(2) By using formulas (7.1.1)–(7.1.3) prove the formulas (7.1.4)–(7.1.9).

(3) By using formulas (7.1.7)–(7.1.12) prove the commutation relations (7.1.13)–(7.1.15).

(4) By using the same line of reasoning as in Section 2.4, prove formulas (7.1.33)–(7.1.36).

(5) Derive formula (7.1.43).

(6) Prove that the eigenvalues of $\hat{S}_x$ and $\hat{S}_y$ are $\frac{\hbar}{2}$ and $-\frac{\hbar}{2}$, while the corresponding eigenvectors are given by formulas (7.1.53) and (7.1.54).

(7) Show that $\hat{\sigma}_x$, $\hat{\sigma}_y$ and $\hat{\sigma}_z$ are unitary matrices.

(8) Prove formula (7.1.61) for states $|1,0\rangle$ and $|1,-1\rangle$.

(9) Prove formula (7.1.62).

(10) Prove the equivalence of the Hamiltonians given by formulas (7.1.85) and (7.1.88).

(11) Prove that the Pauli equation (7.1.87) is invariant with respect to the local change of phase of the wave function.

(12) Derive equations (7.1.93) and (7.1.94).

(13) Derive formulas (7.1.105) and (7.1.106).

(14) Prove formulas (7.1.114) and (7.1.115).

(15) Prove that the solution of equation (7.1.121) is given by formulas (7.1.122)–(7.1.125).

(16) Explain why the number of terms in the sum in formula (7.2.25) is given by the expression (7.2.27).

(17) Explain why $\frac{1}{\sqrt{N!}}$ is the normalization coefficient in formula (7.2.28).

(18) Derive formulas (7.2.55)–(7.2.58).

(19) Prove formulas (7.2.62) and (7.2.63).

(20) Prove formula (7.2.61).

(21) Verify the validity of formulas (7.3.9)–(7.3.13) for annihilation and creation operators.

(22) Consider the particular case of four identical boson-type particles and confirm formulas (7.3.23) and (7.3.24).

(23) Prove formula (7.3.45).

(24) Prove formulas (7.3.46) and (7.3.47).
(25) Prove formulas (7.3.57)–(7.3.61).
(26) Derive formula (7.3.73).
(27) Prove formula (7.4.15).
(28) Prove formulas (7.4.25)–(7.4.27).
(29) Prove formula (7.4.37).
(30) Prove the commutation relations (7.4.53)–(7.4.55).
(31) Derive formula (7.4.63) for the momentum of electromagnetic field.
(32) Derive formulas (7.4.66).

Bibliography

[1] A. Aharoni, *Introduction to the Theory of Ferromagnetism*, Clarendon Press, Oxford, 1996.

[2] L. E. Ballentine, *Quantum Mechanics, A Modern Development*, World Scientific, 2010.

[3] H. Bethe, *Intermediate Quantum Mechanics*, W. A. Benjamin, Inc., 1964.

[4] F. Capasso (Ed.), *Physics of Quantum Electron Devices*, Springer-Verlag, 1990.

[5] C. Cohen-Tannoudji, B. Diu, F. Laloë, *Quantum Mechanics*, vol. 1, Wiley-VCH, 2005.

[6] S. Datta, *Quantum Phenomena*, Addison-Wesley, 1989.

[7] P. A. M. Dirac, *The Principles of Quantum Mechanics*, Clarendon Press, Oxford, 1958.

[8] R. P. Feynman, *Statistical Mechanics*, W. A. Benjamin, Inc., 1972.

[9] R. P. Feynman, A. R. Hibbs, *Quantum Mechanics and Path Integrals*, (Emended by D. F. Styer), Dover, 2005.

[10] D. J. Griffiths, *Introduction to Quantum Mechanics*, Pearson Prentice Hall, 2005.

[11] R. B. Griffiths, *Consistent Quantum Theory*, Cambridge University Press, 2002.

[12] C. Kittel, *Introduction to Solid State Physics*, Wiley, 2005.

[13] L. D. Landau, E. M. Lifshitz, *Quantum Mechanics, Nonrelativistic Theory*, Addison-Wesley, Reading, 1958.

[14] R. L. Liboff, *Introductory Quantum Mechanics*, Addison-Wesley, 2003.

[15] M. Lundstrom, *Fundamentals of Carrier Transport*, Addison-Wesley, 1990.

[16] G. Ya. Lyubarskii, *The Application of Group Theory in Physics*, Pergamon Press Inc., 1960.

[17] N. D. Mermin, *Quantum Computer Science, An Introduction*, Cambridge University Press, 2007.

[18] A. Messiah, *Quantum Mechanics*, vol. 1 and 2, Dover, 2014.

[19] J. von Neumann, *Mathematical Foundation of Quantum Mechanics*, Princeton University Press, 1996.

[20] J. Rammer, *Quantum Transport Theory*, Perseus Books, Reading, 1998.

[21] J. J. Sakurai, *Advanced Quantum Mechanics*, Addison-Wesley, 1967.

[22] R. Shankar, *Principles of Quantum Mechanics*, Kluwer Academic/Plenum Publishers, 1994.

[23] L. I. Schiff, *Quantum Mechanics*, McGraw Hill, 1968.

[24] M. Shur, *Physics of Semiconductor Devices*, Prentice Hall, 1990.

[25] M. Tinkham, *Introduction to Superconductivity*, McGraw Hill, 1996.

[26] D. J. Thouless, *Topological Quantum Numbers in Nonrelativistic Physics*, World Scientific, 1998.

[27] S. Weinberg, *Lectures on Quantum Mechanics*, Cambridge University Press, 2013.

[28] S. Weinberg, *The Quantum Theory of Fields*, vol. 1, Cambridge University Press, 1995.

[29] J. Weiner, P.-T. Ho, *Light-Matter Interaction, Fundamentals and Applications*, vol. 1, Wiley, 2002.

[30] A. Yariv, *Introduction to Theory and Applications of Quantum Mechanics*, Wiley, 1982.

Index

297

Printed in the United States
By Bookmasters